PHILOSOPHY AND THE ORIGIN AND EVOLUTION OF THE UNIVERSE

SYNTHESE LIBRARY

STUDIES IN EPISTEMOLOGY,

LOGIC, METHODOLOGY, AND PHILOSOPHY OF SCIENCE

VOLUME 217

PHILOSOPHY AND THE ORIGIN AND EVOLUTION OF THE UNIVERSE

Edited by

EVANDRO AGAZZI

University of Fribourg, Switzerland,
University of Genova, Italy and
President of the International Academy of Philosophy of Science

and

ALBERTO CORDERO

Department of Philosophy, Queen's College,
City University of New York, U.S.A.

KLUWER ACADEMIC PUBLISHERS

DORDRECHT / BOSTON / LONDON

Library of Congress Cataloging-in-Publication Data

Philosophy and the origin and evolution of the universe / edited by
Evandro Agazzi and Alberto Cordero.
p. cm. -- (Synthese library ; v. 217)
Papers from the Annual Meeting of the International Academy of
Philosophy of Science, held in Lima, Peru, Aug. 1988.
Includes index.
ISBN 0-7923-1322-4 (hb)
1. Cosmology. 2. Astronomy--Philosophy. 3. Science--Philosophy.
I. Agazzi, Evandro. II. Cordero, Alberto. III. International
Academy of Philosophy of Science. Meeting (1988 : Lima, Peru)
IV. Series.
QB981.P562 1991
523.1--dc20 91-21391

ISBN 0-7923-1322-4

Published by Kluwer Academic Publishers,
P.O. Box 17, 3300 AA Dordrecht, The Netherlands.

Kluwer Academic Publishers incorporates
the publishing programmes of
D. Reidel, Martinus Nijhoff, Dr W. Junk and MTP Press.

Sold and distributed in the U.S.A. and Canada
by Kluwer Academic Publishers,
101 Philip Drive, Norwell, MA 02061, U.S.A.

In all other countries, sold and distributed
by Kluwer Academic Publishers Group,
P.O. Box 322, 3300 AH Dordrecht, The Netherlands.

Printed on acid-free paper

Printed in the Netherlands

TABLE OF CONTENTS

INTRODUCTION

It has often been noted that a kind of double dynamics characterizes the development of science. On the one hand the progress in every discipline appears as the consequence of an increasing specialization, implying the restriction of the inquiry to very partial fields or aspects of a given domain. On the other hand, an opposite (but one might better say a complementary) trend points towards the construction of theoretical frameworks of great generality, the aim of which seems to correspond not so much to the need of providing «explanations» for the details accumulated through partial investigation, as to the desire of attaining an horizon of global comprehension of the whole field. This intellectual dialectics is perceivable in every discipline, from mathematics, to physics, to biology, to history, to economics, to sociology, and it is not difficult to recognize there the presence of the two main attitudes according to which human beings try to make «intelligible» the world surrounding them (including themselves), attitudes which are sometimes called analysis and synthesis. They correspond respectively to the spontaneous inclination which pushes us to try to understand things by seeing «how they are made», in the sense of «looking into them» and breaking them into their constitutive parts, or rather to encompass things in a global picture, where they are accounted for as occupying a place, or playing a role, which are understandable from the point of view of the whole.

This dynamics is to be found in the patterns of the cognitive activity of human individuals, but is equally active since times immemorial in the cultural development of mankind. In particular

E. Agazzi and A. Cordero (eds.), Philosophy and the Origin and Evolution of the Universe, vii–xii.

it is bound to the idea of causality in a double sense. On the one side, this causality appears as the expression of the patterns, of the links which bound the elements or constituents of things together (and this leads to the formulation of principles or laws having a static character). On the other side, this causality expresses the notion of a production of things, and is therefore deeply related with the idea of time, and very naturally leads to the problem of the «origin». Mythical or religious cosmologies and cosmogonies (as well as the corresponding conceptions regarding the origin and the destiny of men, or the sense of history) clearly correspond to this picture, but also the doctrines of many philosophers, from ancient Greece up to modern times. In the first case, the universal laws and patterns of reality are seen as the expression of certain supernatural designs, while the origin of things is traced back to the action of supernatural agents. In the second case, certain original elements are introduced, while the general patterns of reality are expressed in the form of universal metaphysical principles, which usually also account for the origin of things. However, the difficulty resides precisely in the way of combining the static meaning of the first sense of causality, with the dynamic meaning of the second.

The rise of modern science in 17th century marked a decisive primacy of the analytic approach, but at the same time, the requirement of generality was not dismissed: simply, instead of trying to establish universal principles valid for the whole, the path followed was that of finding principles of increasing generality for a definite kind of phenomena (think e.g. of the elaboration of the great variational principles of analytic mechanics). These principles were meant to account for the behaviour of any «mechanical system», but could they be taken as principles for the intelligibility of all things? Such a step was possible only by postulating that whatever existing entity (at least in the material world) is a mechanical system and nothing else. This is a clear metaphysical postulate, which was actually adopted in the «mechanistic worldview» which was advocated by many people in the past century. The crisis of classical mechanics (which can certainly be interpreted in several ways) may be seen as the admission that, already in physics, we must recognize the existence of several kinds of interactions, so that not whatever physical system is a mechanical system.

However, the discovery of the differences did not discourage the search for unity, which took another direction: instead of trying to *reduce* all interactions to an alleged fundamental one (as it was essentially the case with the mechanistic worldview), the efforts pointed towards *unifying* these interactions within a suitable general theory. The success of this enterprise is not yet complete, but this is the direction in which contemporary physics is moving. The interesting consequence is that a discourse on the *whole* appears again legitimate in science, since it is no longer based upon a more or less arbitrary generalization of certain partial patterns of physical systems, but on the elaboration of a view which respects differences. This approach has been made possible not only thanks to the existence and development of very powerful theories, such as relativity and quantum physics (no less than by the development of an extremely sophisticated mathematical instrumentation), but also - and in a decisive measure - by the gradual acceptance of an *evolutionistic* point of view.

Evolution has made its first appearance in the sciences of life, and in spite of the possibility of interpreting it in a reductionistic sense (a possibility which is certainly controversial), its most positive contribution consists in being a unifying theory, in which the differences among living beings are not eliminated, but rather connected in a genetic way. What we see in contemporary cosmology is precisely something of the same kind: cosmology is characteristically concerned nowadays not so much with the problem of the structure or constitution of the world, as with the *origin* and *evolution* of the universe, and the interesting thing is that in this way it offers a model for combining the static (or structural) aspect of causality, with its dynamic essential sense. In this way the *whole* appears as a *result*, but not as a result of the application of some rigid patterns to a postulated initial state of affairs (as it was the case, e.g. with the Laplacean cosmology): the whole itself was present at the origin, but in a form which immediately (i.e. practically without any elapsing of time) started evolving, and there is no reason for believing that this evolution will stop (cosmology does not concentrate very much on the question of the future evolution of the universe yet, but does not totally disregard this question either).

What we have said does not mean that everything is clear, unproblemetic or uncontroversial in this picture, it simply wants to stress the novelty and the challenging force of this approach, which has already brought science to adopt attitudes that were hardly thinkable some decades ago. This approach has also influenced epistemology, which has been obliged to pass from the «foundationalist» attitude, which was typical of the first half of this century, to a much more flexible attitude, in which basic elements of «evolutionary» nature are incorporated in the very idea of rationality, in the way of appreciating theory construction and theory change, no less than concepts formation and methodological requirements. This is in a way obvious: if epistemology wants to be a critical reflection on knowledge, with a special concern for scientific knowledge, it was to be expected that the remarkable developments in evolutionary thinking, which have occurred in the last thirty years, from elementary particle physics and cosmology to naturalized anthropology, should revitalize the debate that surrounds the foundations of science. The philosophical consequences of contemporary evolutionary ideas raise a whole series of intriguing questions which thorougly invest science and philosophy: yet they still lie open and with regard to them (i.e. to their merits and their limits) the struggle will probably continue for several years. What seems certain is that owing to this debate we may be on the verge of developments of signficance to both philosophers and scientists.

The present book does not have the pretension of illustrating this problematic in full. It simply wants to single out certain important factors and aspects of the rich thematic evocated above, with an effort of bringing together its scientific and philosophical sides. The paper by E.Agazzi, *The Universe as a Scientific and Philosophical Problem* tries to show how the investigations of contemporary cosmology vindicate the intellectual interest of a classical philosophical problem, without dissolving its specific philosophical connotations, but rather encouraging a renewed dialogue between science and philosophy. After this introductory paper, the scientific aspects of the theme are considered in a short series of contributions having a certain degree of technicality, but still contained within the limits of an accessible generality. Their titles are self-explanatory: *The Geometric Structure of the Universe* by R.Torretti; *Superstring Unification and the Existence of Gravity* by B. Zwiebach; *The Universe of Modern Science and its Philosophical Exploration* by D. Shapere; *From Molecules to Life* by R. Engel; *Meta-Neuroanatomy: The Myth of the Unbounded Mind/Brain* by C. Cherniak. The trajectory from relativity theory, to unification theories, to elementary particles physics, to biology and psychology is outlined in these papers, while the contribution of Shapere already develops the picture of the evolutionist epistemology which is suggested by these developments. A bridge towards more philosophical considerations is represented by the papers of M. Artigas, *Emergence and Reduction in Morphogenetic Theories*, and of J. Mosterín, *What can we Know about the Universe?* Almost all the salient epistemological problems related to contemporary cosmology are discussed in the following three papers: *The Universe as a Scientific Object* by M. Pauri (who points out several conceptual difficulties in assuming the universe as a scientific object proper); *General Laws of Nature and the Uniqueness of the Universe* by E. Scheibe (who through a strict logical analysis tries to reduce the gap between ordinary physics and cosmology, due to the fact that this is confined to the study of a single system); *The Anthropic Principle and its Epistemological Status in Modern Physical Cosmology* by B. Kanitscheider (who analyzes the different options in the formulations of the Anthropic Principle, showing how they are determined by certain meta-scientific criteria of rationality). The paper of A. Cordero, *Evolutionary Ideas and Contemporary Naturalism* gives a detailed account of the recent trends in philosophy of science and tries to show how classical philosophical constraints on the conception of science and nature hinder rather than help to advance naturalism as a contemporary philosophy. Finally the paper of F. Miró

Quesada, *Origin and Evolution of the Universe and Mankind* is a survey of the general thematic discussed in the book.

Some words now as to the «origin» of this volume: it contains selected papers from an international colloquium held in Lima in August 1989, as the annual meeting of the International Academy of Philosophy of Science. It was then followed by a two weeks summer school on the natural sciences and their philosophy. These two events were organized under the auspices of the Lima University, the Peruvian National Council for Science and Technology, the National Science Foundation (U.S.A.), and various other international organizations, which also provided a considerable part of the necessary funds. The editors are therefore indebted to the institutions and individuals whose cooperation made the Lima colloquium and summer school possible. In particular, they should like to express their gratitude to Professor Francisco Miró Quesada, Director of the Institute for Philosophical Investigations of the Lima University, and to Professor Carlos del Rio, President of the Peruvian National Council for Science and Technology, for their invaluable support. We also thank the University Council of the University of Fribourg for the financial help granted for the technical preparation of this volume, and Bernard Schumacher and Edgardo Castro for their valuable help in the editorial work, and in the realization of the camera-ready copy.

Evandro AGAZZI
President
International Academy of
Philosophy of Science

Alberto CORDERO
Member
International Academy of
Philosophy of Science

Evandro Agazzi

THE UNIVERSE AS A SCIENTIFIC AND PHILOSOPHICAL PROBLEM

1. THE SIGNIFICANCE OF SCIENTIFIC COSMOLOGY

Cosmology used to be a classical subdomain of traditional philosophy: according to a well known partition of the whole of reality into three great spheres (Man, Nature and God), it became customary at a certain time to articulate the so-called systematic philosophy into three branches: anthropology or psychology (the philosophy of Man), natural philosophy or cosmology (the philosophy of Nature), and theology (the philosophy of God).[1] However, this partition became controversial with the development of modern science, since a certain effect of this development was the conviction that at least some of these domains of inquiry had been removed from the competence of philosophy and passed on to the competence of science. This was the case, in particular, with Nature: in fact, as is well known, the very notion of natural philosophy rapidly changed its meaning, in the sense that what became actually understood under this term was the new physical science initiated by Galileo and Newton. Of course, at least until the end of the 18th century things were not perceived in this light: those people whom we now qualify as scientists were considered - and considered themselves - to be natural philosophers; they were supposed to cultivate philosophy (or at least a part of it), and neither to develop an investigation of Nature which would stand besides (or even at variance with) the philosophical investigation

E. Agazzi and A. Cordero (eds.), Philosophy and the Origin and Evolution of the Universe, 1–51.

of Nature, nor to promote a new kind of inquiry, which should replace or delegitimate the philosophical study of Nature. Only in the 19th century, and in particular with the positivistic movement, did «science» and «philosophy» become charged with very distinct and even almost opposite meanings.[2]

Within this perspective, many people considered it to be very obvious that, when a certain field of investigation comes under the competence of science, it ceases to be of concern to philosophy; and they were convinced that a progressive reduction of the domain of philosophy, as a consequence of a continuous enlargement of the domain of the sciences, corresponds to an intrinsic dynamics of progress for mankind. Other people were less radical: they were convinced that philosophy has its peculiar features, which not only keep it distinct from science, but will never allow the elimination or replacement of philosophy by science. However there was a general tendency, even among these people, to admit that science had actually replaced philosophy in the investigation of Nature (in other words, for these people, the most important distinction between science and philosophy was represented by the fact that science extends its competence to the whole of the material world, while philosophy preserves its competence with respect to the spiritual world, which includes in particular Man and God).

However, this separation of domains (which has a quite transparent Cartesian flavour) did not prove satisfactory for opposite reasons. On the one hand, the typical "scientific" approach soon began to be applied also to the study of man under several different points of view, and even the terms «anthropology» and «psychology» were used to indicate particular disciplines among the scientific approaches to man (be it in the older sense of «physical anthropology», or in the newer sense of «cultural anthropology», or in the sense of the different forms of doing "scientific" psychology).[3] On the other hand, the admission that philosophy had delegated the study of Nature entirely to the sciences was rejected by several philosophers, who claimed the legitimacy of a «philosophy of Nature» which was to have a task and a full meaning independently of the «natural science» which had taken the place of the ancient «natural philosophy». In other words, science and philosophy were distinguished not on the ground of their alleged respective fields of inquiry, but on the basis of the different approaches they adopted regarding any field of investigation whatever. In such a way adjectives such as «rational» or «philosophical» were used to qualify the philosophical approach in contradistinction to other approaches (in particular the scientific ones): «philosophical» anthropology and «philosophical» or «rational» psychology were vindicated as legitimate fields of rigorous inquiry (concerning e.g. problems such as that of the existence and immortality of the soul, the existence of free will, the relation between mind and body, etc.), besides «physical» and «cultural» anthropology, or «experimental» psychology, which were characterized by the use of purely descriptive methods not trespassing the level of the phenomenal ascertainment of empirical evidence. In the same vein, problems such as those of determinism or indeterminism, causality, finalism, order and chance, specificity of life with respect to matter, etc., were considered to constitute the subject matter of a «philosophy of Nature», which was to be independent of the empirical and theoretical investigations of the natural sciences proper. In an analogous spirit, one spoke of a «rational theology» as a purely philosophical investigation concerning God, distinct from a «positive» theology, which accepts a given revelation as a basis for its claims.

What is interesting in the development sketched above is the fact that no double qualification emerged (until very recently) for cosmology: while we find philosophy of Nature standing beside the natural sciences, philosophical psychology beside experimental psychology, rational theology beside positive theology, philosophical anthropology beside different forms of scientific anthropology, we do not find something like a scientific cosmology standing beside a philosophical cosmology. This fact is not accidental, and it also tells us that the identification of cosmology and philosophy of Nature (which we have hinted at in the beginning) in the traditional partition of philosophy was actually not so obvious as it might seem at first glance. Indeed, the idea of a philosophy of Nature strictly indicates a kind of philosophical investigation which concerns what "*belongs to Nature*" and, in a way, may be conceived of as the philosophical study of the «material being» (as it was often said), but this does not mean that Nature is being taken "*as a whole*" and considered as an object of inquiry in itself. In other words, a philosophy of Nature considers those beings which have the *property* of being "natural", and tries to single out those general features which are connected with this property; but this does not imply (though neither does it exclude) that Nature itself be considered as a global entity to be submitted *as such* to an additional specific investigation. When this second step is taken, we pass from philosophy of Nature to cosmology, and this transition may be seen either as constituting a part of philosophy of Nature itself, or as an enlargement of it, or even as a special sector of metaphysics, according to different philosophical approaches.

If we keep this distinction in mind, we can understand why philosophy of Nature and cosmology (both being originally parts of philosophy) did not evolve in the same way as regards their relation to modern science. In fact, modern science could be seen as a continuation of the study of "material beings" (which had been the task of natural philosophy) performed under the adoption of different *methodological* prescriptions, the most basic of which were the delimitation of the scope of the inquiry to restricted fields of observable phenomena, and the refraining from metaphysical generalizations. Cosmology, on the other hand, seemed to be intrinsically such that it could not be developed according to these new methodological criteria, since no observational means can be designed for grasping Nature "as a whole", i.e. for identifying the *Universe* as an object of investigation; and in the absence of this possibility, it also appears unavoidable that only very general concepts of a metaphysical flavour can be used in such a kind of discourse. That is, if we want to continue speaking of cosmology, we must keep it within philosophy (and in particular consider it as a special branch of metaphysics), but we cannot hope to transform it into a scientific discipline. This conclusion may be found in Kant, who (according to the classification current at his time) places cosmology - besides rational psychology and rational theology - within metaphysics, and maintains the impossibility of its attaining a *cognitive* status. This is practically equivalent to excluding a priori any possibility of making cosmology "scientific", while admitting the «idea» of the Universe, and the related questions which emerge in philosophical cosmology, as having a regulative role in the pursuit of our investigations concerning natural phenomena.[4]

Of course, this did not prevent scientists from making generalizations, from asking questions and proposing theories which undoubtedly had a cosmological flavour, but this was done within the framework of certain specific scientific disciplines, and without evoking the notion of the universe. Concretely speaking, it was in astronomy and in celestial mechanics that such perspectives were proposed, but even in those cases the cosmological approach was restricted to the limited question of the origin of the astronomical systems from a previously existent unorganized bulk of matter subject to Newtonian forces. This in particular is the case with the famous "cosmogonical hypothesis" of Kant-Laplace, and it is worth noting that the title of the work in which Kant proposes this hypothesis does not contain the term «cosmology», but is expressed simply as the *General Natural History and Theory of Heaven* (1755).[5]

In spite of all this, we see that cosmology is today being presented, and discussed, developed as a scientific discipline. What is the meaning of this historical fact? And in particular, what does it imply as regards the relations between science and philosophy? We can say that it first implies a refinement of the very concept of science, in the sense of confirming the multidimensional, or «analogical», character of this concept, and of stressing certain of its features that had risked becoming neglected in certain recent trends in philosophy of science. A second implication is the impossibility of maintaining that clear-cut separation between science and philosophy that was too optimistically claimed by the positivistic movement, and is still widespread in contemporary culture, in spite of many symptoms which indicate its untenability. We shall proceed now briefly to consider these two aspects.

2. COSMOLOGY AND THE MODEL OF SCIENCE

2.1 The question of the scientificity of cosmology

It is well known that modern cosmology has been the object of a not negligible debate regarding its scientific status, a debate which is in a way still alive, since the possibility of closing it would depend on the universal acceptance of a unique "model" of science, a condition which is far from being realized, and (as we shall see) is also far from being desirable at all. From this point of view, the discussion whether cosmology does or does not deserve to be qualified as science has several aspects in common with the debate - which began a century ago and continued during several decades of the present century - regarding the scientific status of the *sciences humaines* or *Geisteswissenschaften* (i.e. psychology, sociology, history, philology, linguistics, economics, political science and so on), which we shall call «human sciences» here, in spite of this terminology not being common in English. It was sometimes claimed regarding these disciplines that they were still very far from being scientific, because they made practically no use of the concepts and methods of "the sciences" (i.e. of the natural sciences). Accordingly, certain "tolerant" people retained the hope that these disciplines could progress towards (at least partial) scientific status by their increasingly adopting the concepts and methods of the natural sciences, while other less tolerant people maintained that these disciplines could never reach the status of sciences, owing to the peculiarity of the features characterizing the world of man, in which free will, subjectivity and emotions play an ineliminable role.

However, this was not the opinion of the professionals of these disciplines: they were aware that they had developed many methods for the critical accumulation and assessment of data, for the rational elaboration of conjectures, for the rigorous comparison of hypotheses and theories, which allowed - in their domains - for a no less rigorous and objective kind of knowledge than the natural sciences could attain in their fields of investigation. Of course, these methods and criteria were specific, and were fully in keeping with the different aims, the different objects, the different levels of generality, the different degrees of certainty that it is *reasonable* to expect in those domains of inquiry. For these reasons they felt themselves fully entitled to qualify their disciplines as «sciences», without accepting the imperialistic monopolization of this concept by the natural sciences. After all, this monopolization was simply the consequence of having attributed a "normative" purport to an historically contingent priority (i.e. to the fact that the natural sciences had been the first to reach the degree of objectivity and rigour that had been promoted to the role of being the characterizing mark of "science"); a priority, moreover, which could even be challenged: in fact, the philological and historical work of the «humanists» of the 15th and 16th century had already reached such a high "scientific" level that it could be surpassed only by the more sophisticated tools of the German philology of the 19th century. This means that these human sciences were born at least a century before the natural sciences. As a consequence of this vindication, a very articulated and rich epistemological reflection was developed (whose well known lines we shall not mention here), aiming at making explicit the specificity of the human sciences with respect to the natural sciences, and at the same time the legitimacy of recognizing them as sciences.[6] Nowadays almost no one objects to qualifying these disciplines as sciences, and this not just because of a spirit of tolerance, but because it has become more or less clear that the concept of science cannot be rigidly bound to a unique model, but has an *analogical* nature, in the sense of being applicable in different ways to various fields of inquiry, provided the minimal requirements of objectivity and rigour be safeguarded.[7]

In the case of cosmology the situation is partly similar to and partly different from the one just described. The similarity consists in the fact that modern cosmology (as was the case with the "human sciences") is cultivated, on the one hand, by many specialists who consider themselves to be scientists and consider their discipline to be science, while it is challenged, on the other hand, by certain critics as not really being a science. The difference consists in the fact that cosmology did not develop as a science outside the natural sciences, but rather within their framework, and actually as a kind of logical development of some of the most basic theories of modern physics, such as general relativity theory and quantum theory. This means that those who work professionally in this field have more direct reasons for feeling themselves to be within the domain of "science", and it means at the same time that those who criticize the scientific status of cosmology must do so on much deeper methodological grounds. This is actually the case, but (as we have said) we shall not enter the current discussion on the scientific status of cosmology. We shall be content with noting the fact that this discipline has *concretely* attained this kind of consideration (in spite of still existing criticism), that it is developed and discussed by scientists and philosophers of science in scientific books, journals, congresses and publications, using arguments derived from science. The interesting question becomes therefore that of seeing which kind of consequences derive from the fact that cosmology has installed itself among the sciences in our time.

2.2 Cosmology and the standards of physical science

Let us begin with a seemingly secondary remark: modern "scientific" cosmology is among the most significant contributions to the understanding of the nature of science as a *speculative* enterprise, i.e. as an enterprise which primarily aims at saying, understanding and explaining "how things are". This remark is not meant to deny that the scientific enterprise as a whole may also have some additional aims, but is only intended to stress the most radical intellectual interest of science, an interest which not only accounts for its originally being so deeply connected with philosophy, but also for its once again coming into a no less significant contact with philosophy in our time (as we shall see more explicitly in the sequel). Indeed no predictions, no useful applications are expected to come from the investigations of contemporary cosmology, no better fitting to our environment, not even a more efficient way of organizing our future experience. All these criteria, which are usually advocated by the supporters of a more or less explicit instrumentalistic or conventionalistic view of science, fall short of accounting for the growing interest in scientific cosmology, which not only limits itself to saying "how things are", but also takes seriously the even less utilitarian task of trying to say "how things have been" in a far remote and almost unimaginable past time.

This fact already entails certain modifications in some of the methodological criteria which have been considered as basic to every scientific inquiry, especially in the domain of physics. And since modern cosmology has emerged within physics, it can easily be understood that precisely the attenuation of these criteria appears among the objections that are sometimes addressed against the scientific status of cosmology. The most directly affected among these criteria is probably that of *testability*, understood in the strict sense of the possibility of conceiving of an experimental test designed with the view of submitting a hypothesis, or a theory, to an empirical check (a condition which is common both to a confirmationist and to a falsificationist conception of what science should be). This is so because cosmology not only does not propose experiments for the testing of its hypotheses, but it does not even make predictions which could serve as a test (as e.g. astronomy has been able to do for centuries). One could perhaps say that cosmology is nevertheless able to make correct retrodictions, but this is not really the case either: a genuine rectrodiction should consist in indicating something precisely identifiable which should have happened in the past (and which we do not already know to have occurred), and then checking through the empirical evidence of a particular sort (corresponding to some kind of independent record of this event) to see whether it actually happened. Some sciences are able to do this: astronomy, for example, can retrodict celestial events of which we may have independent historical records; geology can retrodict the formation of certain characteristic features of the earth's crust at given points, which we may then be able actually to check by suitable means; many historical disciplines elaborate conjectures and theories where certain past events are postulated (i.e., retrodicted), which have not yet been determined actually to have occurred, but may be checked by consulting archives, by archeological excavations, and so on. Cosmology does not do anything of the sort in a proper sense: the fact that it "retrodicts" the occurrence of the "big bang" several billion years ago does not in fact possess the character of a retrodiction, since no independent empirical evidence of this "event" is at all possible or even thinkable; and the same must be said of all the details of the history of the universe which have been reconstructed by the cosmological theories. A certain exception to this may be the discovery of the 3°K background radiation in interstellar space which, though not having been a real retrodiction of the big-bang theory, may be considered the most significant independent "*empirical*" confirmation of it; but this is of course too little to allow one to claim that cosmology has the features of a "*testable*" scientific investigation in the full sense of this term.

The lack of testability is not as such an insurmountable obstacle to the possibility of recognizing a certain discipline as a science: all "*descriptive*" sciences are in fact characterized by an abundant dimension of empiricity, which may be accompanied by a remarkable observational rigour, a highly sophisticated methodology for the ascertaining of facts, a very detailed and intelligent system of classification, etc., without implying the capacity to perform any real test. The reason this situation is considered acceptable is that no genuine theory appears in these sciences, i.e. no claims that really overstep the horizon of what has been observed and described, so that there is no real need of an empirical check of something which is already totally displayed at the empirical level. However, we certainly cannot say that cosmology is a "*descriptive*" science: in fact, not only is it true that cosmological observations are very scarce and, in general, far less reliable than astronomical records (which are already far less accurate than the measurements of experimental physics), but it is the very nature of cosmology that even the most superficial analysis reveals it as being an extremely *theoretical* enterprise, so theoretical that it could even be qualified as *speculative*. But, if this is so, how can we accept that a theoretical investigation be a science, in spite of having such a poor degree of testability? Why can cosmology be considered a part of science?

2.3 Cosmology and the features of the historical sciences.

The answer is that cosmology may be considered a part of science only if one enlarges upon the criteria of scientificity so as explicitly to include those criteria which characterize the historical sciences. This does not simply mean that, within the large domain of "the sciences", we admit a special sector particularly for the historical sciences, which, as we have seen above, are themselves a subdomain of the "human" sciences (this enlargement has already been accepted for several decades). What is meant here is that the very category of historicity must also be extended so as to include the domain of Nature, with all that is implicit in this methodological decision, a decision which involves much more than the simple superficially evident admission of a process of "becoming" from the past to the present and on to the future. To the extent that we seriously admit this enlargement, we must be ready to accept not only that historiography is (at least in certain respects) a science, but also that science in general is (at least in certain respects) historiography. This admission is not too paradoxical, since the idea of extending a certain kind of historical consideration to Nature is not really new: the idea of a history of the Earth has actually been quite clear at least since the work of Buffon in the 18th century, while any theory of biological evolution can be considered scientific only by including among its conditions of scientificity those of the historical disciplines, as could easily be shown. However, if the introduction of an historical outlook in the field of Nature is not really new, one may say that not all the consequences of the adoption of this approach seem to have been completely analyzed yet.

What seems typical of the historical form of investigation (understood in a very broad sense), is that it cannot avail itself of that kind of "enlargement" of reality which is available to the experimental sciences. In fact, the latter are not confined to describing, understanding and explaining "*available facts*", but can (and in a way must) support their explanations by showing that they can account for "*produced facts*", that are artificially brought to existence through experiments. In this regard we can see what distinguishes the simply *empirical* from the *experimental* sciences: historical sciences belong therefore to the empirical sciences (as do e.g. several descriptive natural sciences), and the feature characteristic of these sciences is that of assigning a very special function to time, namely that of constituting a kind of barrier which in a certain sense makes the objects of inquiry inaccessible, or perhaps better, "non-manipulable". For other empirical but non-experimental sciences, this "barrier" may be constituted by space (e.g. we cannot perform experiments in astronomy because we cannot manipulate celestial objects, already owing to their distance from us, and quite independently of other reasons), or by size and complexity (this being the reason that we cannot perform experiments in sociology or economics). Let us note, incidentally, that this condition of not being able to enlarge the domain of available facts, though being typical of the "descriptive" sciences, does not characterize these sciences alone. In fact, astronomy, sociology and economics (as well as the historical sciences), which we have just mentioned, are not purely descriptive sciences, since they also contain interpretations, explanations and theories. This remark allows us to see that a discipline which is not experimental (because it cannot manipulate its objects and enlarge their domain at will), is not reduced by this to being simply descriptive: it can also well be theoretical.

We are now in a better position to evaluate the real meaning of the untestability of cosmological hyptheses of which we have spoken above. This untestability is undeniable if it is understood in the sense of physics, i.e. as the possibility of performing *expe-*

riments. But cosmological hypotheses and theories may be testable in a weaker sense, one which is in keeping with the much less demanding rigour that is customary in the non-experimental empirical sciences, and in the historical sciences in particular, in which hypotheses and theories must be testable in some way, though on the basis of empirical evidence that is not experimental in a proper sense.

2.4 The question of nomological explanations.

These considerations lead us easily to understand why, in particular, contrary-to-fact statements are (correctly) regarded with much diffidence in history, and are strictly excluded from any "scientific" historical account, while they may have a place in experimental science. The reason usually alleged is that one cannot change what has already happened, and test what other outcome might have occurred, while in experimental science it is often possible to produce new conditions, different (i.e. "contrary" in this special sense) from those already realized in some given concrete case, and to see what actually follows. But is this the real reason? It is so only to a certain extent. Indeed, contrary-to-fact conditionals are the typical way of expressing nomological claims and distinguishing them from purely accidental generalizations. Therefore, when we say that historical sciences refrain, or even must refrain, from uttering contrary-to-fact statements, we are left with two possibilities. One is that we say this because we think that there are no laws governing historical facts, or (to put it differently) that historical events are so complex that we must consider them as being practically contingent and accidental, so that historiography should actually reduce to a descriptive science. The second possibility is that we believe in the existence of historical laws, or at least in the possibility of giving causal explanations of historical events, but consider it methodologically unsound to translate our explanations or theories into contrary-to-fact historical statements, simply because they are *in principle* excluded from being testable, since their refer to an inaccessible past. It seems reasonable to think that all the historians who conceive of their discipline as not being purely descriptive, more or less tacitly subscribe to this second position.

The question that we have hinted at here concerning historiography has its expected parallel in the case of cosmology. In fact, an aspect of this discipline which has sometimes been attacked is that it cannot discriminate between nomological and contingent features of the universe, since to do this it should be necessary to have at one's disposal different universes for testing contrary-to-fact statements about them. (Of course the objection is not formulated in exactly these terms, but it may be suitably reformulated in this way). This criticism is rather strong, since it is intended to show that cosmology may at best be a descriptive discipline, in which therefore no real explanations are possible, and for that reason it would not deserve the qualification of being a science, since nomological statements (and moreover the possibility of discriminating what is nomological from what is accidental) are considered to be basic conditions for scientificity. How can this objection be met? A first possibility might be to stress the affinity between cosmology and historiography, while at the same time relying upon the admission that the historical sciences really are sciences, without their having to be nomological (a well known position which has been widely advocated by eminent scholars). In this spirit cosmology might remain scientific without being reduced to a purely descriptive discipline, in a way similar to the historical sciences, which are more than just descriptive and can provide explanations which are not nomological in the strict sense of physics. However this position is not easy to accept for cosmology, since it has grown out of physics, and in this sense its paradigm of scientificity remains basically that of physics, including nomological explanation and theorizing as characteristic marks. In fact, all our cosmological theories share the ambition of presenting themselves as coherent applications of the best among the existing strictly nomological physical theories, so that in a way they may claim to be nomological in character. However physical theories never concern *the universe*, but only certain well delimited kinds of phenomena occurring *in the universe*, so that the methodological question at issue here is, first, whether nomological reasonings concerning separate kinds of phenomena occurring *in* the universe may lead to nomological statements regarding *the* universe itself, conceived as a whole. Secondly, the question has to be asked whether the very concept of the universe, which does not occur in any of the physical theories used as tools by cosmology, itself has a scientifically definable status.

Indeed, the first of these questions might receive a positive answer within the analytic-reductionistic way of thinking, which is still widespread nowadays, according to which knowledge of the "parts" or "components" is sufficient ("in principle"!) for explaining the properties of the "whole". This position is very weak (since it overlooks the fact that, at best, the analysis of the combination of the parts in the whole remains at a level of *description* and does not reach that of an *explanation*, let alone that of a really nomological explanation); however we shall not criticize this position now but admit it for the sake of the argument. Now, in the usual cases of reduction, the "whole", whose alleged explanation would be possible starting from some nomological theory of its parts, is independently given a meaning, and also an empirically identifiable reference, within some other science which, moreover, is able to bring to light certain laws or regularities concerning it. For example, one may believe that the *biological* properties of an organism can be explained through a chemical or physical theory of its cells, but the organism, the cells, and their biological properties (though not corresponding to any concept available in chemistry) have a precise configuration and are satisfactorily described in biology. The same may be repeated e.g. for the empirical laws and properties of a gas, when it constitutes the "whole" with respect to which the kinetic theory of gases tries to give a nomological explanation starting from the hypothetical model of the molecular disorder of its "parts". Now, in the case of cosmology, we hardly find any conceptualization of the basic concept of the universe; even less can we rely upon empirical criteria for assigning a referent to this concept. Therefore we cannot say which are the properties, regularities or possible laws of this object, of which we could eventually provide a nomological explanation using the physical theories which concern what happens "in the universe".

Does all this deprive cosmology of any sense? Not necessarily, but it indicates that this sense cannot be entirely justified on the basis of those criteria which are customary in the physical sciences. At this point the analogy with history may again be useful: it is perhaps not impossible to identify regularities or "laws" of several partial aspects of historical events (e.g. of economic, social, psychological, military aspects or "factors" of human history); however this would not imply either that there are "laws of history" in general, or that these laws are themselves obtainable starting from those of the different aspects involved in the historical events. Moreover, the very concept of history cannot be reconstructed from those of the partial sciences used in historical work (such as economics, sociology, psychology, etc.) but is rather presupposed as a general framework which is fairly understandable, because it corresponds to one of the basic elements in the human approach to reality (the approach consisting in regarding the course of human events in time "as a whole"), an element which also determines rather well the kind of intellectual interest and the criteria of investigation that permit the determination of an "historical science".

For cosmology we can repeat something similar: it is possible to identify regularities and laws regarding many aspects of what happens "in the universe", and to submit them to nomological explanation, but this does not imply that there are "laws of the universe", or that these laws, if they exist at all, are obtainable starting from those of these different aspects. Moreover, the concept of the universe, far from being reconstructed from concepts which characterize the nomological theories on which cosmology relies, is rather presupposed as a general conceptual framework, which is sufficiently understandable since it corresponds to the spontaneous human approach consisting in considering the "external world" as a whole, and in trying to locate man himself with respect to this "external world". This attitude determines the criteria of investigation of any possible cosmology (such criteria must consist in the effort to push all considerations to the widest "boundaries" possible); but this very fact already indicates that the tools offered by the usual empirical sciences may not be sufficient for completely characterizing cosmological concepts.

If, after all our reflections, we consider what really happens in cosmology, we must say that little awareness appears there of the peculiar features we have outlined. In fact the participation of cosmology in the field of the historical sciences remains practically unperceived, as is demonstrated by several considerations. For example, the preoccupation of extending nomological explanation to all issues having a real cosmological relevance is very strong, and it characteristically leads to the more or less explicit emergence of contrary-to-fact conditional statements. Such statements are even explicitly formulated in many discussions concerning the «anthropic principle,» where it is very customary to read: «If the values of such and such physical constants had been even slightly different from what they are, then the universe would have been so different that no life and no conscious beings would exist in it.» This statement, taken by itself, is really unproblematic, since it simply expresses a (more or less reliable) nomological conclusion from the existing physico-chemical-biological theories, according to which the existence of living organisms is possible only under certain physical conditions, that in turn are possible only if the values of certain physical constants remain within given very restricted intervals. Already at this stage we find a difference with respect to the ordinary historical sciences, since in these sciences, as we have already noted, it is not considered to be serious to say, e.g.: «if Cleopatra's nose had been just a little longer, the course of history would have been different», or: «if Hitler had not been born, the history of our century would have been totally different». In the usual historical sciences such contrary-to-fact conditionals are discarded for essentially two reasons: the first is that we do not rely upon nomological relations that would deterministically connect Cleopatra or Hitler to the course of history, so that we can still believe (as many actually do) that the course of history would have been "essentially" the same even if these two individuals had not existed (only "accidental details" would have been different). The second reason is that we cannot actually have at our disposal another world, in which these two individuals did not exist, since we must be content with the study of the contingent world in which human events have occurred and continue to occur.

2.5 The problem of contingency.

Compared with this situation, cosmology does not suffer from the first objection, since it works within a conceptual framework in which the course of cosmological events, so to speak, is nomologically connected with those initial conditions which may be constituted e.g. by the value of certain physical constants; and in this way it would not make sense to claim that «life would have appeared *anyway* in the universe.» This is why contrary-to-fact conditionals taken in this first sense are perfectly legitimate in cosmology . But now what about the second objection? This objection, as we have seen, depends on the fact that it makes no sense to try to find an explanation of something which is in itself contingent. Does cosmology accept this attitude? Apparently it does not, since there is a general tendency toward explaining why the said values of the physical constants (or a lot of other features which conceptually play the same role) are exactly what they are. In other words, nomological reasoning, besides being applied *within* the framework of the universe for explaining certain particular features of it (e.g. life) starting from *given* suitable initial conditions, is applied to explain these very conditions, i.e. to explain the general framework and hence, in a way, the universe as a whole.

But one must be conscious, then, that this actually means removing *contingency* from the universe as a whole. In fact, nomological reasoning is tantamount to proving some kind of necessity within a certain framework (the framework of a given scientific theory), and this necessity does not normally receive a metaphysical meaning because, first of all, it does not imply that the scientific framework adopted be in itself necessary, i.e. the only possible or correct one, and because the initial conditions to which the framework is applied are not thought of as being necessary either. But if we now try to explain the boundary conditions of the universe, i.e. to eliminate their contingency, how could we do so without relying upon a non-contingent ultimate framework, i.e. without relying upon what is practically speaking metaphysical knowledge in the proper sense of this term?

A possible way of escaping this consequence is to admit the real contingency of our universe, while at the same time trying to include it within a more general framework of necessity. This seems to be the idea behind the recent theories in cosmology which assume the existence of a plurality of worlds or universes besides ours: each of these worlds would realize the physical situation corresponding to one of the possible combinations of values of the basic physical constants (to put it briefly), and among these infinite worlds there also exists ours, which in such a way would no longer constitute a contingent and in this sense unexplicable realization of an extremely improbable structure, but simply one among the necessary realizations of what physical laws impose to be also actually realized. It is clear that this interpretation of the plurality of universes avoids the "metaphysical" assumption that one particular set of boundary conditions is so to speak "necessary in itself", but it does this by making the no less metaphysical assumption that whatever is logically compatible with our present physical theories did actually happen, even though we can have real information only about that particular universe in which conscious beings could appear.

The situation is very different when a plurality of universes is considered only in the sense of a logical display of many different alternatives, either for developing the logical consequences of admitting certain hypotheses, or for making statistical evaluations, or for making computer simulations of different "models" of the universe. In all these cases we are in the presence of something which is the continuation of very familiar scientific patterns of investigation, such as hypothetical constructions, thought experiments and the elaboration of models. But in these cases the problem of giving a nomological explanation of the actual form of the universe as a whole (including therefore its origin and evolution) cannot really be solved, and the historical flavour of cosmology surfaces again, in contrast to its being rooted in physical theories. In particular, the special role of time in the historical approach appears to be of particular importance.

3. THE HERMENEUTIC DIMENSION OF COSMOLOGY

As we have already remarked (and as is very widely recognized), modern cosmology did not develop (contrary to what is commonly admitted to be the case for other empirical sciences) in an effort to provide an acceptable explanation of a set of specific empirical facts (empirical evidence regarding the nature of the universe as a whole is indeed hardly thinkable), but rather as the desire to explore certain remote consequences of general relativity theory. It was only because Einstein's equations left different alternatives open that certain empirical evidence was also taken into consideration (besides other theoretical factors), precisely in order to make a choice among the alternatives. In this sense cosmology has, in common with history, the characteristic of being a *hermeneutic,* more than an *explanatory*, enterprise. It tries to "interpret" the sense of certain great physical theories, extrapolating their consequences so as to suggest a global scenario and even a set of particular events which are completely beyond any possible empirical test, for the good reason that they belong to the *past*. Of course, since cosmology performs its hermeneutic task within the intellectual framework of natural science (and of physics in particular), the resulting picture is judged according to the criteria of acceptability of physical theories. This implies two things: first that what is unsatisfactory from the point of view of physics is also considered unsatisfactory in cosmological theories; second, that all the resources provided by physical theories (besides specific moves of a more theoretical nature) are used for removing difficulties or improving the picture in cosmology. This in particular accounts for the spectacular symbiosis of cosmology with high-energy physics which began about 1970.

The meaning of all this is that the ground for the "interpretations" which constitute the core of cosmological theories is offered by the knowledge available at a given time (particularly, but not only, in physics), and this is as it should be. However, as we have already stressed, this knowledge is constituted not so much by empirical evidence as by certain physical theories, or even certain more specialized and advanced parts of them. These theories, or parts of theories, are accepted because they are regarded as reliable due to their being empirically well confirmed; and their use consists in their results being extrapolated back into the past in an untestable way. This passage requires critical evaluation for different reasons.

4. HOW COSMOLOGY AVAILS ITSELF OF PHYSICAL THEORIES.

On the one hand, one must recognize that the said procedure is fully in keeping with the most customary way of using existing scientific theories in pursuing new lines of investigation. However, in the case of cosmology this common way of proceeding seems to be employed under very special conditions. In fact, in the "normal" cases where existing theories are used to investigate new domains, they are certainly "assumed", but at the same time (even when they are assumed with the highest degree of confidence) it is also understood that they could be "falsified" in the course of the new investigation, which therefore also constitutes an additional test of them. In cosmology, on the other hand, they are rather "taken for granted", in the sense that cosmological theories develop as kinds of "intellectual applications" of such very general physical theories, affording no real opportunity to reject or modify them. We are not unaware that in certain recent proposals advanced in cosmology the attitude has surfaced to accept at least a relaxation of certain basic principles of ordinary physics, in order to solve particular theoretical difficulties (think e.g. of the possibility of avoiding gravitational collapse - which is one of the main difficulties of gravitational physics no less than of the standard big bang model - by relaxing the energy condition and admitting antigravity). Still these seem to be *ad hoc* modifications of a rather speculative nature, that do not rely upon independent empirical and theoretical support.

This does not at all mean that *cosmological* theories are not subject to criticism or "falsification". It simply means that these theories must *explicitly* accept as an untouchable basis not only something which we could call "astrophysical evidence" of an *empirical* nature, but also a "a priori knowledge" of a *theoretical* nature. In such a way cosmological theories may be compared and discussed on the basis of their ability consistently to cope not only with their specific empirical evidence (as is the case with many ordinary scientific theories), but also with the theoretical content of *other* theories.

This situation is not uncommon in science, and the kinetic theory of gases may be seen as an example where classical particle mechanics was no less "taken for granted" than where the experimental laws of the specific field (i.e. the gas laws) in order to develop acceptable theories of gases. One may also say (as people such as Hübner, Shapere and myself have stressed several times) that this reliance upon historically available knowledge (including theoretical knowledge) is the normal condition for the development of science. However, in the case of cosmology this merging of different available theories seems more problematic, since it proceeds counter to the recognition of the specificity, partiality and relativity of the scientific concepts and principles which has been the most decisive mark of modern science since the time of Galileo. In fact it may be fascinating to think of a "marriage" of general relativity and quantum theory, but for the moment we do not have a relativistic quantum theory of gravity, and the concepts we use have at least operationally and theoretically different meanings depending on the theory from which they are taken, so that their precise meaning in cosmology risks being lost. What sense can it really make to speak of time intervals which may be of the order of 10^{-30} sec at the occasion of the big bang, and of several billions of years when we speak of the formation of galaxies and stars? What sense can it make to speak of distances which may be of the order of 10^{-30} cm (like the original coherent microscopic region from which our visible universe arose according to the "inflationary" cosmology), and of the almost infinite distances of galactic and extragalactic systems, which should separate the parts of our present universe?

A similar discourse may obviously be repeated for all the physical magnitudes occurring in cosmological theories. In other words, these concepts are used as if they had a uniform meaning throughout cosmology, while in fact there are parts of cosmology where they have, let us say, the microcontextual meaning of quantum theory, parts where they have the macrocontextual meaning of general relativity, while at the same time the cosmological empirical evidence is expressed in terms of classical mechanics, or optics, or electrodynamics. This feature is not really comparable with that of those cases in which we start with concepts of a given theory and then try to define operationally and theoretically homonymous concepts (with a more or less different meaning) of another theory. In the case of cosmology, the fundamental concepts seem to have a sort of "general" meaning, which is not exactly that used in the specific partial theories which cosmology uses, and at the same time seems to be such as to encompass these particular meanings. Is this a methodological flow? Probably not, but only on the condition that we are ready to grant a status to cosmology which is not exactly that of the other physical sciences, but is closer to the status attributed to philosophy. There concepts are allowed a degree of generality which is practically unrestricted, which is not "constructed" by a formal definition or a definite set of operational procedures, but emerges from the different meanings present in human discourse as a kind of common connotation that must be "interpreted" from them.

The legitimacy of this way of proceeding (i.e. of again adopting a hermeneutic attitude) may be granted to cosmology due to the fact that it takes the universe as its field of investigation; on the one hand the universe cannot have the characteristics of any scientific object proper, and on the other hand it cannot even be considered as the set of all scientific objects. The reason is that scientific objects are not "things", but are the referents of certain specific sets of predicates (and this is why they can be precisely defined), so that speaking of the totality of scientific objects would actually mean speaking of the totality of the scientifically definable predicates or properties. Now it is certainly impossible to define or delimit an object having at the same time all these properties (simply because this totality is indefinite and cannot therefore lead to any definition), but it is not absurd to envisage an open horizon in which all these properties may be intellectually included or rather "available", and be used when they prove useful and applicable to the question at issue. This is what happens in cosmology, and it is the reason for cosmology's being related to philosophy no less than to physics and (as we have seen) to historiography.

Let us return now to the reasoning outlined above, where we remarked that, in the "normal" cases, when we assume already established theories for investigating a new field, we are open, on the one hand, to recognizing possible falsifications of these theories in the new field (as happened e.g. when it was realized that classical mechanics was unable to cope with classical electrodynamics, when efforts were made to construct mechanical models of the electromagnetic field). But, on the other hand, even if we take these theories for granted (which means that they are considered to be immune from falsification), this does not immunize the new theory which is built using them as a basis (as was the case, e.g., when particular theories of gases were discarded on the ground of specific falsifications in spite of being constructed on the basis of an "assumed" classical particle dynamics). Now, in the case of cosmology, there is at least some suspicion that this immunization of the auxiliary theories may also pass over to the cosmological theories themselves, perhaps (as certain philosophers such as Hübner have maintained) owing to the intervention of implicit a priori presuppositions which are involved in them. In order to investigate this question, we must look more closely at the sense in which the extrapolation of the results of existing physical theories into the past (which is typical of cosmology as we have already remarked) is untestable, since this does not concern the further testability of these "auxiliary" theories (which we have accepted to take for granted), but of the cosmological theories themselves.

5. THE TESTABILITY OF COSMOLOGICAL THEORIES

The untestability in question depends in the first place (as we have already discussed) on the fact that we cannot experience past events, while very many claims of cosmological theories concern just such events. In this regard we have indicated an affinity of cosmology with historiography, but now we must say something more: this affinity has its roots in the fact that the time of cosmology is not the time of physics, but rather the time of history (or, to be more exact, the concept of time has several meanings in cosmology: sometimes it is the time of microphysics, sometimes it is the space-time of relativity, sometimes it is the time of history). The most direct indication of this fact is that in cosmology we speak of past, present and future: these are by no means connotations of physical time, but of psychological and existential time, that are then extended to the historical time. It is only if we take this into consideration that we can understand why the past is not "available" for the performing of experiments. However we shall return to the problem of time in cosmology at a later point.

A second no less significant reason for this untestability of cosmological claims concerning past events is that, owing to the impossibility of constructing experimental set-ups in the past, we cannot take possible combinations of accidental occurrences into account, and *remove* such sources of uncertainty, as we can in normal experimental practice. To see this point more clearly, we can consider "projection into the future" (i.e. prediction), rather than "projection into the past" (retrodiction). When we use existing physical theories to make predictions, we may hope to be successful (and also to be entitled to use such predictions as testing procedures for our theories or hypotheses) only if we can realize a sophisticated experimental set-up, thanks to which a very well delimited and idealized situation is artificially constructed, which corresponds to the neutralization of all possible accidental disturbances, and can therefore put the theory to a test. But everyone knows that we can predict "normal" more complex facts, which depend mainly (but not totally) on our intervention, only within a certain margin of error, and that this margin becomes even wider when we are dealing with "normal" facts on which we cannot intervene, especially when they become more complex. Weather forecasting is an obvious example, but other examples may be mentioned: e.g. we are extremely interested in the possibility of predicting earthquakes, volcanic eruptions, droughts or inundations, but we feel almost impotent, not because we think we do not have sufficiently sophisticated physical theories, but because we know that too many accidental circumstances intervene in such complex phenomena. Even more problematic, and perhaps nearly meaningless, are such questions as whether the Sahara will be transformed into a tropical forest in 2000 years, or whether new vertebrates besides mammals will be produced by biological evolution in the coming two billion years. This shows that the difficulty does not disappear when we pass, say, from isolated or individual events to more global or general features, or when we pass from short to large time scales: contingency operates at all levels.

What is suprising, in view of what we have noted, is that, while forecasting the future appears to us very hazardous and rather close to science fiction (in spite of being grounded on solid empirical evidence of the present situation and on scientific theories which we believe to be very sound), we are rather inclined to take much more seriously reconstructions of the past history of the world (such as those of the theory of evolution or cosmology) based on the same grounds. But this attitude is by no means more rational: telling "how things have been" is not at all easier than telling "how things will be", if doing so requires essentially theoretical work. We may have a different impression, because we know that the past has already been, and we may also have some records of it. But unfortunately what we want to describe is exactly what does not appear in those records, and therefore we must simply conjecture how it "might have been", no less than when we conjecture what "might one day be".

6. THE A PRIORI ASSUMPTIONS OF COSMOLOGY

Let us now consider other *a priori* assumptions of a different kind, which are indispensable when we try to speak of the past history of the universe, and in particular of its origins. The first of these assumptions is that we implicitly postulate that physical laws and principles, which we consider valid at present, were also valid in the past. At a first glance, this presupposition seems to be the same as is made in usual science, when we tacitly assume the invariance with respect to time of our laws and principles (in fact no scientific inquiry would be possible without such an epistemological presupposition). However we want to lay stress on the different kind of foundation this presuppostion receives in the two cases. In our "usual" practice - as we have been told by a long empiricist tradition - we are led to be confident that a certain law or principle "having been found valid in the past" should continue to be valid also in the future. But in the case of retrodicting the history of the universe, we cannot repeat this claim, since that these laws and principles "have been valid in the past" is just the question at issue. Therefore all the weakness of the empiricist foundation emerges here: the real situation is that we make such claims not on the basis of an experience of a necessarily very short duration (which can actually be qualified as a "present experience"), but on some other *a priori ground*, the most common of which has traditionally been the belief in the *uniformity of Nature*. In cosmology this presupposition of the uniformity of Nature is disguised in the form of different isotropy and homogeneity postulates, of which it would certainly be arbitrary to say that they are empirically grounded, since the empirical evidence is always very limited and, if we should take it seriously, it would rather speak against isotropy. Only by introducing very strong idealizations we can postulate these isotropies.

This is not illegitimate, provided we are aware that we are introducing in this way certain "conditions of intelligibility" which are of a philosophical nature and cannot be justified on purely physical grounds. Indeed, they move in a way in a direction contrary to that of modern physical science, which has been characterized by an ever-greater stressing of the *locality* of definitions and laws, owing to their operational anchorage. The explicit recovery of these requirements of isotropy and homogeneity (which were implicitly admitted in traditional physics) is an interesting confirmation of the philosophical horizon of generality which comes, in cosmology, to complement purely physical ways of reasoning.

7. TIME IN COSMOLOGY

Discussing even the most fundamental of the said isotropy or homogeneity postulates would lead us too far afield, while it may be useful briefly to discuss the question of time, which shows many features in common with this kind of problem. It is nowadays uncontroversial that the *physical definition* of time is necessarily local, and this because time itself, no less than space, may become a physical magnitude, and this depends on the possibility of performing certain concrete measurement operations. Practically speaking, the determination of a temporal congruence always requires the singling out of certain relations in a natural process which are endowed with some kind of periodicity, so that no possibility exists of measuring a "pure" time. The importance of this fact is so well known, that it has been considered as the deepest epistemological lesson of Einstein's relativity, and as the clearest foundation of the operationalistic conceptual approach of contemporary science (as P.W. Bridgman has especially emphasized[8]). But not only does contemporary physics discard the notion of "pure" time. It has also given up the idea of a "universal" time, i.e. of a time that could be defined by enlarging or generalizing the local time of the original physical definition. It is again the theory of relativity which has shown that "global simultaneities" do not exist, and therefore that no universal time exists either: in fact this would be the equivalent of the Newtonian absolute time, the elimination of which is usually presented as one of the most astonishing and revolutionary conquests of contemporary science, whose merit goes to Einstein's theory. But now it is clear that without such a universal time, any discourse on the past, present and future becomes meaningless (and this is also a well known consequence).

In the face of all that we must however recognize that cosmology makes use of the concept of *cosmic time*, which is essentially an equivalent of the old absolute time, and in particular it allows one to speak of the past, present and future of the universe. How is this possible for a science which takes its roots in general relativity theory? The difficulty may disappear, or at least be drastically reduced, if we admit that this cosmic time is not introduced by means of a very problematic "extension" of local time, which would go against GRT, but is rather the expression of the fact that local time is itself an operational "cutting" which is performed within a *general* time that is already "known" in some way, and which constitutes the intellectual presupposition for understanding our experience and making a representation of it. In other words, physical time is the result of a scientific specification of a more general *philosophical* concept of time, and this specification may give rise to a local time when it is operationalized, and to cosmic time when it is assumed "heuristically" in cosmology in order to speak of the universe and its evolution. Let us remark that something very similar happens with the "substratum" of the universe, which has an important role in cosmology, but at the same time is in striking contrast with one of the central results of GTR, i.e. Einstein's celebrated dissolution of every privileged observer, which entails the dissolution of any chronogeometric substratum.

As we have already remarked above, the complications with time in cosmology are increased by the fact that not only relativistic, but also quantum theoretical treatments of time are involved in it. In particular, this leads to certain speculations concerning the "origin" of time itself, and even the pretension that the universe "was" a unique micro-object "when" (i.e. "at the time that") no physically conceivable time "yet" existed. To say that these ways of speaking are metaphorical is the least one should do, but this is not enough, since one should specify whether we can dismiss the metaphor and speak clearly and explicitly. This does not seem to be the case if we pretend to rely upon a scientifically circumscribed definition of time, while it is possible if we accept a general philosophical notion of time, in which homogeneity and directionality are included.

One might object that the existence of cosmic time is an acceptable scientific *hypothesis*, if we admit the existence of Robertson-Walker spaces (which in turn seem to fit in well with the scientifically detectable features of our universe). However this is not a convincing objection, since the existence of cosmic time was admitted even when such a possibility of hypothetically justifying it had not been discovered, and this simply because without admitting cosmic time it is plainly meaningless to speak of any kind of cosmic evolution. For this reason we can maintain that cosmologists would continue to use this notion even if it should turn out that there is some error in the abstract reasonings which permit one to hypothetically justify something which may be interpreted as cosmic time from a special model of space (the Robertson-Walker model). And the important aspect is just this *interpretation*: indeed it is only if we already have this notion of a cosmic time that we may be able to "recognize" it under the formal conditions expressed by the perhaps abstruse results of certain mathematical requirements (i.e. as the existence of a unique foliation of spacetime by Cauchy hypersurfaces of constant mean extrinsic curvature!)

Let us now briefly consider how this notion of a "general" time functions intrinsically within those model in which the uniformity of Nature is presupposed, of which we have spoken above. One reason that the presupposition of the uniformity of Nature is problematic is that our physical laws and principles do not concern Nature as a whole, but something which we could qualify as particular aspects or constituents of Nature, such as mass, energy, charge, fields, etc., so that what we should assume to be invariant are the laws and principles concerning these aspects or constituents. But the difficulty here is that, when we come to the "origins", these laws and principles should in a way be valid before the coming into existence of those entities to which they apply, if we really were to infer the origin of these entities from the validity of these laws and principles. A possible way of avoiding this difficulty is that of admitting that these laws and principles have an atemporal status, in the sense that they regulate the course of events occurring in time, without being time-dependent themselves. This is a possible move, but one should be aware that in so doing one is actually thinking of an absolute time with respect to which these laws and principles are invariant, and which is different from the time which appears as a parameter in the formulation of these very laws, and is "interpreted" by them as being related in certain ways to space, matter, energy, etc. We seem to require an atemporal space-time background, in order to make a theory of "physical" space-time. In order to eliminate the contradiction from this way of reasoning, we must admit that these "times" belong to two different levels of cognition, that one of them is in a way metatheoretical with respect to the other. One way of providing these claims with a sufficiently understandable meaning is to recognize that the general meta-discourse in which the most fundamental time occurs is that of the philosophical background which is presupposed, as a condition of intelligibility, by every more specific and partial scientific discourse.

8. THE PHILOSOPHICAL FEATURES OF COSMOLOGY

It would not be difficult to mention other issues of a similar nature, but what we have said is enough for the purpose of our discussion. Let us then ask: does all this mean that cosmological investigations, and in particular investigations regarding the origin of the universe, are self-defeating, circular or incorrect? Not at all. It simply corresponds to the fact that when tackling such questions we are confronted with the particular problems bound to the "point of view of the whole", a point of view that science usually tends to remove, but which has an almost invincible tendency to reappear, simply because the whole inevitably constitutes the background and framework of all our knowledge. In fact, we can never have a single experience without conceiving it as being part of a whole: we have an awareness of this experience as constituting a certain focussing on a detail which belongs to a much broader structure, within which it occupies a certain place (in a very broad sense of this term). It is the presence of this global horizon which provides us with the possibility of making the single items of our experience intelligible.

Modern science has been characterized, among other ways, by the fact that it puts limitations on this aspiration towards the whole, through a double attitude. Firstly, by claiming that a correct and satisfactory knowledge of the details, of the parts, of the particular aspects, may be attained without presupposing an understanding of the whole. We may call this the "Galilean prerequisite", since it was Galileo who explicitly proposed the delimitation of natural investigation to a restricted set of empirically ascertainable features or properties of things and, moreover, introduced a kind of tripartition among an observed physical system, the observer, and the rest of the world. He assumed that a satisfactory and objective knowledge of a physical system was possible without taking into account either the observer, or the rest of the world (which in this context may mean the "whole" which is considered to be irrelevant to the scientific investigation of an isolated system). Here, by the way, it is possible to see the origin of the principle of locality, which remained basic to physics until the advent of quantum theory.

Secondly, modern science assumed that, when the consideration of complex systems is needed (systems which in this sense may be considered as being "wholes" with respect to their "parts" or constituents), an understanding and an explanation of the whole simply consists in an understanding of the functioning of its parts or constituents: it will "result" from their combination, according to the laws and principles regulating them.

It is probably correct to say that the combination of these two attitudes constitutes the substance of the analytic approach. Under the first point of view it makes the intellectual problematization of the whole completely illegitimate, while under the second point of view it gives to this enterprise a limited legitimacy by indicating the particular conditions for its pursuit. Cosmology is in a way the expression of this kind of legitimation, and of the conditions under which it is admitted to be scientifically acceptable. In particular, cosmology is an expression of that methodological choice which we have already stressed above, i.e. that of availing itself of existing physical theories (which are necessarily "partial" and even "local" in the already specified sense), and trying to draw from them a satisfactory theory of that "whole" which is the universe. However (as we have remarked on that occasion) this strategy may at best provide a certain description and hermeneutic "comprehension" of this whole, without leading to a genuine nomological theory of it.

But the problems which we have just discussed above indicate that even this enterprise cannot be pursued without the intervention of some *a priori* element, and this is a symptom of the fact just mentioned, that the whole is "already there" when we engage in acquiring knowledge, and that we cannot help developing our knowledge according to our conceptions of this whole, which are implicit even in our analytic work.

Since philosophy is typically the intellectual attitude which aims at coping with this effort to understand the whole, it is not difficult to see that the presuppostions we have indicated are typical philosophical presuppostions, and that they play an especially important role when science is willing to confront itself with the problem of understanding the whole. For that reason we must say that the very concept of the *Universe* is a typical philosophical concept, and the fact that science has brought it under its scrutiny necessarily brings science to that interplay with philosophy, which it had known at the beginning, but which has been thought to have been dismissed in more recent times.

A confirmation of the above general remarks comes from several particular considerations. The first is the well known difficulty of defining the universe as an "object", a condition which seems indispensable in order for cosmology to be a "science of something", but one which is hard to fulfill. If we take a definition of the sort introduced e.g. by Bondi,[9] according to which «the Universe...is the largest set of all physically significant objects,» we meet logical problems and paradoxical consequences. If we try to escape them by defining the universe as the «reference set of a certain cosmological model»,[10] we avoid formal inconsistencies, but are immediately confronted with the problem of knowing whether our model really captures the whole, i.e. the universe, or whether it is only the model of a very large and complex structure of objects belonging to the universe, but not exhausting the universe itself. In other words the idea of unity and totality which is the truly formal characteristic of the concept of the universe (in the classical sense of "formal") seems to be too primitive to be captured by explicit scientific definitions. We have already hinted at this fact, and have also hinted at what seems to be the fundamental reason for it, i.e. that a very special kind of "infinity" is involved here. In fact one's spontaneous impression could be that the logical difficulty in conceiving of the universe as the "totality of objects" resides in the impossibility of even theoretically encompassing such a totality, which is at least indefinite, if not actually infinite. However this is not the major problem: when we speak of an object of inquiry, we mean a "thing" considered under a specific "point of view",[11] and since the points of view from which a thing may be considered are potentially infinite, we see that if the universe were to contain the totality of "objects", it would already be potentially infinite even if it contained just one thing. Of course, if it is meant to contain at the same time all things and all objects, the problem of grasping it becomes truly desperate.

Moreover, the consideration of this "whole" implies the violation of the "Galilean prerequisite" indicated above, since not only the single physical systems, but also the observer and the rest of the world must necessarily be included in the universe if it is really to be "the whole". This seems to be actually fulfilled by contemporary cosmology, which (tacitly) includes in particular a theory of the origin of life and the usual evolutionary theories of the origin of man under its umbrella. However this is no answer to our difficulty, since this amounts to including other people among the objects of inquiry, but does not eliminate the peculiarity of the "observer" and of its subjectivity, which remains hidden in the most delicate epistemological features of cosmology. The most notable of these features is the use of the concept of historical time, involving the notions of past, present and future, which are not objective in the sense of being intrinsic to the objects, but have a meaning only with reference to a conscious subject who says «now.». If one considers this point attentively, one might think that the anthropic principle, which introduces the consideration of human consciousness into cosmology in a way which seems gratuitous and extrinsic to many scholars, is perhaps far from being so alien to the epistemological background of this discipline.

9. BOUNDARIES WITH METAPHYSICS

The problems we have tackled lead us to ask the general question: «How can one try to understand the "whole"?» Since the whole is obviously not the content of any possible experience, but, as we have already seen, the implicit background which makes all experience possible, it cannot be the object of any direct empirical investigation. This remark seems definitely to condemn cosmology to providing pseudo-knowledge. However if understanding of the whole were to be conceived this way, not only cosmology, but also philosophy would be condemned as an illusory enterprise. But this is not so, simply because the whole is not to be understood as the "totality of the real", i.e. as the actual cumulative set of all existing individuals, with all their features and properties. It is obvious that if philosophy were to pretend to know the totality of the real, it would deserve the scorning irony with which its enemies often consider it. But philosophers are not unintelligent, ambitious people who believe themselves to be able to know all things and all properties of things by simply speculating or reading other philosophers books. Philosophy tries rather to understand the "whole" in the sense of elaborating and analyzing those most general concepts and principles of reality which could help us to understand it in the diversity of all its aspects, and this hopefully by finding *a few* concepts and principles of *great generality*, which is not a foolish programme, after all.

How this task can be carried out is among the most debated issues of philosophy, and we shall certainly not take it up here. Rather, we shall content ourselves with saying that a solution of this issue which we are inclined to accept, and which is also well in keeping with scientific practice, is that this investigation of the whole is realized as an ideal reconstruction which starts from actual experience, and generalizes principles and criteria of intelligibility which prove successful in some restricted field, but at the same time seem to be successful *because* they have a universal purport. (Whether these principles are abstracted from experience in an Aristotelian manner, or are Platonically inborn, or a priori in a formal Kantian sense, may be left undecided here).

What is important in this procedure is that the consequences of applying these general principles may lead one also to include into the whole (understood this time as the "totality of the real") entities which are not included in the empirical evidence in which the general principles themselves had been used at first, and this because they are meant to be principles of the whole, and therefore be susceptible of indefinite application. This use of principles and concepts we may call the *synthetic use of reason*, to contrast it with the *analytic use* of which we have already spoken.

The move just described is typical of metaphysics, but it is also made in science. Every science tries to provide a complete account of its field of inquiry, and this means several things: the possibility or even the necessity of introducing theoretical entities, besides those which are empirically discovered and described using the empirical means admitted by that science; the unrestricted application of these general principles, which implies not only their indifference with respect to particular or accidental properties of the individual entities to which they apply, but also the possibility of applying them to cover the *whole* field of inquiry both synchronically (which practically means with respect to every region of space at a given time), and diachronically (which means with respect to all possible time instants). In such a way we see that the homogeneity and isotropy of time and space are already implicit in this "point of view of the whole" which we find both in metaphysics and in the most general intellectual attitude of science as well.

Is the consideration of past and future, and in particular the problem of the "origin", included in this approach? Not necessarily. In the case of the sciences, since it is already clear from the beginning that each of them concentrates its investigation on a circumscribed whole, it is tacitly admitted that the "rest of the real" (in many senses which we shall not mention here) remains out of consideration, so that the problem of the origin does not really surface. For example, biology takes life as given without asking from where it comes, chemistry does the same with material substances, physics with matter, energy, particles, fields. However it is not impossible (and it actually happens) that questions about these origins be asked, but then they are not scientific, but philosophical questions and, more precisely, cosmological questions, even though they are not necessarily formulated in cosmology proper. So, for example, the question of the origin of life is of a *cosmological* nature, since it does not properly concern "what life is" (a problem, in a way, for biology), but "what the place of life is in the whole", and this may be seen as a metaphysical question if this whole is understood with the greatest of generality, and as a cosmological question if the whole is understood rather as the whole of Nature.

We have seen therefore that cosmological questions are philosophical questions by their very nature. However, one can try to answer them by resorting to scientific knowledge. So, to take our example, one may try to answer the question of the origin of life by going back to chemistry, and then answer the problem of the constitution of the chemical elements by going back to physics. And then? Is physics the ultimate ground for solving the problem of the origin? The answer is delicate: a problem does not change in *nature* simply because we resort to some particular tools for solving it, and in this sense a cosmological problem remains philosophical in spite of being studied with the aid of a science.[12] On the other hand, one may be convinced that physics does actually provide the solution to this problem, or at least that it is the only discipline competent to provide a possible solution: in this case one simply endows physics with the competence necessary to treat of the *whole* in the fullest sense of this term, and this is tantamount to making of it a metaphysics. This view is not in itself absurd, but it should be supported by showing that physical concepts and physical theories are independent of extra-physical a priori presuppositions, and have an unrestricted and all-embracing validity. But this is just what has been shown to be problematic throughout this paper.

Without repeating now things which we have already hinted at, let us simply stress that all the drastic extrapolations which are usual and indispensable in cosmology have the character of metaphysical generalizations, even though the whole to which they are extended simply is the "whole of Nature". The most instructive example of this is the already discussed fact that local physical laws are assumed to be valid "everywhere and at every time": this seems to rely on the *assumption* of a cosmic uniformity which might be more or less justifiable on the basis of what we empirically know of the accessible universe, but a little reflection is sufficient to see that it is the admission of the unrestricted validity of these laws which actually *defines* the meaning of this cosmic uniformity. For that reason we must say that these laws play the role of non-local principles in a sense that does not coincide with the usual physical meaning of non-locality, but rather corresponds exactly to the classical requirement of *universality and necessity*, which is the distinguishing mark of the metaphysical principles.

That this is so is confirmed by the fact that these laws are assumed to be determined with a practically infinite degree of accuracy, otherwise even an accidental variation in the conditions of application, which could correctly be considered to be negligible from a local point of view, would result in total unreliability from the point of view of the cosmic consequences we should draw from this application (see what we have already said concerning cosmological retrodictions). Now, such an exactness beyond any margin of error is again a typical mark of the metaphysical principles, and is alien to the very concept of a physical law.

Another circumstance that is rather surprising, if we consider it from the point of view of standard scientific methodology, but which is in keeping with the metaphysical way of reasoning, is that the physical laws in question have to be invariant not only, in a generic sense, with respect to time and space, but also with respect to the evolution of the universe. This point is crucial because it shows rather clearly that cosmic uniformity was not really a necessary presupposition for the extrapolation of the local physical laws, as it may have seemed to be when one simply considered space and time. Indeed, these laws appear now to be valid even in spite of a deep non-uniformity, i.e. in spite of their field of application (the concrete universe) becoming radically different in the different stages of its evolution: they are not only "spacetime-invariant", but even "evolution-invariant". Moreover, they are such that they remain independent of the evolution of the universe and at the same time predict such an evolution. From this point of view they share again the nature of the metaphysical principles, which are meant to explain mutation and change, and even to predict it, because they are valid independently of change, because they have an immutable nature. Morever it is essentially because of this feature that they are meta-empirical, since change dominates the empirical realm; and if they are so, no wonder that they are not subject to the conditions of space, time, change, uniformity or non-uniformity, etc. These conditions characterize the empirical, concrete, material world, but not the world of reason. The attribution of the said invariances to the metaphysical principles is therefore a clear indication of their being principles of rationality, and to the extent that physical laws are endowed with similar invariances, they are promoted to the level of being rational, in the sense of being what is specifically meta-empirical.

We shall terminate our discussion of this point by indicating a final consideration, which corroborates the above-sketched view. Even attributing to known physical laws the metaphysical features just mentioned, several difficulties still remain in cosmological theories. How does one try to remove them? Sometimes by introducing new theoretical principles, but even by meta-empirically postulating certain *factual conditions* which are in turn not justifiable within the theory which assumes them. Among the difficulties, or at least, as we have seen in part, among the "open questions" of the standard cosmological model, are zero-curvature, the causality paradox, the horizon structure, isotropy and homogeneity; and a merit of the inflationary model is that it can solve these problems. This model however relies upon two conjectures: that our uniform visible universe is the result of the expansion of a quantity of matter occupying a region of the order of 10^{-30} cm, an expansion which began when the universe was just 10^{-35} sec. old; that on this almost vanishing scale of length no chaotic substructure was present in space, so that this "smoothness" could evolve to the (alleged) smoothness of our observable world, thanks to the action of antigravity. It is easily seen that these two hypotheses are genuinely untestable *factual* assumptions of a *synthetic* character, for which no other justification is provided in the theory than the fact that they solve certain *theoretical* open questions. Here again we find the features of the *synthetic use of reason* which is typical of metaphysics.

A separate, but no less significant, case is constituted by the direct use in cosmology of expressions which are verbally identical to certain metaphysical formulations in the strongest and even most debated sense of metaphysics. We refer here, e.g., to concepts such as those of "genesis" and of "creation out of nothing" which are used in contemporary quantum cosmology for interpreting certain theoretical (or mathematical) results.[13] In order to avoid any genuinely metaphysical interpretations of these expressions one may try to point out, first, that they are not inescapable, since it is not absolutely certain that GTR (within the framework of which they are derived) is the only pertinent theory for tackling the problem of the very initial state of the universe. This is however a very cheap escape, for we cannot rely on GRT for the construction of our cosmological theories, and then remind ourselves that this theory might not be the only or the best basis for cosmology, just at the moment we discover that it may lead to metaphysical interpretations we dislike. A more serious objection seems to be that the mathematical results which are so interpreted are, after all, only certain particular cases of *singularities*, which depend on special mathematical properties of the complicated spacetime manifold which is used for representing a universe satisfying certain cosmological assumptions. In our case, this would amount to recognizing the existence of an *absolute zero of time*, and this should not be more surprising or "metaphysical" than the discovery of the existence of an *absolute zero of temperature*.[14] However things are not so simple. First of all because in every mathematical model of a concrete system we are constrained to interpret *all* the mathematical expressions as expressing properties of that system, so that we cannot dismiss some of them as accidental mathematical by-products having no concrete significance. (Note, by the way, that the tendency in cosmology is exactly the opposite, that all the possibilities disclosed by the mathematical machinery are exploited when they can serve to give plausibility to some welcome assumption: simply think of antigravity, whose supporters stress that it can be justified in the Grand Unified Theories). Now, if this is how things stand, one must recognize that the existence of an absolute zero of time is categorially different from the existence, say, of an absolute zero of temperature, since it poses the problem of an *origin*, which the instance of temperature does not. On the other hand, if one remains within the framework of GRT, it is obvious that no physical action may be envisaged which could causally produce the universe "before" the absolute zero of time. Does this prevent us from asking whether there might be a cause of the universe? Of course not, provided we do not pretend this cause to be a *physical* one, i.e. a cause which is expressed through the highly specialized and restricted features of our present physical theories. In that case such a cause would genuinely deserve to be called metaphysical. Physics certainly does not *oblige* us to admit such a cause, but it is far from preventing us from doing so either. Only if we make of physics a metaphysics (i.e. if we pretend that physically definable causality is the *only* causality) can we make this claim.[15] In view of these remarks it seems arbitrary to say that the situation considered here, if correctly appreciated, shows that the universe is itself an uncaused cause or, in traditional terms, a «*causa sui*», so that «Anyone, who can live with the concept of the Deity as an *uncaused cause*, can surely live with the Universe itself as the uncaused cause.»[16] In fact we must say that the physical concept of causality does indeed exclude the possibility of an uncaused cause (otherwise every causal explanation would be intrinsically arbitrary, since we might always suggest of any phenomenon that it is simply caused by itself), while admitting that the concept of causality at its highest level of generality (i.e. at a metaphysical level) does not exclude the possibility of an uncaused cause. Now, it is certain that if this concept appears to be logically sound in itself, we cannot refuse to apply it, provided we remain conscious that the *way* in which it is applied defines its domain of application: this means that if we apply it in cosmology, we are *ipso facto* making a claim of "metaphysical cosmology". It is true that we can very well live with the idea of the Universe as the uncaused cause, but this simply means (unconsciously) accepting an immanentist, rather than a transcendentist, metaphysics.

Quite analogous reasonings could be provided regarding other ways of "eliminating" the problem of accounting for the initial or boundary conditions, e.g. within quantum gravity:[17] these eliminations are meaningful only to the extent that physics is essentially endowed with the role and power of a metaphysics.

10. CONCLUSIONS

What we have seen in this paper is that, with the development of cosmology, contemporary science has again found many links with philosophy, which seemed to have been discarded for more than a century. In fact, not only was the tenet largely accepted that scientific investigation must carefully avoid philosophical commitments or dependences, but it was also tacitly assumed (as we have seen at the beginning) that the progress of science would gradually make philosophy superfluous. At most it was conceded that philosophy could perform useful work by analyzing science (especially the language and the methodology of science) and by bringing the current world-view up to date by integrating in it the advancements of scientific knowledge. This positivistic outlook is clearly overcome by the most recent trends in science, in the sense that it becomes ever more clear that scientific research *incorporates* philosophical concepts and principles, and really contributes to refining them and making them fruitful. Cosmology, after all, is not the only example: present investigations on the mind-body problem constitute another interesting field where philosophical conceptualization and scientific theories come into very close contact, a contact whose meaning is not that we finally bring the light of science to illuminate philosophical obscurities, but that we derive from science a great deal of information and knowledge for improving our philosophical understanding of this fundamental issue.

In fact, every time we have been obliged - in our preceding reflections - to recognize that certain cosmological theories or assumptions are not fully justified, or overstep the current standards of physical science, we have never drawn the conclusion that this has condemned cosmology. Rather, we have been led to see that the adoption of such assumptions was peculiar to its way of being scientific, a way which not only implies the "use" of certain general philosophical principles (as is also the case in other sciences), but is further characterized by the adoption of several intellectual attitudes (typical of which is the adoption of the "point of view of the whole") which are among the salient traits of philosophy. All this does not mean (as certain positivistically minded scholars seem to believe) a risk of corrupting the purity of scientific rigour, but a step towards recovering that unity of knowledge which we have lost for too long a time. This unity comes not so much from the use of a *unique* type of language, of conceptualization, of methodology (the unity of reductionism), but rather from the exploitation of different intellectual approaches in a mutual interrelation and feed-back regarding certain common *problems*. Now, as we have seen, the universe is hardly conceivable as a scientific "object"; it is also difficult to define it as a precise "concept" (it is rather something like an "idea" in a Kantian sense); but it has certainly posed one of the most challenging intellectual *problems* for mankind since its origin. Philosophy, mythology, religion, art and science have each tried to tackle this problem, and it would be arbitrary to exclude one or the other of these approaches, since all help us to "understand" the universe. Therefore it is certain that the most promising advances in the intellectual understanding of the universe may come from recognizing it to constitute at the same time both a scientific and a philosophical problem.

Department of Philosophy
University of Fribourg (Switzerland)

NOTES

1. More precisely: while the expression «natural philosophy» has been in use for many centuries, the term «cosmology» was formally introduced to indicate a special subdomain of philosophy by Christian Wolff, whose systematic partition of the philosophical disciplines - and the corresponding terminology - became standard. He proposed to divide traditional metaphysics into two parts: general

metaphysics or «ontology,» and special metaphysics, which in turn he subdivided into «theology,» «psychology,» and «cosmology» (see Wolff 1728,#77). In this way, cosmology became the new denomination of the traditional natural philosophy, however with certain new shades of meaning which we shall consider later. This classification, which passed soon into the textbooks, is taken as obvious also by Kant, who discusses these domains of traditional metaphysics in the Transcendental Dialectics of his *Critique of Pure Reason*. In particular, in the 1st Book, 3rd Section of the Transcendental Dialectics, he explains: «The thinking subject is the object of *psychology*, the content of all phenomena (the world) is the object of *cosmology*, and the thing, which includes the highest condition of the possibility of anything thinkable (the being of all beings) is the object of *theology*. Therefore pure reason provides us with the ideas for a transcendental doctrine of the soul (*psychologia rationalis*), for a transcendental science of the world (*cosmologia rationalis*), and finally also for a transcendental knowledge of God (*theologia transcendentalis*)» (Kant 1776, A 334, B 391).

In spite of the strong Kantian objections to the pursuit of "rational cosmology"- developed in the *Critique of Pure Reason* - this term has continued to enjoy a standard use by a variety of philosophers, who did not accept Kant's arguments against the possibility of a rational metaphysics. We find it in such German philosophers as Hegel and many others, in practically all the manuals of scholastic philosophy (which mostly adhered to the Wolffian scheme of classification of the sectors of metaphysics), and also in philosophers not belonging to the scholastic school, at least up to the first decades of our century.

2. The fact that natural science and natural philosophy were not considered to be separate disciplines even in the 18th century may result from the consideration of the partition of "cosmology" itself proposed by Wolff and accepted by his followers. In fact he used to distinguish a «general cosmology» (which is a branch of ontology and has a speculative character) from an «experimental cosmology» (which has an observational and descriptive character): «General cosmology is the science of the world or universe in general, i.e. in that it is a being and is a composite and modifiable being» (Wolff 1731, # 1). Moreover «General scientific cosmology is that, which demonstrates a general theory of the world [*theoriam generalem de mundo*] starting from the principles of ontology; on the contrary experimental [*experimentalis*] cosmology is that which derives from observations a theory [*theoriam*] which is established or will be established in scientific [*scientifica*] cosmology» (Wolff 1731, # 4). In the subsequent literature the term «rational» was often introduced in place of the Wolffian «general», and this is

what we find in Kant. What is worth noting here is the fact that the truly "scientific" cosmology was maintained to be the more speculative one, while the "experimental" cosmology was considered as a descriptive basis from which a fully fledged cosmological theory, anchored in the ontological framework of "general" cosmology, could be developed. It is easily seen that the Wolffian conception of cosmology implicitly entails a splitting of this discipline into two separate parts: one metaphysical and one "scientific" (where "scientific" must be understood now in the new sense of modern science). When Kant developed his fundamental criticism of "rational cosmology", he concretely deprived it of any pretension to being a "science", and what was left was the possibility of doing cognitive cosmology only at the level of the empirical descriptive cosmology. As is well known, Kant himself does not deny the possibility of a metaphysics of Nature, but in a very new sense, i.e. as the determination of the a priori conditions of the cognitive construction of physical objects (as it is presented, e.g. in his *Metaphysical Foundations of Natural Science*; see Kant 1786).

3. For example, scientific anthropology, understood in the sense of a purely biological study of man, was explicitly established towards the middle of the 19th century. Simply as one of the many steps of this transformation, let us mention that the old chair of anatomy existing at the *Jardin du Roi* in Paris since 1635, became a chair of «Natural history of man» in 1832, and finally the first French chair of *Anthropology* in 1855 under the teaching of Quatrefages. For an excellent historical survey of the evolution of the term «anthropology» in the history of European civilization, see the third chapter of Part IV: «Natural history of man and the origins of modern anthropology» in Gusdorf 1960.

As to psychology, it is interesting that Wolff had also distinguished an «empirical psychology» (see Wolff 1732), and a «rational psychology» (see Wolff 1734), more or less in the spirit of the similar distinction he had proposed for cosmology. The scientific approaches to psychology which started in the 19th century were in a way a development of the Wolffian approach of «empirical psychology», while philosophical psychology was left aside and totally disconnected from the scientific approach. For an interesting account of these developments see again Gusdorf 1960, especially the last part of chapter four of Part IV: «The empiricist theory of knowledge and the origins of psychology».

4. These well known developments of Kant's position concerning rational cosmology are presented especially in his discussion of the «cosmological antinomies» (*Critique of Pure Reason*, A 408 ff., B 434 ff.).

5. What we are maintaining here is that while natural science (let us call it «physics» for brevity) investigates particular "natural phenomena", cosmology investigates "the world" as a whole (according to the different definitions of cosmology we have found in the literature), and for this reason it cannot be a domain of physics. However one may object that an explicit consideration of the "world" was not excluded from the field of investigation of the "new" physical science. For example, already in Newton's *Principia* we find a whole book (the 3rd and final book) devoted to the topic «The system of the world» (*De mundi systemate*), but it is very interesting to note that, in the short introduction, Newton himself stresses the «philosophical» character of this part: «In the preceding books I have laid down the principles of philosophy; principles not philosophical but mathematical: such, namely, as we may build our reasonings upon in philosophical inquiries. These principles are the laws and conditions of certain motions, and powers or forces, which chiefly have respect to philosophy; but, lest they should have appeared of themselves dry and barren, I have illustrated them here and there with some philosophical scholiums.... It remains that, from the same principles, I now demonstrate the frame of the System of the World». What follows is then a general "philosophical" discussion of a methodological character (*Regulae philosophandi*) and an application of the theory of gravity to the study of the solar system. The conclusion of the work is the famous *Scholium generale*, in which we really find the perspective of the "whole", trespassing the limited scope of the "world" understood simply as the solar system, and it is certainly not accidental that in this perspective Newton finds it necessary to evoke explicit metaphysical principles, including the existence of God as an ultimate explanation of the cosmic order. This was by no means a kind of extrinsic addition, but is a clear indication of the general spirit of "natural philosophy", which continued to inspire scientists. A confirmation of this thesis comes from a celebrated scientific work in which the newly introduced term «cosmology» explicitly appears: the book *Essai de cosmologie* (Essay on Cosmology) of Maupertuis (1750). In this work the author tries to apply the "physical" principle of least action (which he had enunciated in 1744) to unify the laws of the universe and also to prove the existence of God, giving rise to a great philosophical and scientific debate, in which persons such as Euler and Voltaire took part.

We find a different approach in the already mentioned hypotheses on the origin of astronomical objects formulated by Kant in 1755 (see Kant 1755), by Lambert in his *Cosmologische Briefe* (Cosmological letters) of 1761 (see Lambert 1761), and by Laplace in the "note VII" of his *Exposition du système du monde*

(Expostion of the system of the world) of 1796 (see Laplace 1796). It is not interesting here to analyze the differences among these proposals: their common feature is that the Newtonian theory of gravitation is assumed as the unique and sufficient framework for explaining the present state of the astronomically describable world (Kant and Lambert formulate hypotheses which are applicable to all stellar systems, while Laplace's is restricted to only the solar system). However, while in Laplace this construction is purely scientific in the sense that he is unaware of the philosophical presuppositions involved, in the case of Kant (as we have already seen), and also in the case of Lambert, this construction is not disconnected from more general philosophical and metaphysical considerations concerning cosmology, which they had developed in other works. All this seems to confirm our claim, that the term «cosmology» (which by the way was not used by Laplace) tends to remain in use within philosophy (works with this title have been frequent until recent years), but had no circulation in science. It is not by chance, perhaps, that Duhem's work *Le système du monde. Histoire des doctrines cosmologiques de Platon à Copernic* (The system of the World. A History of Cosmological Doctrines from Plato to Copernicus) restricts its attention to the period preceding the birth of modern science, while a subtitle «An Essay in Cosmology» appears on such a thoroughly philosophical work as Whitehead's *Process and Reality* (1929).

6. For a good account of these epistemological debates one may profitably consult Gusdorf 1960, especially the section on «The historical sciences» in the fourth chapter of the fifth part of the book. In the other parts of this chapter one also finds reliable accounts of the debates concerning other "human sciences".

7. On this point see Agazzi 1979.

8. See in particular Bridgman 1927.

9. See Bondi 1960.

10. See Kanitscheider 1990, p.339.

11. It would lead us too far afield to justify this claim here. I allow myself to refer to what I have published on this subject on several occasions. Some titles are given in the references.

12. This is by no means a polemical remark. In fact it fully corresponds to the most widespread way of conceiving of science to say that the scientific approach consists primarily in a certain way of formulating cognitive problems, a way which, according to many scholars, has even had the effect of "eliminating" as pseudo-problems several problems of philosophy. Therefore, it follows that certain problems cannot be seriously taken into consideration in science. Now, in the case of cosmology, we find that many problems, which are identical with those of philosophical cosmology, are seriously taken into consideration. Hence, they remain philosophical as problems but, on the other hand, we see finally that this fact cannot "eliminate" them and does not prevent them from also being considered with the aid of science. As a confirmation of this let us simply take a brief quotation from a very "speculative" philosopher, namely Hegel, who gives the following indication of the (philosophical) questions of cosmology in # 35 of his *Encyclopedia of the Philosophical Sciences*: «The third part [of the traditional metaphysics], i.e. cosmology, was concerned with the world, its contingency, necessity, eternity, limitation in space and time, with the formal laws in its mutation, and finally with the liberty of man and the origin of evil.» It is certainly difficult to deny that most of the questions listed above are among these which are deeply debated in contemporary "scientific" cosmology, but this is exactly the reason for recognizing that this cosmology cannot be just scientific.

13. See for example Wheeler 1977, and Vilenkin 1982.

14. See Kanitscheider 1990, pp.344-345 for this discussion.

15. Let us remark that already at the level of common experience we are familiar with forms of causation that are not "physical" in the sense of not being accounted for by physics. All intentional acts, which are caused by voluntary decisions, are of this kind. The claim that these decisions "in principle" or "in the last analysis" also depend on physical causality is a dogmatic tenet of a bad materialistic metaphysics.

16. See Barrow 1988, p.227.

17. See again Kanitscheider 1990, pp. 348-349.

REFERENCES

AGAZZI, E.

1969 *Temi e problemi di filosofia della fisica*, Manfredi, Milano,1969; Abete, Roma, 1974 (2nd ed.)

1976 «The Concept of Empirical Data. Proposals for an Intensional Semantics of Empirical Theories», in M. Przelecki et al.(eds.), *Formal Methods in the Methodology of Empirical Sciences*, Reidel, Dordrecht, 1976, pp.143-157

1978 «Eine Deutung der wissenschaftlichen Objektivität», *Allgemeine Zeitschrift für Philosophie*,3 (1978), pp.20-47

1979 «Analogicità del concetto di scienza. Il problema del rigore e dell'oggettività nelle scienze umane», in *Epistemologia e scienze umane*, Milano, Massimo, 1979, pp.57-78

1985 «Commensurability, Incommensurability, and Cumulativity in Scientific Knowledge», *Erkenntnis*, 22 (1985), pp.51-77

BARROW, J.D.

1988 *The World within the World*, Clarendon Press, Oxford, 1988

BONDI, H.

1960 *Cosmology*, Cambridge Univ. Press, Cambridge, 1960

BRIDGMAN, P.W.

1927 *The Logic of Modern Physics*, Macmillan, New-York, 1927; several reprints

DUHEM, P.

1913-1959 *Le système du monde: Histoire des doctrines cosmologiques de Platon à Copernic*, Hermann, Paris,1913,-17 vol.I-V; 1954-59, vol.VI-X

GUSDORF, G.

1960 *Introduction aux sciences humaines. Essai critique sur leurs origines et leurs développements*, Les Belles Lettres, Paris, 1960

HEGEL, G.W.F.

1817 *Encyclopädie der philosophischen Wissenschaften im Grundrisse*, A. Osswald, Heidelberg, 1817; Suhrkamp, Frankfurt a.M., 1970

KANITSCHEIDER, B.
1990 «Does Physical Cosmology Transcend the Limits of Naturalistic Reasoning?», in P. Weingartner and G.J.W. Dorn (eds.), *Studies on Mario Bunge's Treatise*, Rodopi, Amsterdam-Atlanta, 1990

KANT, I.
1755 *Allgemeine Naturgeschichte und Theorie des Himmels*, J.F.Petersen, Königsberg, 1755; H. Fischer, Erlangen, 1988
1781 *Kritik der reinen Vernunft*, Hartknoch, Riga, 1781; F. Meiner, Hamburg, 1976
1786 *Metaphysische Anfangsgründe der Naturwissenschaft*, Hartknoch, Riga, 1786; Fischer, Erlangen, 1984

LAMBERT, J.H.
1761 *Cosmologische Briefe über die Einrichtung des Weltbaues*, Kletts Wittib, Augsburg, 1761; Engl. trans. by S.L. Jaki, Edinbourgh, 1976

LAPLACE, P.S.
1796 *Exposition du système du monde,* Imprimerie du Cercle-Social, An IV de la République française, (1796), Paris; Fayard, Paris, 1984

MAUPERTUIS, P.-L- M. DE
1750 *Essais de cosmologie*, various places, including Berlin, 1750

NEWTON, I.
1687 *Philosophiae Naturalis Pincipia Mathematica*, London, 1687; English transl., *Mathematical Pinciples of Natural Philosophy*, W. Benton: Encyclopaedia Britannarica, Chicago, Geneva (etc.), 1987

WOLFF, CHR.-F.
1728 *Discursus praeliminaris de philosophia in genere.* In *Philosophia rationalis sive logica...*, Officina libr. Rengeriana, Frankfurt-Leipzig, 1728
1731 *Cosmologia generalis*,ibid., 1731
1732 *Psychologia empirica*, ibid., 1732
1734 *Psycologia rationalis*, ibid., 1734

Roberto Torretti

THE GEOMETRIC STRUCTURE OF THE UNIVERSE

Allow me, first of all, to thank the International Academy of Philosophy of Science for inviting me to speak to you and with you on this occasion. My subject is «the geometric structure of the universe». I would never have dared to choose it, but I made no resistance when Evandro Agazzi proposed it to me; so I assume full responsibility for this immodest attempt to cover in one lecture such a broad and intricate subject. Indeed, within the standard framework of contemporary cosmology the *geometric* structure of the universe is the structure of the universe *tout court*; at any rate, it is all there is to its large-scale structure. However, other lecturers will elaborate further on its more significant aspects. Thus my talk is mainly intended to provide a necessary, but necessarily schematic conceptual background for the fuller view of present-day physical cosmology you will be given later.

A few preliminary remarks will help bring my subject into focus. The phrase «the geometric structure of the universe» owes its current meaning to Einstein's theory of gravity, better known as General Relativity. This was the first theory of mathematical physics which could handle the universe as a physical object. The cosmological solutions to Einstein's gravitational field equations found by Einstein himself,[1] Willem de Sitter,[2] Alexander Friedmann,[3] the Abbé Lemaître[4] and Kurt Gödel[5] entitle us to ascribe a geometric structure to the universe if certain conditions are met. It is in virtue of this very ascription of a geometric struc-

E. Agazzi and A. Cordero (eds.), Philosophy and the Origin and Evolution of the Universe, 53–73.

ture to the universe, that the latter becomes, in the context of General Relativity, a definite object of scientific study. Geometric structure, in the sense here intended, might more properly be called chronogeometric structure, insofar as it has to do not or at least not primarily with relations between permanent points in space, but rather with instantaneous points or so-called *events* or *possible events* in space-time. General Relativity is a geometrodynamic or, if you wish, chronogeometrodynamic theory of gravity. On the one hand, the simplest kind of freely falling bodies, viz. spinless, uncharged particles, do no more than trace, unprompted and unhindered, the straightest lines in space-time. On the other hand, such lines are not given *a priori*, as in the Minkowskian chronogeometry of Special Relativity, but depend on the actual distribution of matter and non-gravitational energy.

My chief aim in this lecture is to shed some light on these momentous, at first blush baffling ideas. They stem from radical innovations in geometry introduced during the 19th century, in particular by Bernhard Riemann. I must therefore talk a little about his work, as untechnically as I can. You have all heard of the Non-Euclidian geometry, which Lobachevski and Bolyai developed about 1830 by tampering with Euclid's 5th Postulate. They showed that if one assumes that through a point outside a given straight line there goes more than just one straight which lies on the same plane as the former but does not meet it, one obtains a system of geometry which displays some surprising features, but is no less consistent than the Euclidian system. The difference between Euclidian and Lobachevskian geometry turns on the behavior of coplanar straights at infinity; but it shows up in interesting properties of figures near at hand. Thus, it is a characteristic of Euclidian geometry that two plane or solid figures can have the same shape and yet differ in size. Consequently, in a Lobachevskian space there is not a single cube or square. It might seem, therefore, that Lobachevskian geometry is just an academic exercise, with no relevance to the world we live in. But the cubes and squares we find in our environment are just very nearly cubic or square. Couldn't it be that at astronomical distances it is impossible to construct a quadrilateral which is even approximately square? Lobachevski tested this hypothesis by measuring the angles of the triangle formed by three stars: Sirius, Rigel and Star N.29 of Eridanus. He verified that, within the admissible margin of error, they added up to two right angles, the Euclidian value. But the possibility remained open that a significant discrepancy would be observed at larger distances.

Riemann realized that Euclidian and Lobachevskian geometry were but two very special cases among the countless geometric structures that could conceivably be vested in space. He thought that scientists might eventually find it appropriate to use an unorthodox geometry for the description of physical, especially of microphysical phenomena, and proposed a broad framework for the development of alternative geometries. Riemann's revolutionary work rests squarely on another revolution in geometry, carried out two centuries earlier by Fermat and Descartes. By their method of coordinates points in space are represented by ordered triples of real numbers. Figures such as an ellipse or a sphere are then represented by suitable sets of number triples satisfying certain equations, so that geometry gets to benefit from the advantages of algebra and analysis. The introduction of coordinates naturally suggested the idea of spaces of more than three dimensions, whose points would be represented by ordered real number quadruples, quintuples, etc. The idea came up already in the 17th century. Close to the end of the 18th century, Lagrange represented a mechanical system with *n* degrees of freedom as a point in motion in a «configuration space» of *n* dimensions, and also as a curve in a space of $n + 1$ dimensions the additional dimension corresponding to time. Now, it is important to see that what makes a collection of objects represented by *n*-tuples of real numbers into something one may sensibly regard as «points» in an *n*-dimensional «space» is not just the fact that each is labelled by *n* numerical coordinates. What really does the trick are the neighborhood relations which the set of real number *n*-tuples (the so-called Cartesian product R^n) inherits from the ordered field of real numbers, and induces in turn in the said collection of objects. (Technically speaking, the collection will sport the weakest topology which makes each coordinate function into a continuous function.) Moreover, it is not essential that the entire collection be represented by a single coordinate system in R^n. For such a collection to merit the name of «space» and to become a suitable object of geometrical inquiry it is sufficient (i) that every element or «point» in the collection has a neighborhood which is mapped into R^n by a coordinate system; and (ii) that when the same neighborhood happens to be represented by several coordinate systems the latter are mutually related by smooth one-to-one functions: «coordinate transformations». (Smoothness, rather than mere continuity, is required if geometry is to continue to enjoy the assistance of analysis, as a physicist would surely expect.) The entire «space» is then charted in a piecemeal or «patchwise» fashion in R^n, much like the round earth is charted, region by region, on the flat pages of an atlas. But the «space» as a whole might not be homeomorphic, i.e. topologically equivalent to R^n. An n-dimensional «space» in this sense is roughly what Riemann called an n-fold extended quantity, and we now call a real n-dimensional differentiable manifold or n-manifold, as I shall say for short. Since a flat map can in turn be mapped one-to-one into R^2, the geographical analogy should be taken literally: the surface of the earth is a 2-manifold, and so is indeed any smooth surface.

In the 1820's Gauss had carried out important investigations on the geometry of curved surfaces. There are, of course, definite quantitative relations between, say, the lengths of paths, the sizes of angles, or the areas of figures drawn, say, on the surface of an egg, or a pretzel, or a saddle. By the geometry of a surface I mean the system of such relations. Now, the geometry of an ordinary surface in Euclidian space is of course inherited from the Euclidian geometry of the latter; but, contrary to what one might think at first blush, the geometry of a surface does not depend on the way how it lies in space. For example, when you roll up a flat page printed with geometrical figures to form a cylindrical tube, you do not shorten the lines, or widen the angles, or shrink the polygons on the page. Gauss figured out a way of characterizing the geometry of a surface *intrinsically*, that is, without appealing to any relations between points on the surface and points on the space outside it. To describe his achievement I shall appeal to the following intuitive idea: Think of particles moving on the surface in every direction with every conceivable speed. The velocities of all the particles going through a given point form a vector space, the 2-dimensional space of tangent vectors or tangent plane at that point. Gauss in effect encoded the geometry of a surface in a tensor field of rank 2, that is to say, in a correspondence which assigns, to each point of the surface, a bilinear function on vectors at that point. I need not explain here what is a bilinear function. It is sufficient to say that it will assign a real number to each ordered pair of vectors at the point, in such a way that the numbers assigned to the four pairs which can be formed from two linearly independent vectors determine the number assigned to any other pair of vectors. The tensor field in question is known as the metric tensor, or simply the *metric*. It is symmetric, which means that its action on the pair <v,w> of vectors at a point *P* yields the same real number as its action on the pair <w,v>. It is positive definite, which means that its action on a given *non*-zero vector taken twice yields a positive number. The metric, in turn, determines a scalar field that is, a smooth real-valued function on the surface, which is known as the Gaussian curvature. Again, I do not wish to burden you with technical details; but let me indicate, by way of example, that the Gaussian curvature of a Euclidian plane, and hence of a cylindrical surface, is everywhere equal to 0; that the Gaussian curvature of a surface on which the straightest lines -the so-called geodesics of the surface- behave as Lobachevskian straights is a constant negative number, and that the Gaussian curvature of an egg is everywhere positive and varies smoothly from point to point.

If, with Riemann, we view every smooth surface as a 2-manifold, we can see his own work as a fairly straightforward generalization of Gauss' methods and results to the *n*-dimensional case. Given an *n*-manifold *M*, its geometry, i.e. the system of quantitative relations between lengths, areas, volumes, hypervolumes, etc., on *M*, is to be defined intrinsically, without any consideration of a higher-dimensional manifold in which *M* might be imbedded. The key notion is that of the tangent space of vectors at any given point P of M. This is the n-dimensional analogue of the tangent plane at a point of a surface. To explain this notion, I resort again to the image of particles moving in every direction and with every conceivable speed through P. The idea of a particle's motion in an n-manifold can be made definite by thinking of its description in a coordinate system. The particle's position at a given time t is labelled by coordinates $x_1(t),...,x_n(t)$. Suppose that the particle goes through P at time t_0. The n-tuple of time derivatives $[\frac{dx_1}{dt},...,\frac{dx_n}{dt}]$ taken at $t = t_0$ defines then a vector in R^n, representing the particle's velocity as it goes through P. The tangent space at P is the space generated by all such vectors representing the velocities of particles going through P and is therefore a copy of R^n. As in the 2-dimensional case studied by Gauss, the geometry of M is defined by the metric, a symmetric positive definite tensor field of rank 2.[6] Depending on the metric, a 3-manifold can be Euclidian or Lobachevskian, or sport any one of countless alternative geometries. Any geometry characterized in this way by a rank 2 positive definite symmetric tensor field on an n-manifold is a Riemannian geometry. What all such geometries have in common is that they are, so to speak, almost Euclidian on a small neighborhood of every point. More precisely: the Euclidian metric agrees everywhere with any such metric to the first order in the coordinate differentials. Riemann contemplated other geometries, which would not meet this condition. However, given the known success of Euclidian geometry, he thought that physics would do well to stick for some time to the almost Euclidian Riemannian geometries. Riemann also introduced the n-dimensional analogue of Gaussian curvature. This, however, is not a scalar field (unless, of course, $n = 2$), but a tensor field of rank 4, that is, roughly speaking, a smoothly varying assignment to each point of the manifold, of a 4-linear function on ordered quadruples of vectors at that point (cf. footnote 6). This tensor field, commonly known as the Riemann tensor or curvature tensor of the manifold, is determined by the metric. Let me note, by the way, that when General Relativity ascribes a curvature to space in the presence of gravity, this does not mean that space is curved or bent in some intuitive, or rather, counterintuitive sense, but simply that in the presence of gravity the Riemann tensor assigns non-zero values to some ordered quadruples of spatial directions. The idea that the geometry of physical space might depend on the forces of nature was already expressed by Riemann in 1854. As he pointedly noted: «The empirical concepts on which the metrical determinations of space are grounded, namely, the rigid body and the light ray, have no validity in the very small. Hence it is quite conceivable that the metrical relations of space in the very small do not meet the requirements of [standard] geometry. Indeed, one ought to assume it, if the phenomena are thereby explained in a simpler way.»[7]

In September 1905, in a paper «On the electrodynamics of moving bodies», Albert Einstein published the principles and some important consequences of what is now known as the Special Theory of Relativity. In it he showed how to reconcile the equivalence of all inertial frames, proclaimed by Galileo and Newton, with the laws of classical electrodynamics, which, as understood at the time, apparently distinguished one such frame, viz. the one supposedly defined by the «aether» or medium of transmission of electromagnetic waves, from all the others. This apparent distinction came about because on switching from one inertial frame to another physicists changed the position coordinates to adjust for the relative motion of the frames, but left the time coordinate untouched. Einstein postulated the equivalence of inertial frames and the validity of one simple and well-corroborated principle of electrodynamics viz. that the speed of light *in vacuo* is the same everywhere and in every direction, and does not depend on the velocity of its source, and *derived* a rule for the transformation of coordinates from these two postulates. Transformations that follow this rule are called Lorentz transformations, or (in their most general form) Poincaré transformations, while transformations of the traditional sort are called Galilei transformations. The laws of classical electrodynamics remain invariant under Poincaré transformations and thus comply with the equivalence of frames as understood by Einstein, but the laws of classical mechanics are Galilei invariant and thus clash with it. Einstein therefore introduced some drastic changes in the laws of mechanics, which made them Poincaré invariant. The old and the new laws yield practically the same predictions for the low energy mechanical experiments which supported the former. Their glaring discrepancy at high energy levels has been long since resolved in favor of the latter. Indeed, insofar as Special Relativity is built into Quantum Electrodynamics, it has been confirmed by the most accurate experiments in physics, to a degree of approximation that was unheard of when Einstein proposed the theory in 1905.

At that time, however, the most accurately confirmed theory of physics was Newton's theory of gravity, which was, of course, Galilei invariant. Poincaré noted this difficulty and sketched a Poincaré invariant theory of gravity in a paper published in 1906.[8] Einstein may have looked for some such theory for a short while, but in December 1907 he had already realized «that within the frame of the special theory of relativity there is no room for a satisfactory theory of gravitation».[9] This realization was linked to an idea he had one day while sitting at his desk in the Swiss Patent Office and which he later described as «der glücklichste Gedanke meines Lebens», «the most fortunate idea in my life».[10] It occurred to him then that if he were falling freely in a uniform gravitational field he would not be aware of his own weight. In our own days, we are all acquainted with this fact thanks to the repeated levitation of astronauts in front of television cameras; but unless we study relativistic physics we do not perceive its implications. Einstein did, and concluded that a frame freely falling in a uniform gravitational field is physically equivalent to an inertial frame. Hence, no frame at rest in a uniform gravitational field or moving in any way in a non-uniform gravitational field can be equivalent to an inertial frame. Since in the real world no gravitational field can be said to be uniform, except approximately and over a small region, there is no such thing as a global inertial frame of reference.

Einstein's insight promptly yielded a couple of important predictions on the optical effects of gravity,[11] but it would only come to full bloom when combined with an idea of the mathematician Minkowski, a former teacher of Einstein's at the Zurich Polytechnic. Minkowski showed also in 1907 that the seemingly paradoxical features of Special Relativity looked perfectly natural if the new kinematics was conceived as the geometry of a 4-manifold endowed with a quasi-Riemannian metric.[12] This manifold, which Minkowski called *die Welt*, «the world», is now known as Minkowski spacetime. The metric enables one to distinguish between spatial or «spacelike» and temporal or «timelike» directions in the manifold. They are separated at each point by a hypercone (the 3-dimensional analogue of a two-sheet cone), often called the light-cone because, according to the theory, light *in vacuo* is constrained to it.[13] The life history of a free particle traces a straightest line from past to future, that is, a timelike geodesic. I said the metric is quasi-Riemannian, because it is not positive definite: its action on a particular vector at a spacetime point may yield a positive or a negative number, depending on whether that vector points in a «spacelike» or a «timelike» direction. But apart from this no doubt important difference, the Minkowski metric is like a Riemannian metric. At first Einstein spurned Minkowski's contribution as a mere formalism, devoid of genuine physical significance. But he must have soon been conquered by its simple elegance, for in 1912 he was able to perceive, in its light, the remarkable analogy between his approach to gravity and Gauss's theory of surfaces.[14]

As I said before, the geometry of any smooth surface practically agrees with plane Euclidian geometry on a small neighborhood of each point. Hence, the groundplan of Lima's historic center can be represented virtually without deformation on a piece of squared paper. Likewise, the chronogeometry of the world practically agrees with the Minkowski metric on a small neighborhood of each event. Hence the experiments performed aboard a spaceship can be referred virtually without deformation to an inertial coordinate system. But the whole earth cannot be snugly wrapped up with a single, undeformed sheet of graph paper. Nor can the world, teeming as it is with gravitational fields at loggerheads with each other, be covered by a global inertial frame. Just as the local (approximate) flatness of the earth accounts for the successful use of Euclidian geometry in city planning, so the local (approximate) inertiality of freely falling frames accounts for the success of Special Relativity in laboratory physics.

Guided by this analogy and the mathematical advice of his friend Marcel Grossmann Einstein worked for the next few years on a chronogeometrodynamic theory of gravity, in which the spacetime trajectories of freely falling bodies and light rays *in vacuo* would be governed by the spacetime metric, and the latter would depend on the distribution of matter and light. His efforts came to fruition when he published, on November 28, 1915, the field equations the General Relativity.[15] They express the equality between a tensor field intended to represent the spacetime distribution of non-gravitational energy, and a tensor expression constructed from the spacetime metric and its first and second derivatives. Solving the field equations amounts to extracting the metric from this expression. Such second order differential equations can only be solved under rather strenuous simplifying assumptions. Consider for example the so-called vacuum Schwarzschild solution, which was found in 1916 by Schwarzschild and by Droste, and is the cornerstone of the new celestial mechanics.[16] It is assumed that spacetime is perfectly empty, with a metric which is spherically symmetric in space. Demonstrably, this implies that the metric is static in time and converges to the Minkowski metric at spatial infinity. The solution depends on a parameter m and is defined everywhere except at the spatial center of symmetry and on a spatial sphere of radius $2m$ (if the gravitational constant equals 1). The latter singularity can be eliminated by a suitable coordinate transformation, but the former is inerradicable. The solution aptly represents the gravitational field in the empty space surrounding a spherical body of mass m which stands at the spatial center of symmetry, and is enormously far from the nearest body. Take now a test particle that is, a body too small to make a significant difference in the metric and place it anywhere in the field. Its life history traces a timelike geodesic spiralling about the spatial center. The Schwarzschild solution was applied at once to see whether Einstein's theory can account better than Newton's for the behavior of Mercury. To do this, one equates *m* with the mass of the sun, subtracts from Mercury's observed trajectory the perturbations attributable to the presence of the other planets, and compares the resultant trajectory with the geodesic described by a test particle whose initial distance from the spatial center is equal to the distance between Mercury and the center of the sun at a given time. Their agreement is uncanny, for such a test particle will experience precisely the perihelion advance of 43" per century which had been measured in Mercury's orbit and which Newtonian mechanics was unable to explain.[17]

Einstein published the first cosmological solution of his field equations in 1917.[18] He did not produce it in order to account for any empirical facts. Indeed, one of the two items of astronomical information he adduced in his argument, namely, that heavenly bodies move relatively to one another at fairly low speeds, did not agree with Slipher's measurements of nebular Doppler shifts, which had been reported in scientific meetings and journals since 1913.[19] Einstein's motivation was, so to speak, philosophical. He was convinced that a geometrodynamic theory of gravity must satisfy what he called Mach's Principle, by which «the [metric] field is *exhaustively* (*restlos* Einstein's emphasis) determined by the masses of bodies».[20] He took it to mean that, at a great distance from massive gravitational sources the metric cannot be defined. This openly contradicts one of the conditions of the Schwarzschild solution, namely, that in the vacuum surrounding a spherically symmetric source the metric at spatial infinity is Minkowskian. At that time, this condition was still thought to be an independent assumption, and not a consequence of spherical symmetry. So Einstein together with the mathematician Grommer tried in vain to find «centrally symmetric, static gravitational fields which degenerate <...> at infinity».[21] He then realized that the question of the behavior of the metric at spatial infinity would vanish if there was no spatial infinity to begin with. You will recall that a Riemannian *n*-manifold need not be globally, but only patchwise, homeomorphic (topologically equivalent) to R^n.

For example, the sphere is globally homeomorphic to S^2, the topological subspace of R^3 formed by all real number triples <x,y,z> which meet the condition $x^2 + y^2 + z^2 = k^2$, for some fixed real number k. Likewise we can have a 3-manifold which is homeomorphic to S^3, the topological subspace of R^4 formed by all real number quadruples <x,y,z,w> meeting the condition $x^2 + y^2 + z^2 + w^2 = k^2$, for some k. Let us call such a manifold a supersphere. Now consider a spacetime in which each spatial slice (representing the whole of space at a particular time) is a supersphere. Obviously, superspheres just like ordinary spheres are topologically compact, so there can be no question here of spatial infinity. With some such idea in mind Einstein looked for a new solution of the field equations. He took it for granted that, in the large, the world looks roughly the same in every direction, from any vantage point, at any time, so he made the simplifying assumption that matter is uniformly distributed in space as a motionless presureless fluid. The solution he found required a change in the field equations themselves. This may sound like cheating, and Einstein himself would later describe the revision of the field equations as the greatest mistake of his life. But the thing is not so bad as it seems. The field equations must match the symmetric rank 2 tensor field T_{ik} representative of matter-energy with a tensor field of the same kind constructed from the metric and its first and second derivatives.[22] Energy conservation implies that the covariant derivative of T_{ik} vanishes everywhere ($T^{ik}{}_{;k} = 0$). The original version of Einstein's field equations certainly satisfies these requirements, but they are not the most general set of equations which does so. If we rewrite the original field equations in the form

$$E_{ik} = -T_{ik} \tag{1}$$

(where E_{ik} is a tensor constructed from the metric I shall call the Einstein tensor), the most general version should read

$$E_{ik} + \lambda g_{ik} = -T_{ik} \tag{2}$$

where l is a freely adjustable parameter, known as the cosmological constant, and gik stands for the metric.[23] Eqns. (2) follow at once from eqns. (1) and the condition $T^{ik}{}_{;k} = 0$, for, by definition of the covariant derivative, $g^{ik}{}_{;k} = 0$ everywhere. If $\lambda = 0$, (2) reduces to (1). If λ is sizable, gravitation will show up mostly through phenomena of repulsion, not attraction. Einstein assumed that l has a very small value and obtained a solution of eqns. (2) in which every instantaneous spacial slice of the world (technically speaking: every spacelike hypersurface orthogonal to the timelike flow of statically distributed matter) has everywhere the same constant positive curvature (in the Riemannian metric induced on such slices by the quasi-Riemannian spacetime metric). This immediately suggests that the said spacial slices are superspheres, so that spacetime, endlessly extending backwards and forwards in time, is topologically like the skin of a supercylinder (homeomorphic to $R \times S^3$). Evidently, the question of spatial infinity does not arise in Einstein's static universe; but neither does the question that plagued finitistic cosmologies since Antiquity, viz. if I come to the end of the world, couldn't I reach beyond it simply by stretching my arm?[24] for, though finite, this space has no end. No wonder then that Einstein's cosmological solution of the revised field equations was greeted as a major progress in philosophy.

It did not, however, allay for long his own philosophical worries. In the same year 1917 the Dutch astronomer de Sitter produced a solution of the revised field equations for a totally empty world (which is, of course, another way of having a static and homogeneous distribution of matter).[25] Two test particles placed on the same spatial slice of de Sitter spacetime fly away from each other at a speed which increases with their mutual distance. Now, an empty universe may seem the supreme scientific extravagance, but as an approximation to what is actually seen through the telescope it is in fact no worse than Einstein's plenum. From a sufficiently lofty standpoint the very galaxies could be seen as test particles. Slipher's results, which showed redshifts unexpectedly prevailing among the spectra of distant galaxies, could then be read as a corroboration of de Sitter's solution.

In 1922 and 1924 the German *Zeitschrift für Physik* carried two papers by one Alexander Friedmann, a Russian meteorologist, which showed that the cosmological solutions of Einstein and de Sitter are just two limiting cases of a countless family of exact solutions of the (revised) field equations.[26] Friedmann allowed the cosmological constant λ to take any value, less than, equal to or greater than 0. He also retained Einstein's assumption that matter is a pressureless fluid, but did not require it to be motionless.[27] He made a few geometrical assumptions which we now know to be justified if (i) the overall history of the universe appears the same, in every direction, from the standpoint of any speck of matter, though it may look different at different times; and (ii) the universe is stably causal (which roughly means that one will never meet and shake hands in the future with one's own earlier self, nor could one be enabled to do so through a slight change in the spacetime metric). The Friedmann solutions share some interesting features which I will now summarize. Since, at this level of abstraction non-gravitational forces are not contemplated, all matter is freely falling. Hence, its timelike flow occurs along timelike geodesics.[28] Any such geodesic, extended as far as it will go, may be regarded as the life course of a non-decaying particle. I call it a Friedmann worldline. Except in the special case of the Einstein solution, the Friedmann worldlines either diverge from each other, or converge towards each other, or first diverge and then converge. In all Friedmann solutions the Friedmann worldlines meet orthogonally with a succession of spacelike hypersurfaces, i.e. 3-manifolds, each one of which may naturally be identified with physical space at a given time. Any such space has everywhere the same Riemannian curvature, but the successive spaces of a given Friedmann solution may have different curvatures. However, if one of its momentary spaces has a positive, negative or zero curvature, then all its momentary spaces also have a curvature which is, respectively, positive, negative or zero. Thus a Friedmann spacetime either has the manifold topology of

Minkowski spacetime -if its space curvatures are negative or zero- or that of Einstein's supercylindrical spacetime -if its space curvatures are positive. But if the topology of the Friedmann universes is thus fairly tame, their affine structure- i.e. their system of geodesics -is startling. Suppose we live in a typical Friedmann universe in which the Friedmann worldlines are now diverging- either forever, or until such time as they will begin to converge. Suppose you mentally go backwards in time along an arbitrary worldline γ. Then, as you delve into the past, second by second, year by year, all the other worldlines will converge towards γ, so that matter in every spatial neighborhood of γ will be denser and denser, its average density increasing beyond all bounds as the time you have gone backwards approaches a certain value -the fashionable figure is about 12 billion years. Of course, the density of matter cannot be actually infinite at a given spacetime point, since neither the energy tensor nor the metric would be defined at such a point. But it is clear that, in a Friedmann universe of the said kind, if you wish to label the past points of any Friedmann worldlines with real numbers in agreement with the temporal distance between those points, you cannot use for this purpose the entire field of real numbers R, but only a finite open interval. Thus, if you choose a time unit, assign time 0 to the present and use the negative reals to label the past, the entire past of each Friedmann worldline will be mapped onto some interval (-T, 0). You may then call T, as Friedmann did, presumably with tongue in cheek, *die Zeit seit der Erschaffung der Welt*, «the time since the creation of the world». But you ought not to think that you have succeeded in dating creation at time -T. -T is just a real number which is the greatest lower bound of the range of your time coordinate. It cannot be, however, the date of any event, for events can only occur on Friedmann worldlines, and therefore, under the foregoing assumptions, must carry a date greater than -T. Indeed, for any event in a Friedmann universe whose starting time can be assigned, there must have been some time, however short, that went before it. In this way, thanks to Riemann's wonderful idea of manifolds, relativistic cosmology is able to conceive a world with a finite past, but which does not have a beginning.

Similar remarks can of course be made, *mutatis mutandis*, about the future points of the Friedmann worldlines in a Friedmann universe in which they converge. Technically speaking, in the typical Friedmann universe (not the static Einstein universe or the empty de Sitter universe), the Friedmann worldlines, parametrized by time, are incomplete geodesics, and the spacetime is timelike geodesic incomplete. For some time scientists thought that this remarkable feature of the Friedmann solutions was due to the neat simplifying hypotheses on which they rest, but that in the rougher circumstances of real life such geodesic incompleteness would be avoided. Thus, de Sitter wrote in 1933 that «the conception of a universe shrinking to a mathematical point at one particular moment of time <...> must <...> be replaced by that of a near approach of all galaxies during a short interval of time».[29] But the singularity theorems discovered by Penrose, Hawking and Geroch in the 1960's imply that, under very general, highly plausible physical assumptions, any relativistic universe in which the worldlines of matter converge towards the past or towards the future as in a typical Friedmann universe is timelike geodesic incomplete.[30] Of course, the proper question to ask is on what grounds may someone claim and to what extent is it true that we live in such a universe. On these points I expect to learn much from the lectures that will follow.

Universidad de Puerto Rico

NOTES

1. A. Einstein, «Kosmologische Betrachtungen zur allgemeinen Relativitätstheorie», *K. Preuß. Ak. der Wissenschaften, Sitzungsberichte*, 1917, pp. 142-152.

2. W. de Sitter, «On the relativity of inertia. Remarks concerning Einstein's latest hypothesis», *K. Nederlandse Ak. van Wetenschappen, Proceedings*, 19: 1217-

1225 (1917); «On the curvature of space», *K. Nederlandse Ak. van Wetenschappen, Proceedings,* 20: 229-243 (1917); «On Einstein's theory of gravitation, and its astronomical consequences», *Monthly Notices of the Royal Astronomical Society*, 78: 3-28 (1917).

3. A. Friedmann «Über die Krümmung des Raumes», *Zeitschrift für Physik*, 10: 377-386 (1922); «Über die Möglichkeit einer Welt mit konstanter negativer Krümmung», *Zeitschrift für Physik*, 21: 326-332 (1924).

4. G. Lemaître, «Un univers homogène de masse constant et de rayon croissant, rendant compte de la vitesse radiale des nébuleuses extra-galactiques», *Annales de la Société Scientifique de Bruxelles*, A 47: 49-59 (1927).

5. K. Gödel, «An example of a new type of cosmological solutions of Einstein's field equations of gravitation», *Reviews of Modern Physics*, 21: 447-450 (1949).

6. Let me add a short technical explanation for the sake of conceptual cleanness. The tangent spaces at every point of M can be merged very naturally to form a $2n$-manifold, the tangent bundle over M. The vector spaces of linear, bilinear, or, generally, k-linear functions (for some positive integer k) on the tangent space at each point of M will thereby be merged into further bundles over M. Let Θ stand for one of these bundles. I shall say that an object a I Q is placed at the point $P \in M$ if a is a tangent vector at P or is a k-linear function defined on suitable objects placed at P. A section of Θ is a smooth mapping f: $M \rightarrow \Theta$ such that, for each $P \in M$, $f(P)$ is placed at P. (The epithet 'smooth' applied to a mapping from an n-manifold to an m-manifold makes good sense thanks to the condition of smoothness imposed on coordinate transformations in the characterization of manifolds.) In particular, suppose that Θ is the bundle of bilinear functions on tangent spaces of M and that f assigns to each $P \in M$ a symmetric positive definite bilinear function on tangent vectors at P. Then Ä is a Riemannian metric on M.

7. B. Riemann, «Über die Hypothesen, welche der Geometrie zugrunde liegen», Abhandlungen der K. *Gesellschaft der Wissenschaften zu Göttingen*, 13 (1867), p. 149.

8. H. Poincaré, «Sur la dynamique de l'électron», *Rendiconti del Circolo Matematico di Palermo*, 21: 129-175 (1906). A five-page abstract of this paper

was published in the *Comptes Rendus* of the Académie des Sciences of June 5, 1905 (CR, 140: 1504-8). Einstein's first paper on Special Relativity was submitted on June 30th, 1905.

9. A. Einstein, «Autobiographisches»; in P.A. Schilpp, ed., *Albert Einstein, Philosopher-Scientist*, Evanston, The Library of Living Philosophers, 1949, p. 64.

10. From a manuscript, in the Pierpont Morgan Library in New York, as quoted in A. Pais, *Subtle is the Lord The Science and the Life of Albert Einstein*, Oxford, Clarendon Press, 1982, p. 178.

11. Namely, the gravitational redshift and the bending of light-rays in a gravitational field. See A. Einstein, «Über das Relativitätsprinzip und die aus demselben gezogenen Folgerungen». *Jahrbuch der Radioaktivität und Elektronik.* 4: 411-462 (1907).

12. Minkowki's lecture of 1907 was not printed until 8 years later: H. Minkowski, «Das Relativitätsprinzip», *Jahresbericht der Deutschen Mathematiker-Vereinigung*, 24: 372-382 (1915). In 1908 he published «Die Grundgleichungen für die elektromagnetischen Vorgänge in bewegten Körper» (*Nachrichten der K. Gesellschaft der Wissenschaften zu Göttingen, Math.-phys. Kl.*, 1908, pp. 53-111), and delivered his famous lecture «Raum und Zeit» (*Phys. Zeitschrift*, 10: 104-111 (1909)).

13. Light-cones in Minkowski spacetime are also called null-cones (from Null, the German word for zero), because in the Minkowski metric the spacetime interval or «separation» between any two points joined by a geodesic confined to a light-cone is always equal to 0. Such geodesics are known as null or lightlike geodesics. This terminology can be carried over to General Relativity, where, once again, the spacetime trajectories («worldlines») of massless particles (photons, gravitons *in vacuo*) are, of course, null geodesics. However, the null geodesics through a spacetime point generally do not form a hypercone in the metric sense of this word, and indeed they may not even form a 3-manifold homeomorphic with a hypercone. Hence, the term 'null cone' is normally reserved in General Relativity for the hypercone of *null vectors* at a point, generated by the vectors tangent to null geodesics through that point. Full-blooded physicists would only reluctantly call this hypercone a light cone, for it resides in a purely mathematical space, and though conceptually linked to

physical light, it cannot be filled with it. But this should not be understood to mean that light fills a different hypercone, resident in the same space as the null cone of vectors, as was suggested by one of the sketches that Mário Novello drew in the blackboard to illustrate his lecture at the Conference.

14. Einstein tells of this new insight in the preface to the Czech edition of his «popular exposition» of Relativity. See J. Stachel, «Einstein's Search for General Covariance, 1912-1915»,*Einstein Studies*, 1:63-100 (1989).

15. A. Einstein, «Die Feldgleichungen der Gravitation», *K. Preuß. Ak. der Wissenschaften, Sitzungsberichte*, 1915, pp. 844-847; cf. Ibid., pp. 778-786, 799-801, and ref. 12.

16. K. Schwarzschild, «Über das Gravitationsfeld eines Massenpunktes nach der Einsteinschen Theorie», *K. Preuß. Ak. der Wissenschaften, Sitzungsberichte,* 1916, pp. 189-196; J. Droste, «The field of a single centre in Einstein's theory of gravitation, and the motion of a particle in that field», *K. Nederlandse Ak. van Wetenschappen, Proceedings*, 19: 197-215 (1916)

17. Einstein derived this result using approximation methods in November 1915. See A. Einstein, «Erklärung der Perihelbewegung des Merkur aus der allgemeinen Relativitätstheorie», *K. Preuß. Ak. der Wissenschaften, Sitzungsberichte*, 1915, pp. 831-839.

18. A. Einstein, «Kosmologische Betrachtungen zur allgemeinen Relativitätstheorie», *K. Preuß. Ak. der Wissenschaften, Sitzungsberichte*, 1917, pp. 142-152.

19. V.M. Slipher, «The radial velocity of the Andromeda Nebula», Lowell Observatory Bulletin, N.58 (1913); «Spectrographic observations of nebulae», *Popular Astronomy*, 23: 21-24 (1915); «A spectrographic investigation of spiral nebulae», *Proceedings of the American Philosophical Society,* 56: 403-409 (1917). Einstein's other piece of astronomical lore was that stars are uniformly distributed in space. Again this would soon be disproved by American astronomers wielding the new long-range telescopes; but for the purpose of Einstein's argument it would do just as well to invoke the uniform distribution of galaxy clusters, or, at least, superclusters, which until very recently was uncontestedly supported by observational astronomy.

20. A. Einstein, «Prinzipielles zur allgemeinen Relativitätstheorie», *Annalen der Physik*, 55: 241-244 (1918), p. 241.

21. A. Einstein, ref. 1, p. 146.

22. The field equations should agree approximately with Poisson's equation in the situations well accounted for by the Newtonian theory. Therefore, they must be partial differential equations of the second order, at least. Einstein thought that given the accuracy of the available data it was pointless to look for equations of the third or higher order which, let me add, would have been even more intractable, mathematically speaking, than his second order equations have turned out to be.

23. For greater precision, let g_{ik} and T_{ik} stand as above, respectively, for the spacetime metric and the stress-energy tensor, and also, where appropriate, for their respective (i,k)-th components relative to the chosen coordinate system. I denote by g the determinant of the metric components relative to the chosen coordinate system. Following the Einstein convention, summation is understood over repeated indices. I write $f_{,h}$ for the derivative of the function f with respect to the h-th coordinate, and Γ^j_{ik} for $\frac{1}{2}g^{jh}(g_{hi,k} + g_{kh,i} - g_{ik,h})$, with h,i,j,k ranging over $\{0,1,2,3\}$. The Ricci tensor is given by

$$R_{ik} = \frac{1}{2}(\log g)_{,ik} - \Gamma^s_{ik,s} + \Gamma^\gamma_{is}\Gamma^s_{rk} - \frac{1}{2}\Gamma^s_{ik}(\log g)_{,s}$$

The original field equations should then be written as follows:

$$R_{ik} = -k(T_{ik} - \frac{1}{2}g_{ik}T^r_r) \qquad (3)$$

where the constant k depends on the choice of units. But (3) is equivalent to

$$R_{ik} - \frac{1}{2}g_{ik}R^r_r = -\kappa T_{ik} \qquad (4),$$

which, in turn, if we set k = 1, spells out the field equations (1). In the cosmological paper of 1917, Einstein wrote the revised field equations thus:

$$R_{ik} - \lambda g_{ik} = -\kappa(T_{ik} - \frac{1}{2} g_{ik} T^r_r) \qquad (5).$$

$$R_{ik} - \frac{1}{2} g_{ik} R^r_r + \lambda g_{ik} = -\kappa T_{ik} \qquad (6),$$

which spells out (2).

24. ἐν τῷ ἐσχάτῳ οἷον τῷ ἀπανεῖ οὐρανῷ γενόμενος πότερον ἐκτεῖναμι ἂν τὴν χεῖα ἠ τὴν ῥάβδον εἰσ τό ἔξω ἤ οὐ (Archytas of Tarentum, in Diels-Kranz, Fragmente der Vorsokratiker, 7th ed., 47 A 24).

25. W. de Sitter, reference 2.

26. A. Friedmann, reference 3.

27. The Abbé Lemaître, who independently rediscovered Friedmann's solutions in 1927 (reference 4), did not require matter to be pressureless, but only that its pressure be perfectly isotropic.

28. Technically speaking, the flow which is a vector field, generates a congruence, i.e. a manifold-filling family, of timelike geodesics

29. W. de Sitter, «On the expanding universe and the time-scale», *Monthly Notices of the Royal Astronomical Society*, 93: 628-634 (1933), p. 631.

30. A systematic exposition of the singularity theorems, with proofs and full references to the original papers, is contained in S.W. Hawking and G.F.R. Ellis, *The Large Scale Structure of Space-Time*, Cambridge, Cambridge University Press, 1973. See also R.*M*. Wald, *General Relativity,* Chicago, The University of Chicago Press, 1984, Chapter 9.

Barton Zwiebach

SUPERSTRING UNIFICATION AND THE EXISTENCE OF GRAVITY

Superstring theory is a novel and promising framework for a unified theory of all physical phenomena. It also seems to provide the first example of a theory of gravity compatible with the principles of quantum mechanics. In this paper I want to review the basic ingredients of superstring theory and discuss recent attempts at elucidating its underlying principles.

Superstring theory appears to provide a framework for the unification of all forces of nature. At the present time we know of the existence of four fundamental forces. From the strongest to the weakest they are: the strong force, responsible for the binding of quarks in hadrons, the weak force, which gives rise to radioactivity, the electromagnetic force, and finally the force of gravity. It should be explained first what is meant by unification of the fundamental forces. Let us do so. In physics theories are usually defined by action principles. From an action principle one can deduce equations of motion for the theory. The dynamical variables are the so called quantum fields and an action principle defines specific calculational rules for quantum mechanical observables. A non unified theory of the four interactions would correspond to having separate and independent action principles for each of the four interactions. It is said many times that this is not esthetically pleasant. While this may be the case, the truth is that it actually appears to be inconsistent to have independent action principles. For example, elementary particles in general

E. Agazzi and A. Cordero (eds.), Philosophy and the Origin and Evolution of the Universe, 75–86.

feel several of the forces; an electron feels the electromagnetic force, the weak force and the gravity force; quarks feel all four forces. In a completely unified theory one has a single action principle.

The first unification in physics was the unification of electricity and magnetism by Maxwell in the last century. Maxwell had no choice but to consider a unified system of equations for both the electric field E(x) and the magnetic field B(x) since it was clear from the experiments of Faraday that these two fields interacted with each other. The next step in the unification was taken by Glashow, Weinberg and Salam in the late 1960s. It had become clear experimentally that the electromagnetic and weak forces were closely related. A single action principle, an electroweak quantum field theory was proposed and shown to describe correctly experimental results. It predicted the existence of the intermediate weak bosons $W^{+,-}$ and Z^0, whose existence was dramatically verified a few years ago.

This electroweak theory used a novel idea, that of spontaneously broken symmetry. The theory is written with a symmetry that is not observed in nature. Specifically, for this theory the symmetry was chosen to be based on the Lie algebra of SU(2) x U(1). The experimentally observed symmetry for this theory is just U(1). But the theory itself breaks its initial symmetry spontaneaously down to the observed symmetry. In order to calculate with any theory one must first find a vacuum state, a state of lowest energy. It happens that the vacuum state in this theory breaks the original symmetry and reduces it to the observed U(1). Again, one may ask, what reasons are there for this large symmetry to begin with? Is it just because it is esthetically pleasant to have a large symmetry? The answer is that it is actually impossible to calculate quantum mechanical observables in a theory of weak interactions and electromagnetism unless one starts with this large symmetry. While the choice of a large symmetry for the electroweak theory is based on pragmatic considerations, it is probably no accident that nature uses beautiful mathematical theories with large symmetries. In fact, nice mathematics has sometimes been a useful guide for the finding of new physical theories. This was the case for the equation of the electron proposed by Dirac.

The standard model of particle physics consists of a single action principle for the electroweak and the strong interactions. As we saw above, the electroweak theory requires an SU(2) x U(1) symmetry group. The strong interactions happen to require the symmetry group SU(3). The standard model is a single quantum field theory based on the symmetry SU(3) x SU(2) x U(1). It contains quantum fields associated with the particles that mediate the forces, and quantum fields associated with the known elementary matter particles. This theory is, so far, in complete agreement with experimental observation. It is also a theory that is calculable using the rules of quantum mechanics. Why isn't this theory called a grand unified theory of the electromagnetic weak and strong interactions? The name «Grand Unified Theory» has been reserved for speculative theories in which the symmetry SU(3) x SU(2) x U(1) of the standard model is replaced by a larger symmetry. The choice of this larger symmetry is not unique, and it leads to several candidate theories. Some popular choices are SU(5), SO(10) and E_6. The theory with a larger symmetry is nicer, in the mathematical sense, and also in the pragmatic sense since it has fewer adjustable parameters. This large symmetry is also assumed to be broken by the vacuum down to the observable symmetry SU(3) x SU(2) x U(1).

It is important to emphasize that the introduction of a larger symmetry in the electroweak theory was a necessity for calculability, while in the case of the grand unified theories it is not. Nature does not seem to vindicate the idea of grand unification. Most grand unified models lead to the physical prediction of proton decay at some observable rate, and experiments have not observed any proton decay so far. While some sort of grand unified theory may be correct, the simplest and most compelling models have been ruled out.

Are we satisfied with the standard model? Not really. It has many adjustable parameters, not predicted by the theory, but that have to be measured. It does not explain why the symmetry of nature is SU(3) x SU(2) x U(1), or why there seems to be three generations of quarks and leptons. At a deeper level, it is not really a complete theory since it does not include gravity. Again, it is not consistent to ignore gravity, since gravity couples to all matter particles and all gauge bosons. It is, however, a useful approximation, given that the strength of gravity is so weak.

We now come to a crucial point. Einstein's theory of gravity is a classical field theory. The theories of electroweak and strong interactions, however, are quantum field theories. Nobody has been able to come up with a consistent framework in which one has quantum fields and a classical gravity field interacting with each other. Given the extraordinary success of the standard model it was natural to try to interpret Einstein's action as a quantum field theory. Unfortunately (or perhaps, fortunately) it turned out that Einstein's theory, as a quantum field theory is incalculable. This difficulty is the reason why it is said that Einstein's general relativity is incompatible with quantum mechanics. Technically speaking, gravity is non-renormalizable. Let us explain this.

Usually quantum field theory observables, such as probabilities of scattering of particles, are calculated using perturbation theory. The interactions between particles are assumed to be small and the observable O is expanded in a power series in terms of Planck's constant h, namely $O = O_0 + \hbar O_1 + \hbar^2 O_2 + \ldots$ In perturbation theory, in order to calculate O, one calculates successively O_0, O_1, O_2, and so on. In this process of calculation, one almost always finds infinities. Infinities can be of different types. Renormalizable quantum field theories correspond to theories in which one only finds a finite number of different kinds of infinities. Each kind of infinity gives rise to an ambiguity that must be taken care of by a measurement. For example, in a quantum field theory describing the interactions of photons and electrons (called quantum electrodynamics) one finds two kinds of infinities. They correspond to the measurements that must be made of the charge and the mass of the electron. Gravity is non-renormalizable, this means that in any calculation of an observable there appear an infinite number of different types of infinities. Since this can only be taken care of by doing an infinite number of measurements, the theory has a lack of predictive ability.

Many people thought, at first, that the problem may not be in Einstein's theory, but just that the perturbation method is not applicable, and that other calculational schemes would have no difficulties. Nevertheless, no such schemes have been found and we now believe that the difficulties found in perturbation theory are to be taken seriously. In fact, a little reflection shows that it should not be easy to have a quantum theory of gravity. In any quantum field theory of matter, the matter fields are quantized, but they live on the smooth four dimensional spacetime manifold. In gravity, the field variable is the metric tensor g, which defines distances in spacetime. If this field is quantized, one is doing something very drastic to the smooth manifold. It is not very clear what would happen to the concepts of distances or time intervals. This shows that quantizing the gravity field is in principle very different from quantizing matter fields, since the gravity field essentially describes the arena in which phenomena take place. Most people believe that the problem of quantizing gravity is so difficult, both technically and conceptually that its resolution would have to await many years, possibly well into the next century. This theory of quantum gravity would be based on a new and deep geometrical principle which would be impossible for us to guess at the present stage of development of both physics and mathematics.

Amazingly, we might have accidentally stumbled upon the theory of quantum gravity. We are very far from understanding the principles it is based upon, or what implications it has for our understanding of the nature of space and time. Nevertheless all we know about it shows that it seems to reconcile quantum mechanics and gravity, at least at the pragmatic level of perturbation theory. This theory seems to have no perturbative infinities. The theory we are talking about is superstring theory, and we now will explain what it seems to be about.

Superstring theory is a unified theory. It is based on an extended object, a string, or a loop, rather than a particle. Roughly speaking the modes of oscillation of this underlying string are seen by us as our particles. The extraordinary thing about superstrings is that one of its particles is always the quantum of gravity, called the graviton, so superstrings are theories of gravity. In addition, they include all the necessary ingredients to be a theory of everything. Superstrings are formulated in a spacetime of ten dimensions, rather that the usual four-dimensional spacetime we are all familiar with. It also has an almost unique symmetry group, that based on the exceptional Lie algebra E_8xE_8. Finally, another important ingredient is supersymmetry, which is a symmetry that relates bosons and fermions and appears to be an essential factor in eliminating the troublesome infinities of quantum theories of gravity. Spontaneous symmetry breaking must take place if superstring theory has something to do with the real world. In a first stage, the ten dimensional symmetry would be broken down to four dimensions by spontaneous curling up of six of the spatial dimensions. While this happens it is possible that the symmetry group E_8xE_8 will break down to a familiar grand unified group such as E_6. Finally, supersymmetry must be broken (since it is not observed in nature) and E_6 must be broken down to the symmetry of the standard model SU(3) x SU(2) x U(1). A good deal of work will be necessary to learn if all this takes place for sure, and whether the result is in complete agreement with observation. The hardest point to understand appears to be the breaking of supersymmetry, it is not yet clear how it happens.

In the remaining of this article we will not discuss the connection with observation but rather we will try to explain what are the fundamental theoretical issues in superstring theory. We will not take a historical approach. Our treatment will not be elementary or completely self contained. We hope, however, to give a feeling of how physicists see string theory at the present time. In particular, and for simplicity, we will be taking about bosonic strings (strings whose excitations correspond to spacetime bosons) rather than about superstrings (strings whose excitations include spacetime fermions, in addition to bosons). Realistic string theories can only be obtained from superstrings, but most of the relevant conceptual issues can be discussed in the simpler context of bosonic strings.

A (bosonic) string is a loop in spacetime. If we call X^μ the spacetime coordinates, with $\mu = 1,2,...D$, and D being the number of spacetime dimensions, the string is described by the set of functions $X^\mu(\sigma)$ where σ is a variable in the interval [0,1]. As σ varies $X^\mu(\sigma)$ traces a curve in spacetime. This curve is the string, and it is a closed string if the initial point of the curve $X^\mu(\sigma=0)$ coincides with the final point of the curve $X^\mu(\sigma=1)$. As the string moves it describes a surface in spacetime, this surface is parameterized by functions $X^\mu(\sigma,\tau)$ where σ and τ play now the role of coordinates on the surface traced by the moving string. An action principle for the moving string was proposed by Nambu and Goto. The action S is given by

$$S = \frac{1}{2\,\pi\,\alpha'} \int d\sigma\, d\tau \sqrt{-\left(\frac{\partial x}{\partial \tau}\right)^2 \left(\frac{\partial x}{\partial \sigma}\right)^2 + \left(\frac{\partial x}{\partial \tau}\frac{\partial x}{\partial \sigma}\right)^2}$$

In here the constant α' has units of length squared. It is the only adjustable constant of the theory and it determines the scale. It is therefore taken to be equal to $(\ell_p)^2$, where ℓ_p denotes the Planck length, which is approximately 10^{-33} cm. We have denoted $a^2 = \eta_{\mu\nu}a^\mu a^\nu$, where $\eta_{\mu\nu}$ is the D-dimensional Minkowski tensor. The above is an action principle for the *classical mechanics* of the string. The mathematically inclined reader may verify that the above integral computes the area of the surface traced in spacetime by the moving string. Note that the action is written as an integral over a two-dimensional space parameterized by (σ,τ). It is useful to think of the (σ,τ) coordinates as coordinates in a reference two-dimensional space R. The dynamical variables, X^μ are just fields on this two-dimensional space. Mathematically, the X^μ are maps from R to the spacetime since given a point (σ,τ), on R, the functions X^μ (σ,τ) give us a point in spacetime. Since the X's are the coordinates of spacetime, in the above action the coordinates of our spacetime are just two-dimensional fields. Thus classical string mechanics is actually two dimensional field theory. Much was learned from studying the above action. String quantum mechanics was studied using standard methods of quantization. It was established that the quantum mechanical states of oscillation of the string could be thought of as particle states in spacetime. Among these particle states a spin two massless state was always present. But a spin two massless state is the quantum that transmits the force of gravity. This showed that a closed string theory would always include gravity. Moreover, the theory only works in some specified number of dimensions D. For string theories D = 26 and for superstring theories D = 10.

The action we discussed above can be shown to be equivalent to another action, mathematically nicer and easier to work with

$$S = \frac{1}{2\pi\alpha'} \int d\sigma\, d\tau \sqrt{-g}\; g^{ab}\, \partial_a X^\mu\, \partial_b X_\mu$$

In this action the distinction between the reference two-dimensional spacetime R parameterized by (σ,τ) and the surface traced by the string in spacetime is emphasized. A metric tensor g is introduced on the surface R, and the indices a and b run over two values, since the surface R is two dimensional. The two-dimensional fields are now the g_{ab}'s and the X's. This action is still an action for classical string mechanics, but as a two-dimensional action it is far more elegant and useful than the previous one. The action has been written using a metric tensor, thus it has the symmetries of general relativity in two dimensions. The action is invariant under any two dimensional coordinate transformation. Two dimensional coordinate transformations form an infinite dimensional symmetry group. Moreover, the above action is also invariant under the transformation g_{ab} ---> Wg_{ab} where W is an arbitrary function of the two dimensional coordinates This is called Weyl invariance, and it holds only in two dimensions. Coordinate invariance and Weyl invariance make the above action tractable. Quantum mechanical observables can be calculated with the above action using the path integral approach to quantization pioneered by Feynman. In this approach any quantum mechanical amplitude A is obtained as the sum of the amplitudes obtained with all possible reference surfaces R. One therefore has to integrate over all possible surfaces R:

$$A = \int dR \text{ Amplitude } (R)$$

It is only because of coordinate and Weyl invariance that these integrals can be done. In integrating over surfaces one should not include surfaces that differ by either a coordinate transformation or a Weyl rescaling, since both metrics give the same action (because of the invariance under such symmetries). Therefore one should not integrate over all possible surfaces R, but only over those that cannot be related by either a coordinate transformation or a Weyl rescaling. Such surfaces are called Riemann surfaces (surfaces with complex analytic coordinates). Two different Riemann surfaces cannot be related by the action of coordinate transformations and Weyl rescalings. The different Riemann surfaces form finite dimensional spaces, called moduli spaces. In a moduli space M each point x corresponds to a different Riemann surface. The sum over the various surfaces in the above equation should therefore be thought of as an integral over some moduli space. Thus quantum mechanical amplitudes are calculated as

$$A = \int_M d_x \text{ Amplitude } (\chi)$$

Moduli spaces are still the object of intensive mathematical studies. Perturbation theory in string theory is studied by evaluating the above integrals explicitly. All studies to date indicate that no infinities arise from these integrals. Integrals over moduli spaces are well behaved. Even though there seem to be no problems in sight, there is, as yet, no complete rigorous proof of the perturbative finiteness of string theory.

In trying to understand the true meaning of string theory essentially two types of ideas have been tried. One of them emphasizes the fact that string theory is a special two-dimensional field theory, it is actually a conformal field theory, which essentially means that the theory is defined on Riemann surfaces. There are an infinite number of conformal field theories and each of them seems to correspond to a different vacuum for string theory, or in other words, different classical solutions of the fundamental theory of strings. This is not so strange because any theory containing gravity has many possible vacuum solutions. In this approach the dynamical variable is a general two-dimensional field theory described by a set of fields and interactions. The hypothetical dynamical principle would be such that the solutions are conformal field theories. Progress in this approach has been slow since it turns out to be very difficult to parameterize the space of two-dimensional field theories (a space in which every point represents a field theory).

A second approach aims at constructing a field theory of strings. Recall that the actions we examined were actions for string mechanics, even though they were field theories in two dimensions. This is clear, because even though we know that the string describes gravitons, nowhere in these actions we find the D-dimensional graviton field $g_{\mu\nu}(x)$. We can proceed by analogy. For string mechanics the dynamical variable was $X^{\mu}(\sigma,\tau)$, for particle mechanics the appropriate variable would be $x^{\mu}(\tau)$. Particle field theory is done by introducing fields $\phi(x^{\mu})$, that is by making the field depend on the particle coordinate, which was the dynamical variable in particle mechanics, and deleting the τ dependence. For strings we can therefore introduce a string field

$\Phi(X^n(\sigma))$, which is a functional of parameterized strings. It is not hard to see that such field will contain an infinite number of ordinary spacetime fields. A parameterized string can be described as:

$$X^{\mu}(\sigma) = x_{\mu} + \sum_{m=1}^{\infty} x^{\mu}_{m} \operatorname{Cos}(2\pi m \sigma) + \tilde{x}^{\mu}_{m} \operatorname{Sin}(2\pi m \sigma)$$

Here x^{μ} represents essentially the center of mass of the string. When the x^{μ}'s (both varieties) are small, the string is just some little loop near x^n. Thus in the limit of small strings x^{μ} can be thought of as the spacetime point where the string is. The string field can then be expanded in a Taylor series around $x^n = 0$ as follows

$$\Phi(x^{\mu}(\sigma)) = \Phi(x^{\mu}, x^{\mu}_{m}; \tilde{x}^{\mu}_{m}) = \sum \Phi_i(x^{\mu}) f_i(x^{\mu}_{m}; \tilde{x}^{\mu}_{m})$$

where the functions f_i are a suitable set of linearly independent functions of the x^{μ}'s (both varieties). Note that the coefficients Φ_i of the expansion must depend on x^{μ}, which is the only variable that is not taken to be small. Therefore the Φ_i represent spacetime fields! Since there are an infinite number of functions f_i there are an infinite number of spacetime fields. Having defined string fields, the next question is what should the action be. An action for open string fields was discovered in 1986, and an action for closed string fields was found in 1989. This action for closed string fields resembles the Einstein's action for the graviton field in that both actions are completely nonlinear. While in the case of the Einstein action we know that the underlying geometrical principle is general coordinate invariance, it is still very unclear what is the symmetry principle behind the closed string action. Its symmetry group is definitely much larger than that of any theory we know. Proper understanding of the closed string action may open up new branches of mathematics and finally give us the desired physical understanding of what gravity and spacetime really are.

Center for Theoretical Physics of Massachussetts Institute of Technology

REFERENCES

D'Hoker, E. and D.H. Phong (1988). *Reviews of Modern Physics* 60:917.

Green, M., J. Schwarz and E. Witten (1987). *Superstring Theory* Cambridge: U.P.

Schwarz, J.H. (1987). *International Journal of Modern Physics A*,Vol. 2, No. 3: 593

Dudley Shapere

THE UNIVERSE OF MODERN SCIENCE AND ITS PHILOSOPHICAL EXPLORATION[1]

I SCIENCE AS OBJECT AND FRAMWORK OF PHILOSOPHICAL INQUIRY

Our ultimate concern is with some larger aspects of the interpretation and significance of the world-view of science: questions about the nature of the knowledge that has been acquired, about the extent to which we can consider it to be knowledge, about the methods by which it has been acquired, about what constitute adequate reasons for belief, about what it is for something to exist and to have properties, about humanity's place in the universe, about humanity's values and their basis, and about the means and possibilities of fulfilling them. These are questions which have traditionally been assigned to the philosophy of science or to philosophy more generally. What is the relevance of the current scientific picture of the universe to such questions?

A wide range of scientific approaches and results - biological, anthropological, archaeological, psychological, historical -converge to provide at least the beginnings of insight into the primitive origins of human thought about nature and knowledge.[4] Those fields, and especially evolutionary biology, provide a framework, a body of background knowledge, which sets the broad terms within which we must see the roots of belief about the

E. Agazzi and A. Cordero (eds.), Philosophy and the Origin and Evolution of the Universe, 87–202.

world in which we find derselves. That background framework instructs us that the species *homo sapiens* has descended from ancestral hominoid and hominid species. Though the exact details of the descent of our species from its immediate ancestors are far from being completely clear, the general concept is established clearly and definitely enough to inform us that *the sources of original human (and pre-human) views of nature must have lain in the everyday experiences of those early human and prehuman beings* - in the practicalities of obtaining food, of staying alive, of nurturing and defending kin, of fear and awe and hope and delight in the face of their surroundings, and all that went with those experiences and coping with them.

Out of those attempts came not only specific practical recipes for coping, but also a plethora of broader mythical and religious beliefs and philosophical speculations that we can perceive only dimly before the dawn of recorded history. Among them must have been the ancestors of the first ideas we find in written history about the universe in which we live, the thoughts of the presocratic philosophers - those ideas themselves being ancestors of much later science and philosophy. At least in general terms compatible with our framework knowledge about the evolution of the human species, we can understand that such ideas must have arisen in earlier, more concrete, and more intuitive (pre-oral, prelinguistic) dealings with the world. They had their origins, that is, in the efforts of emerging mind to deal with the circumstances of life, to adjust to an environment that was at once hostile and awe-inspiring, and ultimately to make sense of the world into which human beings had been thrust.

This background framework stems from one of the best-established parts of our present scientific picture, the fact of human evolution. In its light, we are forced to accept the further fact that what we take to be requirements for understanding, and for how to attain it, are not necessary and inviolable requirements after all, but are subject to modification or rejection in the light of deeper and more careful study. For all their later sophistication, such ideas had their ultimate roots in everyday, primitive experience, and must be seen as possibly being subject to whatever limitations we may find that experience to have. Further, since, according to our best scientific understanding, our brains - as parts of us - are also part of an evolutionary process, we must take as part of the framework of our investigation the principle that we have no access to any necessary requirements of inquiry, existence, or understanding. Human experience, extended by instrumentation based on prior knowledge, is the only access we have to knowledge and understanding, including understanding of what knowledge and understanding themselves should consist in.

But the fact of human evolution is not the only piece of the scientific picture that functions in the framework of our investigations. The biological aspect of the framework only informs us that requirements which we place on inquiry, existence, and understanding *may* not hold as necessary - that they *may* have to be violated, transformed, or replaced. For, like all else about human beings, our ideas concerning knowledge, existence, causal influence, explanation, and so forth, must originally have arisen in the context of dealing with everyday (and, originally at least, primitive) experience, and it is therefore at least possible that those ideas will have to be qualified or even rejected when our inquiries go beyond the everyday. That much we learn from the evolutionary framework of our inquiries into those concepts. It does not tell us that they *are* so violated, much less that they must be. That they indeed have been, and how they have had to be transformed and in many cases replaced, is something that we have learned, and continue to learn ever more deeply, from physics and cosmology.

Those subjects have taught us that, when our investigations go beyond the level of everyday experience, to the realms of the very large and the very small, departures from the everyday and its intellectual descendants are *as a matter of discovered fact* needed if we are to make sense of the broader and more fundamental nature of the universe in which we find ourselves. In physics and cosmology, everyday ideas of what objects are like, of space and time, of causal interactions or forces, have all been repeatedly violated.[5] Even our understanding of the relations between these have been severely altered. Further, deep revisions in our understanding of what it is to "make sense of," to understand, that universe have accompanied such alterations: profound revisions of our ideas of what understanding - explanation - consists in. So too have our ideas of what it is for something to "exist" been changed,[6] though those alterations, like those in our ideas of what understanding consists in, have at best received only lip service, and not the emphasis they call for. (Thus, without realizing the origin of ideas in everyday "middle-sized" experience, and therefore failing to grasp the contingency of the claims involved, classical metaphysics saw them as necessary truths transcending the processes and outcome of scientific inquiry.) The pace and magnitude of such changes have increased enormously in the contemporary field-theoretic descendants of quantum theory and its cosmological applications. Thus the physical and cosmological components of the contemporary scientific world-picture add a second major ingredient to the framework within which the philosophical exploration of science must proceed.

Modifications and rejections of prior expectations are only to be expected in the light of what we know about the history of the human species and its ancestors. Such rejections and modifications have occurred and must be taken seriously. How to go about taking them seriously can be sketched briefly through the following example. A common view of quantum mechanics is that there are deep problems regarding its interpretation: how are we to understand that theory? Indeed there are such problems; but "taking the theory seriously" has to do with how we are to approach them. In many approaches, certain antecedent criteria are brought to bear, implicitly or explicitly: criteria, for instance, of what can be meaningful (*e.g.*, operationalist or verificationist criteria), or of what it is for a theory to be a "realist" one, or for certain theoretical concepts to describe or correspond to something that "exists." But a background framework that tells us that we must learn what it is to understand, and for something to be "real," demands that, rather than considering quantum mechanics and its successors to be in need of "interpretation," we need to ask *Whether quantum mechanics and its field-theoretic descendants may be telling us something about how we are to understand the universe at a fundamental level, what it is to be "meaningful," to be a "realistic theory," to "exist," and so forth*; or in other words, *Whether our views of what it is to understand (and so forth) need to be altered in the light of what we have learned through these theories.* The problem is not whether quantum theories can be taken as giving a "realistic" picture of the way things are; for that way of posing the issue presupposes an understanding of what it is for a theory to be "realistic." Rather, what it is for a theory to be realistic - and what it is for something to be real, or exist - is something we need to be learning, not presupposing.

To ask such questions is to turn the philosophical tradition upside down. It is to refuse to take for granted the possibility of human access to truths about existence, explanation, *etc.*, which science must obey, and in terms of which it is to be interpreted. Instead, it is to admit the possibility that even such ideas are human constructions, originating in experience of an everyday "middle-sized" world - ideas which, like any human ideas, may be subject to alteration or rejection in the light of what we learn through investigation. We need to ask not what we can do for quantum theory, but rather, what quantum theory can do for us. This is what it is to "take seriously" the departures from traditional preconceptions which quantum theories and other scientific developments impose.

The picture given us by current science which will be sketched in this essay provides the *framework* of philosophical inquiry in the following sense. Some parts of the picture (including some but not all aspects of the problems and directions of research involved in it) are well-established, and *it is those parts that must be taken as background information usable in framing the uses and the conduct of our inquiry concerning the conclusions of science and the methods by which they have been attained.* Naturally, such background is to be used with a judiciousness appropriate to the evidence available, the recognized problems involved, and the general possibility that we might be wrong despite the available evidence and our assessment of the known problems.

There is of course more to the background framework of philosophical inquiry than just the two aspects I have mentioned, the fact of human evolution and the departures of physics and cosmology from everyday explanatory expectations. These two are given prominence here because of what they imply about the necessary and radical revision of the philosophical enterprise. For they demand a reinterpretation, and in some cases a replacement, of the questions asked in that enterprise, of the methods by which those questions are to be answered, and of the sorts of answers for which we are to seek. This threefold revision is one whose necessity we have learned in the course of the most successful form of inquiry in which human beings have engaged: the scientific enterprise.

More generally, a framework for an inquiry - any inquiry, as we have learned - consists of a body of background knowledge which provides, to at least some degree, a specification of the kinds of problems that need to be investigated, the methods by which those problems are to be approached, and the characteristics of the kind of answers to be sought.[7] Such a guiding framework will not, in general, consist of all background information available, but only of a subset of it. If the problems are solved, the framework and the topic being explored are unified, the framework giving an explanation of the area of exploration. On the other hand, the investigation can lead to modification or even rejection of at least some elements of the framework.

But the very fact that our best science provides the framework for our inquiry requires that the contemporary scientific picture be the *object* of our inquiry also. Here more is required than merely the best of current science. It is the full picture, not merely the best established parts of it, that are relevant to our study. It includes the framework as described above, but also much more: it includes also the less well-established loopholes and problems as far as we can discern and assess them, the directions of research it suggests, and the alternatives still remaining open. In other words, we must include as parts of the object of our study all of the following: an understanding of the theories and the reasons for believing them, or for seeing them as most promising, and for rejecting their rivals; major questions facing the theories and their compatibility with one another; alternatives open (including different interpretations of the major views and the problems facing them); research directions and why they are promising; assessment of the kind and degree of seriousness and importance of the problems; and in short, all aspects of the current (and changing) scientific view of the universe.

That full picture constitutes the object of inquiry of the philosophy of science in the sense that that discipline is concerned to understand what this picture tells us about such questions as the following: the character of knowledge and its limitations; how we go about getting knowledge (and in particular what constitute good reasons for belief); the nature of existence; what it is to understand; and the implications of the picture for our understanding of ourselves, our place in the universe, our values, and the means and possibilities of fulfilling them. These and related questions have traditionally been the domain of philosophy. However, as we have seen, our approach to them must be radically different from that taken in much of traditional philosophy. Our object of inquiry is science itself, and its aim is to learn *from* the conclusions, problems, methods, and standards of explanation gleaned in inquiry, not to apply antecedent criteria thereto. For as we have learned in the long and difficult process of scientific inquiry which has led to our present scientific picture, such antecedent criteria have no other ultimate basis than human experience and the modifications which such experience and its instrumental extensions force upon us. All other supposed grounds of knowledge ("the hardness of the logical must"[8]) are mere imaginings.

So far, I have taken it for granted that there is something that can be identified as "the present scientific picture." But is there such a thing? Is it possible to point to a body of beliefs which is generally agreed, among authorities, to be well-supported by evidence? And if so, does that body of beliefs have the kind or degree of unity that forms a coherent "picture"? Can we, *within* that framework, describe the problems, controversies, and alternatives open (problems *within* the picture)? Can we reasonably assess the problems *confronting* that picture, and the alternatives which remain reasonable in the light of those problems (problems *about* the picture)? Can we give an account of the major lines of research open in the light of both sorts of problems?

It will be the purpose of Part II of this essay to show that there is indeed such a "picture," and that, though much remains to be done and many problems and alternatives remain open, the picture does possess a great deal of unity. Indeed, as will be shown, the unity achieved and promised (on the basis of excellent reasons) is far greater than it has been in the past. That very achievement is one of the things that has made the past two decades in physics and cosmology one of the greatest eras in the history of science. (And that very greatness, and the unity achieved, make an assessment of its implications even more important, for the purposes mentioned in the opening paragraph of this paper, than would otherwise be the case.) In accordance with the conclusions of Part I of this essay, Part II will examine the insight and unification achieved. Part III will survey some of the major problems in the current theories, and Part IV will outline some of the major directions of current research toward meeting those problems. Finally, Part V will show that, despite the many outstanding problems, the current scientific picture of the universe and its origin and evolution is a remarkably successful, coherent, and promising one, and that it constitutes, in specific ways, a framework (in the sense defined above) within which astronomical, chemical, and biological theories must at present be constructed and situated. In particular, it encompasses theories of the formation of galaxies and clusters of galaxies, theories of the origin and evolution of the chemical elements and their abundances, theories of the origin and evolution of stars and planetary systems, and theories of the origin and evolution of life. An assessment will be made of the kinds and degrees of seriousness of the problems in this total picture (both within and about it), particularly in the light of claims that the next major step of development in physics may well be a "theory of everything" and the "end of theoretical physics."

The purpose of this investigation (apart from its intrinsic interest and importance) will be to back up these considerations, with details. The focus of the present paper will be on the science rather than the philosophical investigation: on surveying the picture which forms the framework and the object of philosophical investigation. However, to show how certain traditional problems are to be approached, a number of philosophical issues will be touched on, in varying degrees of detail and explicitness. Among them are the following: the problem of reduction and emergence, that is, of whether and in what sense the methods and results of one field can be "reduced" to those of another; the question of the nature of scientific explanation, and particularly the question of the effects of new scientific theories on scientific method; the nature of reasons for accepting or doubting a scientific hypothesis the nature of scientific problems; the sense in which a theory can be said to be "complete"; the nature of space and time; and the role of verifiability and observability in science.

The discussion given in this paper is by no means intended to be complete. As far as biological evolution is concerned, there is a cutoff at the origin of life, omitting (except for the brief remarks in the final paragraph of this Part) questions about higher organisms and intelligence. The latter especially would, of course, be an important topic in a fuller study; but little is yet understood about this subject, and there is much disagreement among workers in the field; here we are concerned with a background of things that are more fully understood. But even within the limits of the present paper, many important topics are omitted from discussion, and some that are included are only touched on briefly. It would have been impossible to survey all the details and problems concerning each topic, though I have tried to indicate the importance of such details, and in every case they lie behind the conclusions drawn. The present paper aims only at giving some of the essential features of the larger picture of modern science, with brief views onto the landscape of related philosophical issues.

At each stage of discussion, I have given references of a variety of sorts, among which the following should be mentioned. In all cases, I have tried to cite a few of the seminal works involved in the introduction of the ideas discussed. In some cases I have listed what I consider to be excellent surveys of the developments concerned, particularly where those surveys are readily accessible, as in the *Annual Review* volumes. In a few cases, where I believe the recent historical development to be especially illuminating about the developments and about changes in scientific method and perspective in general, I have given references to a series of works over a time period, particularly conference reports.

Somewhere, sometime, creatures in whom selection for a larger cerebral cortex had taken place were able to put that trait to new uses which were different from those (perhaps as allowing more efficient tree swinging, or perhaps for something else) for which that trait had originally been selected. Instead of having to react to immediate contingencies as they arose, the trait permitted the creatures possessing to anticipate events, to an unprecedented extent, and thus to apply past experience more effectively in dealing with unforeseen dangers. It was a classic case of preadaptation: of the putting to use of a trait, originally selected for other purposes, to a new use which then proved of a new adaptive value, so that improvements in the effectiveness of the trait for the new use became selected for in turn. The development should not be surprising: nature, acting in its blind way to select traits of adaptive value, would have promoted a trait which allowed the creatures possessing it to anticipate events, to apply past experience to dealing with present contingencies. But that trait, in turn, freeing the creatures from dependency on immediate experience, at length served as a preadaptation resulting ultimately in the ability to ask questions about the deeper nature of the world in which the creatures found themselves. That long and gradual process has constituted the emergence of mind, whose ultimate products we are concerned to understand.

II THE UNIVERSE OF CONTEMPORARY SCIENCE[9]

SECTION 1. THE GAUGE-THEORETIC REVOLUTION.

By the end of the 1950's, physicists had come to the conclusion that the processes of nature are governed by four fundamental forces or interactions. The gravitational force is responsible for the gathering and cohesion of matter in large systems, and for the phenomena of falling bodies, projectiles, and orbital motions. The electromagnetic force brings about the cohesion of atoms in molecules and of molecules in larger bodies, and is therefore responsible for the everyday phenomena of physics, chemistry, and biology other than those resulting from gravitation. In addition, it controls the transmission and absorption of light. The strong force holds the protons and neutrons in atomic nuclei together, and governs much of the behavior and characteristics of those nuclear particles and various others when not bound in nuclei. (Other behavior of particles is directed by the weak and electromagnetic forces and, to a minute extent that can in all but the most exotic circumstances be ignored, the gravitational.) The weak force accounts for the radioactive decay of elements, and, more profoundly, for the conversion of one chemical element into another. In the quantum theories of the electromagnetic, weak, and strong interactions, the forces are carried by "exchange particles," which are particles of whole-number spin (bosons). For each force, there is one or more such particles whose exchange between other particles (and in some cases between the bosons themselves) constitutes the force.[10]

As understood in the early 1960's, these four forces were quite different from one another. (Clearly they *are* very different in the universe as we find it today.) Despite its dominance on the scale of our ordinary experience, gravitation is by far the weakest of the four, its apparent strength being due to the fact that it is not balanced by any opposite (antigravity) force, and thus that it can be built to indefinitely large strength merely through the accumulation of more and more matter. The electromagnetic force, on the other hand, comes in positive and negative charges. These neutralize one another under most circumstances, thus masking the fact that that force is vastly stronger than the gravitational. (A magnet can pick up a nail, overcoming the gravitational pull of the entire earth.) Relative to the strong force, the strengths of the four forces at ordinary energies are as follows:

Strong force	1
Electromagnetic force	10^{-2}
Weak force	10^{-5}
Gravitational force	10^{-40}.

While the gravitational force acts on all bodies in the same way, the electromagnetic force acts only on electromagnetically-charged bodies. The weak and strong forces are, each in its own way, still more discriminating. The gravitational and electromagnetic forces have infinite range, but the strong force falls off effectively to zero at distances beyond 10^{-13} cm., while the weak does so at distances roughly one one-hundredth of that.

Despite these apparently profound differences between the forces, by the 1970's a successful unification of two of the four had been achieved, and considerable progress had been made toward a unified theory of three of the four.[11] In 1968, Weinberg and Salam independently advanced a theory unifying the electromagnetic and weak forces.[12] This "electroweak theory" had the form of a local gauge theory, a form which had been developed in another context by Yang and Mills in 1954.[13] However, such Yang-Mills gauge theories had been viewed as mere curiosities, since they seemed to permit useful calculations (*i.e.*, in this case, to be renormalizable) only if the fundamental particles of the theory were massless. For three years after its publication, the Weinberg-Salam theory was similarly viewed - in fact largely ignored. However, in 1971 't Hooft presented convincing arguments that the renormalizability of Yang-Mills theories was not destroyed by the appearance of mass in the particles.[14] Thus, Weinberg and Salam had assumed that the electroweak force was carried by four massless exchange particles. However, the unity of the force would exist only at relatively high energies, and below a certain energy, the unity was "broken" by a dynamic process known as the Higgs mechanism , separating the electromagnetic and weak forces that we know in our relatively low-energy universe. In this symmetry breaking, the Higgs mechanism, endowed three of the four electroweak exchange particles with mass. These were the W^+, W^-, and Z^o particles carrying the weak force; the fourth particle, the one which did not gain mass, was the photon, the exchange particle of the electromagnetic force. 't Hooft's arguments, subsequently rigorized by a number of other physicists, made it clear that the breaking of the electroweak symmetry between the electromagnetic and weak forces did not destroy the symmetry, which only became "hidden."

This proof of the utility of Yang-Mills gauge theories was coupled with a series of spectacular experimental successes of the Salam-Weinberg theory over the following few years.[15] The confidence engendered by the theoretical and experimental successes of the theory led to the attempt to apply the gauge-theoretic approach to the strong force. By 1974 a family of gauge theories of that force, referred to as quantum chromodynamics, had been proposed.[16] There were powerful reasons for the appeal of the gauge-theoretic approach beyond its success in the case of the electroweak theory and in the general case of renormalizability. In the gauge-theoretic approach, once one had a gauge field, the characteristics of the associated force were completely determined; no longer did one have a situation like that of Newtonian mechanics, wherein the forces that governed the behavior of matter had to be introduced wholly independently of the characteristics of that matter itself. The gauge theory of the strong force in particular had an additional appeal. Even apart from the issue of non-renormalizability, older theories of that force had bristled with mathematical difficulties; calculation was all but impossible. But the new gauge formulation was shown to imply that the strong force decreased in strength with decreasing distance, asymptotically approaching zero as the distance between the interacting particles approached zero.[17] (In particle physics, decrease in distance can be shown to be equivalent to increase in energy, so that the higher the energy, the weaker the strength of the strong force.) This meant that strongly-interacting particles at high energies could be treated as free of forces: "asymptotic freedom" made it possible to make useful calculations of strong-interaction processes at high energies.

Thus, by the mid-1970's, theories of the electroweak and strong forces had been constructed. Furthermore, those two theories had the same mathematical form, that of a local gauge theory. Even without any deeper unification of the two theories, their similarity was such that they could be conjoined into what has come to be known as the "Standard Model." This Standard Model is highly successful in accounting for the non-gravitational features of the world around us today. Indeed, there is not a single experimental result in any relevant area that is incompatible with it.

The unification energy for the electroweak interaction - the energy above which the electroweak force would be manifested in all its unity, and below which the electromagnetic and weak forces would be distinct - could be determined as on the order of 10^2 GeV (billion electron volts), the corresponding temperature being 10^{15} °K. The question naturally arises as to whether there exist, or ever have existed, conditions under which the electroweak unification would have been realized. As a matter of fact, a theory of the early universe, increasingly successful since 1964 in banishing its rivals, was at hand, and it provided the locale and time at which the energies required for the electroweak unification would have been present.

SECTION 2. PARTICLE PHYSICS MEETS COSMOLOGY.

In the years since Einstein applied his theory of General Relativity to cosmological questions in 1917,[18] the theory that the universe had expanded from an extremely hot initial state, perhaps a singularity, had been developed, first by Lemaitre[19] and in further detail by Gamow[20] and others. However, the theory fell on bad times in the late 1940's and early 1950's. In the light of measures of galactic distances and of estimates of the rate of expansion (Hubble constant), it appeared that the universe would, on that theory, be younger than many of its constituents. As if this were not trouble enough, the theory also foundered on the problem of showing how the chemical elements could have arisen in the postulated hot initial state: although the lightest elements could have been synthesized, the production of heavier ones was blocked by the fact that there are no stable nuclei with atomic masses 5 and 8.[21] A rival cosmological theory arose, the Steady-State theory, to oppose the Lemaitre-Gamow theory, which one Steady-State advocate, Fred Hoyle, christened the "Big-Bang theory."[22] According to the Steady-State theory, the expansion of the universe does not imply that there ever was a high-density state from which it expanded. Rather, the universe has, on a sufficiently large scale, always appeared, and will always continue to appear, very much the same, in average density and other features, as it does now or as it will at any other time. To preserve the average density in the face of the galaxies' moving apart, the Steady-State theory specified that new matter will be created in the spaces between them. The universe is thus eternal in time, and on sufficiently large scales should show no evidence of variety or evolution.

But the Steady-State theory was soon confronted with a series of criticisms based on observations. First, the negative reason for its introduction, the "age problem" which troubled the Big-Bang theory, was resolved by the late 1950's. Use of Cepheid variables as a primary basis of distance-determinations had failed to distinguish two types of such variables, each with its own relation between period of variability and absolute luminosity; correction of that error led to an increase in the relevant galactic distances by a factor of about 2.6.[23] Another distance-indicator, brightest stars in a galaxy, turned out to have suffered from confusion of large glowing clouds of gas (HII regions) with stars, and that realization led to a further increase of the distance-scale, and therefore of the time-scale, by a factor of about 2.2.[24] Through these and other adjustments, the apparent conflict between the age of the universe and the ages of objects in it was thus greatly reduced if not entirely eradicated.[25]

By 1955, counts of radio galaxies were indicating that such objects are more frequent at greater distances. Since the farther galaxies are seen at an earlier time than the nearer ones, this seemed clear evidence that, contrary to the claims of the Steady-State theory, evolution has occurred among such galaxies.[26] Quasars, first discovered in 1961, pointed to the same conclusion: there are more quasars far away from us than nearby, suggesting that the universe does not appear nearly the same at every time. These observations did not necessarily refute the Steady-State theory; ways out of these objections could be contrived.

On the other hand, the discovery of the cosmic background radiation did convince most workers that the Big-Bang scenario was indeed the correct one. Prediction of relic radiation from the origin of the universe had been made on several occasions.[27] Now, in 1964, Penzias and Wilson, searching for the source of noise which was interfering with satellite communication, found radiation which was isotropic in distribution, at close to the temperature predicted, and with the black-body characteristics expected of the relic radiation.[28] Further successes in predicting the abundances of helium-4, helium-3, deuterium (hydrogen-2), and later of lithium-7, that would have been produced in a Big-Bang scenario, were additional evidence convincing most scientists of the validity of at least the broad features of the hot Big-Bang theory.

In physics, temperature and energy are related concepts, and so the high energies present in the early stages of the Big-Bang universe were a natural place for the processes of high-energy unification and symmetry-breaking to have occurred. The marriage of particle physics and Big-Bang cosmology thus followed almost as a matter of course as part of the gauge-theoretic revolution.[29] The application of the electroweak theory and quantum chromodynamics - together forming the Standard Model - to the Big Bang theory led to profound new insights into the history of the universe. Through it, scientists could understand processes occurring as early as 10^{-35} seconds after the Big Bang itself. In Part IV, we shall see how scientists are trying to extend this picture back to the Planck time, 10^{-43} seconds after the Big Bang, when all four fundamental forces would have been unified.

Applications of physical theory to such fractions of the first second of the history of the universe, and with such details, might seem the height of audacity when comparable inferences cannot be made about much more recent periods. However, it must be remembered that conditions at the beginning of the universe were much simpler than they became later. For example, in the early universe the energy - the temperature - was very high, so that asymptotic freedom held to a very high approximation; and as we have seen, strong-interaction calculations under such conditions are very simple, as they are not in the later, colder universe.[30]

SECTION 3. HISTORY OF THE UNIVERSE: THE LARGER PICTURE.

Thus the fusion of particle physics and cosmology gave, or at least promised, a picture of the sequence of events occurring in the early history of the universe. Before 10^{-43} seconds, as we shall see in Part IV below, all was unity. When that time was reached, the universe had cooled sufficiently in its expansion so that the gravitational force became separated from the electroweak/strong force. A very long time later, when the universe had reached the age of 10^{-35} seconds, the strong and electroweak forces separated, presumably through a Higgs mechanism or some analogous process. Much later still, at the age of 10^{-10} seconds, the electroweak force split to form the electromagnetic and weak forces. Thus arose the four fundamental forces with which we, or at least physicists, are familiar today.

However, this picture, provided by the marriage of particle physics and cosmology, is not confined to the opening instants of the history of the universe. For that same combination of physics and Big Bang theory is open to extension to the subsequent history of the universe, and, in the sense I described in Part I, forms a framework, a background of ideas, within which theories of later processes and events must be constructed. After approximately

100 seconds, the universe had cooled to the point where deuterium, helium, and some lithium nuclei could form. For the first hundred thousand years or so, radiation dominated all processes, and free electrons scattered radiation after less than a centimeter of travel. But then cooling allowed electrons to be captured by atomic nuclei, and radiation became free to travel over great distances, as it does today. Later still, perhaps at about a billion years after the Big Bang, galaxy - or galaxy cluster, or cluster-of-cluster - formation began, an event or sequence of events to which I will return. Collapsing clouds of gas led to the birth of stars (perhaps during or even preceding galaxy formation). It having been demonstrated that the chemical elements beyond helium (with traces of lithium) could not have been produced in the Big Bang, scientists had earlier turned to working out the details of how the heavier elements could be produced elsewhere, namely in the interiors of stars, in the process of stellar evolution. In the late 1930's, nuclear reactions in the cores of stars had been established as the source of stellar energy,[31] and that theory, worked out in far greater detail over the following decades,[32] was gradually extended to account for the evolution of stars of various masses. In the process, depending on whether the star was massive enough to produce the relevant nuclear reactions, energy would be released by nuclear conversions of elements up to iron. Still heavier elements than iron (these heavier elements requiring input of energy for their synthesis) would be produced in supernovae explosions of dying massive stars, and through a few other processes which have by now been investigated in considerable detail. (Agreement of predicted with observed abundances is in general quite good, especially considering the difficulties involved.) Through the process of supernova explosions, and by gentler processes of convection raising nuclear-processed material from the deeper interior to upper layers of the star, followed by shedding of outer layers by old, less massive stars, the "cooked" elements would be mixed with the primordial hydrogen and helium of the interstellar medium, there to become available for new generations of stars. That increasing availability of heavier elements would ultimately make possible the formation of solid

objects, including planetary systems and planetary atmospheres. The evolution of relatively simple organic compounds would proceed, ultimately, it was expected, giving rise to self-replicating molecules and the evolution of life.

A picture such as this is sure to suggest to some the doctrine of "reductionism": that, for example, life is to be understood in terms of chemistry, and that in turn in terms of physics. Some see such reduction as "eliminative," as suggesting that life (and, for that matter, chemistry, stellar and galactic astronomy, and atomic physics) as being "nothing but" fundamental physics. This view has been encouraged by philosophical analyses of the concept of reduction. Such analyses, of course, vary considerably in the details of their formulations; however, many of them hold, roughly, that a "reduction" shows that (1) the entities and their properties spoken of in the reduced theory or field are composed of or wholly understandable in terms of the entities and properties in the reducing one, and (2) that the laws of the reduced theory or field are deducible from those of the reducing one.[33] Such formulations of alleged conditions of reduction are seriously flawed, even with regard to specific versions of these two conditions. Most importantly for present purposes, however, such doctrines a crucial point beyond these two conditions, one which removes any suggestion that scientific unification necessarily does away with the need for a diversity of research approaches in dealing with the various segments of the picture. For example, explanation in terms of context rather than simply in terms of composition can be and remain quite different from explanation in fundamental physics.[34] Nevertheless, the employment of special modes of biological explanation, should they be and remain necessary, would not contradict the possibility of explanation in deeper physical terms. For in view of the success of the current physical picture of the evolution of the universe, it must be possible to understand in those deeper terms not only the ultimate origins and nature of the materials of life, but *also why the special modes of explanation are needed*: that is, *(1) how the context (environment), which leads to special modes of explanation being required in biology, itself can be understood as having arisen from antecedent physical (and astronomical) circumstances, and (2) how and why the existence of such an environment requires the special form of explanation.*

With such understanding available, this view of "reduction" goes beyond the older, more familiar sorts of views of reduction, and furthermore is far more in accord with the achievements of modern science.[35] For the addition of these two conditions (coupled with necessary alterations of the conditions focussed on in the older views) makes it possible to see how, simultaneously, a field like biology or chemistry or even atomic physics can be understood in terms of a deeper one while yet still retaining the autonomy of its explanatory, and also its investigative, procedures.

* * *

The picture of the universe of modern science does not stop with the present. It also provides a context in which the far-distant future, indeed the ultimate fate, of the universe can be comprehended. As we shall see in the next section, it is still not known whether the universe is open or closed - whether, that is, it will continue to expand forever or slow to a halt and then contract to a Big Crunch. (However, as we shall see, evidence is heavily against its being closed.) In either case, the physical theories I have described permit the calculation of a sequence of future milestones in the history of the universe, just as they do for its earlier events.[36] Though details vary, the following is a typical account, assuming that the universe has not undergone contraction in the meantime. (The assumption is a reasonable one: it has not done so yet, and, as was just mentioned and will be borne out later in this essay, evidence is heavily in favor of its continuing to expand in the future.) After 10^{12} years - on the order of one hundred times the present age of the universe - star formation will cease. By 10^{14} years, even the longest-lived stars, having used up all the nuclear fuel by which they had previously produced energy, will have become white or black dwarfs. When the age of the universe has reached 10^{15} years, close encounters of stars will have stripped all planets, anywhere, from stars. After 10^{17} years, white dwarfs will have cooled to become black dwarfs. At the age of 10^{19} years, ninety to ninety-nine percent of stars will have evaporated from galaxies, the remainder having been swallowed in black holes at the galactic centers. All this time, the cosmic background radiation would have been cooling with the continued expansion of the universe, so that by 10^{23} years, its temperature would be 10^{-13} °K. If protons do indeed decay, they would have done so by 10^{33} years or somewhat later; the only remaining particles would be leptons, perhaps forming "atoms" consisting of a positron and an electron orbiting around one another at distances comparable to that of the present observable universe. The 10^{117}th birthday of the universe will be marked by the completion of the process of evaporation of the last remaining black holes (by a process known as Hawking radiation). If neutron stars and white dwarfs still survive, they will, by the time $(10^{10})^{76}$ years have elapsed, they too will have decayed to black holes, themselves in turn ultimately to decay.

III PROBLEMS IN THE PICTURE

SECTION 1. LINKS IN A CHAIN.

So far I have given a survey of "the modern scientific picture." However, there is a further aspect of that picture which must also be included as part of the total viewpoint which constitutes the object and framework of our philosophical inquiries, in the sense laid down in Section I of this essay. That has to do with the *problems* of the current scientific picture, including problems relating to those portions which are accepted and problems relating to the directions of inquiry which presently-accepted ideas suggest.

The problems of this "picture" are best seen in terms of a different metaphor. Instead of speaking of a "picture," let us think of the present scientific world-view as a chain, made up of various links. Some of the major links which we have surveyed are the following.

(1) A beginning in a Big Bang.

(2) An era of unification of all forces (to the Planck time, 10^{-43} seconds after the Big Bang).

(3) Following the Planck time, separation of the gravitational force from the other forces.

(4) Separation of the strong force from the electroweak. (10^{-35} seconds after the Big Bang.)

(5) Separation of the electromagnetic force from the weak. (10^{-10} seconds.)

(6) Formation of the light elements, deuterium, tritium, helium-3, and helium-4, and lithium. (The first few minutes after the Big Bang.)

(7) "Recombination"[37] of electrons with nuclei to form atoms as we know them. (A few times 10^5 years, when the universe has cooled sufficiently for this to occur. Associated with this event is the end of the era of radiation-dominance, this latter having been an era when interaction of photons with free electrons made the mean free path of light extremely short.)

(8) Formation of galaxies (or clusters of galaxies). (Time uncertain - see Section 4 of this Part.)

(9) Formation of stars and planetary systems. (A continuing process once galaxies and the requisite elements have formed.)

(10) Formation of the chemical elements (apart from primordial ones) as part of the evolution of stars, and their dispersal into the interstellar medium to be available for the formation of later-generation stars and solid bodies.

(11) Formation of planetary systems.

(12) The origin of life. (On earth, approximately 3.85 billion years ago.)

A fuller discussion than that given here would also include subsequent stages of the evolution of life and the universe. Those aspects will not be treated here beyond what has been said at the end of Part I.

All of these links with the exception of (2) and (8) are quite strong, though for different reasons. Discussion of (2), and of the state of things up to the split of gravity from the other three forces (that is, preceding (3), and the processes involved in (3) itself), will be given in Part IV, below. (8) and the larger problems associated with it will be discussed at the end of the present Part, and further in Part IV.

SECTION 2. STRONG LINKS.

The era between (3), the separation of gravity from the rest of the forces, and (4), the separation of the strong force from the electroweak, is the era of the Standard Model. This theory, together with the Big Bang theory to which it is applied, is the bedrock upon which the modern scientific picture is built. The components of the Standard Model, the electroweak theory and quantum chromodynamics, have successfully passed a series of experimental tests, among them the following: the successful prediction of weak neutral currents, and of the masses of the W and Z bosons (the foregoing have to do with the electroweak theory; the following are concerned with quantum chromodynamics); deep inelastic scattering which reveal the quark structure of nucleons; hadron jets; and an impressive array of other successes. As I remarked earlier, there is at present no observation or experiment which is incompatible with the Standard Model, from the smallest observed events (on the scale of 10^{-15} centimeters) to the largest scales observable, not including gravitational phenomena. As we have seen, when applied to the Big Bang theory, it provides the basis of a history of the universe from approximately 10^{-35} seconds after the Big Bang to times many orders of magnitude longer than the present age of the universe. As to the Big Bang theory, though doubts about it are still occasionally raised,[38] it too accounts for a great deal of observational evidence as no rival theory can. It is based on General Relativity and the host of supporting evidence for it and against its rivals, plus other telling information.[39] Evidence pertaining specifically to the Big Bang theory includes, as we have seen, the fact that predictions regarding primordial nucleosynthesis of helium-4 give excellent results, and that those regarding deuterium, helium-3, and lithium, though more difficult, are consistent with Big-Bang predictions. (This is Link (6).) The black-body character of the cosmic background radiation, now accepted as established, also supports the Big Bang theory in the form of a "hot" origin, and gives very strong support to Link (7), the recombination epoch, in which the background radiation originates. Other problems regarding the Big Bang theory will be discussed and assessed later.

The most successful part of the Standard Model is the electroweak theory. Its unification of the electromagnetic force (quantum electrodynamics) and the weak force has proved highly successful. It rests on the application of the (local or Yang-Mills) gauge-theoretic approach to the formulation of these theories, the renormalizability of such theories, and the breaking of the symmetry between the two forces. (This latter is Link (5); Link (4), the separation of the strong from the electroweak force, is generally assumed to be of the same general type.) The theory has triumphantly passed a succession of experimental tests, as indicated above in connection with the Standard Model. Resting on the concept of symmetry breaking of a non-Abelian gauge field, the theory is compatible with any of several mechanisms for accomplishing that breaking, the present favorite being spontaneous symmetry breaking in the form of the Higgs mechanism. Though the predicted Higgs particle has not been found at energies explored thus far, empirical evidence concerning the mechanism of symmetry breaking is expected from the next generation of particle accelerators. Quantum chromodynamics also has been highly successful, again as indicated above (deep inelastic scattering, jets, *etc.*).

In spite of their successes, however, the Standard Model and General Relativity are incomplete in certain respects. The Standard Model contains a number of undetermined parameters, while General Relativity breaks down, because of inevitable singularities, at the level of quantum theory. Though neither of these sorts of problems has to do with a disagreement with experience, they do suggest the need for a deeper theory. These problems, and directions being explored for their resolution, will be discussed in Part IV, below, along with a fuller treatment of the Big Bang theory. We shall see that, as far as we have reason to believe at present, they will prove to be problems indicating the *incompleteness* of the theory rather than its *incorrectness*.

Problems (9), (10), (11), and (12) are also justifiably regarded as problems of incompleteness. Problems (9) and (11) concern the formation of stars and planetary systems.[40] [(9) and (11) are separated here because planetary systems will be accompaniments of star formation only when sufficient heavier elements have been cooked in the interiors of early-generation stars and the requisite elements dispersed to the interstellar medium for later generations of stars.] In the Milky Way galaxy, it is estimated that perhaps 50%-70% of all stars are double or multiple. However, though hypotheses are not lacking, the conditions under which stars form as multiple systems or as single systems are by no means well understood. Possibly, but by no means certainly, the formation of planetary systems may be associated with the processes by which single stars are formed (perhaps having to do with the rotational velocity of the cloud from which the objects are formed.) The material that otherwise would have gone into a partner star or stars would presumably be available for formation of lesser objects. But whether single-star status is either necessary or sufficient for formation of planetary systems, and extent of influence on that possibility of such factors as the mass of the protostellar cloud, its rotation and its magnetic field, are as yet poorly understood. Even given the existence of a solar nebula, there are still a number of competing theories of the formation of planetary systems, though much has been learned here too. Despite these difficulties, hope for increased understanding comes from recent and projected infrared satellite observations penetrating the thick cold molecular clouds which have hitherto hidden the process of stellar birth, and from recent discoveries of solid matter (at least rock-sized) surrounding a number of stars. As to Link 10, that having to do with the nucleosynthesis of chemical elements in general, we have seen that application of physics to the production of light elements in the early moments following the Big Bang give quite successful predictions of the abundances of those elements. As to the rest, we have a highly successful theory that most of them are cooked (by nucleosynthesis) in the centers of stars, in supernovae explosions, and in a few other less important but understood processes. They are spread into the interstellar medium by a variety of processes (stellar winds, especially in red giant stars, and by nova and supernova explosions). Predictions of theory are in fairly good agreement with observed abundances in the solar system, stars, the interstellar medium, and other galaxies, and many exceptions can be accounted for (*e.g.*, the deficiency, as compared to abundances elsewhere in the universe, of volatile elements in the four terrestrial planets). This is therefore taken as the right approach, despite there being many difficulties of detail. These will be discussed shortly.

Problem 12 concerns the origin of life (self-replicating molecules).[41] The most general problem in this connection is this: How can a molecule as huge and complex as DNA have evolved from pre-existing smaller molecules? While organic molecules are known to form in interstellar space, and are found in meteorites, and while more complex organic molecules are known through laboratory experiments to be formed in a number of processes which would have been readily available on the primitive earth, the jump from those molecules to self-replicating macromolecules is enormous. Besides the large size of present-day DNA, the fact that it can be synthesized only by protein catalysts (enzymes) which can presumably only be of its own making leads to a chicken-and-egg problem: which came first, the DNA or the proteins? The discovery by Cech and Altman that RNA, even short segments of it, can be self-catalyzing as well as self-replicating[42] has led to hope that this problem can be circumvented. Perhaps short strands of self-replicating RNA were precursors of DNA, and RNA has been found to be capable of self-catalysis. However, there are serious difficulties in maintaining that RNA itself was the *earliest* form of life or self-replicating molecule.[43] Probably some still more primitive type of molecule or other structure preceded RNA, which may have taken over the function of self-replication from that primitive form, just as DNA would later take over from RNA. Many candidates have been proposed for the more primitive mechanism, and, based on the considerations described here and many others, the great majority of workers in the field are confident that a solution will ultimately be found which will place the origin of life squarely in the larger context of molecular evolution in primitive earth conditions. (Those primitive conditions - *e.g.*, the character of the Hadean and Archaean atmosphere or succession of atmospheres - are themselves not well-understood, even though the possibilities are.) Prebiotic organic molecular evolution, and, to the extent described above, that of the planet and planetary system in which it occurs, is already understood in terms of the still larger picture of the evolution of elements surveyed in this essay. Within this context, it remains an open question whether (i) the rise of self-replicating molecules, though its possibility is understandable in chemical terms, is highly improbable, its occurrence on Earth being a sheer and possibly unique accident; or (ii) the rise of life is inevitable given some range of proper conditions. In either case, it is widely accepted that the rise of self-replicating molecules can be understood in terms of chemistry and the conditions prevailing on the planet concerned, the rise of those conditions making the accident or inevitability of life possible. More complex environments after the origin of life make possible the rise of eukaryotic cells, multicellular organisms, and intelligent life, again whether those steps have a low or high probability given prior evolution of organisms and environments. As I have remarked, those phases of evolutionary history will not be discussed in this essay.

The problem of the origin of life involves far more specific detail than I have been able to indicate. Some flavor of the detail that must be entered into can perhaps be seen from the following. Where did the elemental constituents of biogenic molecules come from? (There is evidence that the early Earth may have lost must of its water and biogenic elements as a result of the catastrophic impact which, on recent theories, may have led to the formation of the Moon. Was the Earth replenished in these materials by subsequent cometary and meteoritic impacts?) Given those elements, what were the prebiotic chemical routes and physical conditions by which each of the relevant bases, sugars, amino acids, nucleotides, coenzymes, *etc.*, were synthesized? How could they become segregated from other molecules (*e.g.*, those of different chirality) which would be biologically irrelevant? How could they accumulate in the relevant environments and be protected from disintegration (for example by heat or ultraviolet radiation) under the harsh conditions prevailing on Hadean and Archaean Earth? No single energy source appears capable of producing all the components, and it is difficult to see how some of them (*e.g.*, ribose) could have accumulated at all. Again, how did the genetic code arise, and a mechanism for translating this code into protein synthesis? In short, a very great deal of detail must be accounted for by a theory of the origin of life. There is no room here for mere facile "handwaving," mere "scenario-creating." *Exactly how* the process, in all its aspects, takes place, in all its chemical and physical detail, must be understood.

SECTION 3. SCIENCE AND THE CONCERN WITH DETAIL: AN EXAMPLE.

This discussion of what is involved in the problem of the origin of life brings out a feature that is common to *all* our links: the truly detailed character of the problems, and the insistence on detailed - and, where appropriate, quantitative - agreement between prediction and detailed and quantitative empirical data. This feature is therefore instructive as to the nature of the process in general, and for this purpose, let us look at another example from among our links.

Consider Link (10), the theory of nucleosynthesis. This vast and difficult subject requires *detailed quantitative study* of, among other things, all of the following:

(a) Cross sections and reaction rates of hundreds of nuclear reactions involving the many chemical elements and their isotopes (along with binding energies in general, half-lives for radioactive elements and isotopes, more particular conditions under which photodissociation and other destructive processes that can occur, and a host of other data. Experimental results for many of these are unavailable, and must be calculated from theory.

(b) The variety of processes by which nucleosynthesis can take place, the conditions under which each of these processes will take place, and the possible sites where those conditions will be realized.[44] Eight basic processes are known, each of which (the nucleosynthetic pathways involved) must be understood in detail: hydrogen "cooking" or "burning" (in the nuclear sense) of helium; helium cooking of higher elements; the _-process (carbon and oxygen burning), the equilibrium process (silicon burning), the s-process (slow neutron capture to produce heavier elements from "seed nuclei" produced in the preceding processes), the r-process (fast neutron capture, responsible for much of the abundances of elements beyond iron), the p-process (proton capture), and the x-process (post-Big Bang synthesis of the light elements deuterium, helium-3, lithium, beryllium, and boron, all of which are destroyed by thermonuclear reactions in stars). Locating possible conditions and sites for these processes to take place is often more difficult than it sounds. For example, supernova explosions of young stars (*i.e.*, Type II supernovae, insofar as that is a well-defined category - which it is not) are one obvious choice for rapid neutron capture (the r-process). But exactly where and when in those explosions?

(c) Conditions in the environment of the nucleosynthetic site. These conditions include temperature and pressure, chemical composition (which varies with the age of the star and abundances at the time and locale of the star's formation), existence of convection currents (which can dredge up the newly-synthesized elements to outer layers of the star), mechanisms by which the elements can be dispersed into the interstellar medium, and knowledge of what that site is. Later stages of stellar evolution - not yet always well-understood in detail - must be taken into account, including cooking in various layers surrounding the stellar core. Calculations for each element depend on prior production of lighter (and sometimes, as with elements undergoing beta decay, heavier) elements.

(d) A knowledge of various mechanisms of dispersal of the newly-cooked elements into the stellar medium, by stellar winds, especially in the case of red giants and supergiants, and by more explosive events. (In particular, for instance, much remains to be understood about supernova explosions.[45])

(e) Painstaking determinations of galactic and cosmic abundances of the elements, based on careful study of the interstellar medium, integrated spectra of galaxies other than the Milky Way, solar system abundances, including especially meteorites and in particular carbonaceous chondrites (which are believed on other grounds to be unaltered relics of the primitive nebula which became the solar system). Anomalous abundances must be explained (*e.g.*, as was noted above, the inner planets of the solar system have comparatively little hydrogen and other volatile elements; ^{26}Mg is anomalously abundant in carbonaceous chondrites - supposedly relics of the presolar nebula - and many of these differ from one another and from the sun in relative abundances). The relative abundances so determined from such a variety of sources must then be integrated to give a set of cosmic abundances.

All the results of theory (nuclear, stellar evolution, nucleosynthetic), for each element and every isotope, must in principle be in quantitative accord with the results of experiment. To repeat: there can be no mere hand-waving here: the goal must be comprehensive, precise quantitative agreement *in all details*. The task is obviously massive. What I want to emphasize is the detail - the quantitative detail - which must be achieved. It is no wonder that the process of determination has taken the efforts of a host of researchers during all of the decades since the 1930's, and is by no means completed, there still being discrepancies in particular areas between prediction and observed abundances. However, viewed from another perspective, it is a wonder that so much has been accomplished in so short a time. The relevant subjects - most generally, nuclear physics and the theory of stellar structure and evolution, but a great many other and more specific subjects - date only from the 1930's, and had to be developed over the succeeding decades. In view of the complexity and difficulty of the task, it is remarkable that in many specific cases there is excellent agreement with observations and broad agreement in most. This success makes following up this line of research a reasonable one. That reasonableness is increased by the fact that all other possible sources of stellar energy and element production - meteoritic impacts on stars, chemical burning, *etc.* - have been shown to be inadequate for explaining even the most general features of these processes, much less the vast amount of detail assembled regarding them.

The claim that a certain picture is correct, that certain problems concerning that picture are problems of mere incompleteness (rather than of incorrectness), and that certain lines of research are reasonable, all depend, ultimately, on such details and not on generalities. It must be kept in mind that *the same level of detail is present in connection with other aspects of the picture I have drawn, and of the links between portions of the picture and problems concerning the picture. This will remain true of the discussion of lines of current research in Part IV.* More generally, insofar as it is reasonable to take a theory of a certain body of information (observation) seriously, and insofar as it is reasonable to follow up its suggestions (taking its problems as ones of incompleteness, and taking its indications for further research seriously), that reasonableness rests on the following considerations: (a) that all known alternative ways to account for that body of information have been explored *in detail* and have been found unacceptable; (b) that one of these alternatives has been found able to account *in (quantitative) detail* for a certain portion of that body of information; and (c) that that alternative is consistent with accounts of related bodies of information, both theoretical and observational - as the theory of nucleosynthesis of chemical elements is consistent with the Big Bang theory in cosmology and with the theory of stellar evolution. ((a) and (b) encapsulate the element of what I have called "success" in science: the detailed study of a body of related information, and the ability to account for it in detail. (c) has to do with the coherence of that successful theory with other theories.[46]) These criteria are no doubt somewhat flexible, and must be applied judiciously. Nevertheless, they are not the sort of criteria which have been emphasized by many philosophers. These latter insist that *all possible* alternatives be ruled out, that *all possible* problems be solved, and *all possible* lines of research should be investigated, or otherwise the theory cannot be taken seriously.[47] Such criteria cannot be satisfied, even in principle. However, that fact does not mean that the scientific picture is unacceptable as knowledge. It is this attention to detail - this lack of handwaving - that distinguishes the scientific enterprise, at its best, from popular and philosophical attacks on it and on the possibility of its success.

In an article on the formation of the earth and the solar system, George Wetherill, summarizing the state of affairs in his field, remarks that

> [The standard model for the formation of the sun and planets] tells us little that can be considered at all certain. Some specific problems have been explored relatively thoroughly and quantitatively, but most are in the qualitative "scenario" stage. At present the model can be regarded as a very useful working hypothesis, one that can provide a focus for workers of varying backgrounds and experience, facilitate communication between them, and permit a rational division of labor.[48]

In many respects, this summary could equally well be applied to many subareas of our present scientific "picture" which have been discussed so far. Except for the Big Bang theory, which will be discussed more fully in Part IV, in all but one of the links discussed up to this point, for example, there is at present a "standard model," of greater or less acceptance, specificity, and detailed working-out. At one extreme is the Standard Model combining the electroweak theory with quantum chromodynamics, which, despite many variations and open questions, is very widely supported on the basis of impressive evidence, both experimental and theoretical. At the other extreme, the one area where there can be said to be no "standard model," is the problem of the origin of life, where there remains intensive debate about mechanisms and conditions. Between these extremes, links (6) and (7) are also considered to be well under control, though the uncertainties and alternatives are perhaps greater here than in the case of the Standard Model.[49] A still greater tangle of detailed problems remain for links (8), (9), (10), and (11); nevertheless, there is in each of those cases a theory, or at least an approach, which is widely considered to be the best.

However, even in those cases where many problems and alternatives remain open, or where there is nothing like a "standard model," calling the work a "scenario," or "scenario-building," can be a misleading understatement of what is involved. It might suggest that those scenarios are simply hypotheses, conjectures, that are pulled out of thin air, without basis.[50] But there is, for all the above links - even that of the origin of life - a very large "background framework," a body of knowledge, heterogenous in some respects and more or less systematic in others, in terms of which the construction of alternative possibilities takes place. That background framework, built on long and detailed investigation, has come to be generally accepted on the basis of a multitude of good reasons. It is the base on which new problems are raised, new theories constructed and tested, new lines of research pursued. Thus, even in the case of the problem of the origin of life, the various competing possibilities are all *reasonable* ones in the light of our best information about the processes of star formation, the Hadean and Archaean era on earth, comparative planetary studies, and a host of other relevant information. Hypotheses are not proposed and considered arbitrarily, without reason; conversely, it is not unreasonable to ignore speculations which are, in the light of what we have learned, unpromising.[51]

The totality of these components, together with those which will be discussed in Part IV, constitute what I am calling the "modern scientific picture." Note that not all the parts of this picture are equally well-established; and thus the problems regarding each component must also be part of the "framework" within which, and the "object" about which, philosophical discussion must take place. The same is true of reasonable alternatives to the "standard model" of a theory, and the variety of alternative reasonable lines of research. (A theory or line of research, or statement of a problem, is "reasonable" here if it makes sense in terms of the background framework, and if it makes sense, or holds promise of making sense, of the detailed subject-matter that has been investigated. For theories, "making sense" should be interpreted in terms of the "success" of the theory with respect to its subject-matter - how well it accounts for the details of that subject-matter - and its coherence with the background.)

The various individual components of the picture not need to be equally well-established; but also, in the case of some components, there may not even be a "standard model." Even where the component is a "standard model" theory, it need not be necessary to the picture. The electroweak theory is compatible with other mechanisms of symmetry-breaking than the present "standard" view, the Higgs mechanism. Again, various models of star and planetary formation are compatible with the "background" Big Bang theory with particle physics, and perhaps (within limits) with alternative cosmological theories. Many possibilities are open as to where and when the r-process occurs. More distantly from the standard pictures, as we shall see, are other possibilities; but even in those cases, the "non-standard" possibilities are framed against a background which in most respects is the same as that of the "standard" ones, and if and when they are taken seriously, they are so considered for good reasons. The point is that many of the details - the components, the links - in the picture drawn (or chain constructed) in this paper may yet be replaced, without the rest of the picture being affected in any drastic way. Nevertheless, for all links but one considered so far, there is a "standard model," and with regard to the one exception, we have good reason to pursue the avenues now being pursued. At present we have reason to believe that at least in broad outlines (and often more), the problems in present approaches are, as considered so far, problems of incompleteness rather than of incorrectness.

SECTION 4. A WEAKER LINK: GALAXY FORMATION AND WIDER PROBLEMS ABOUT THE NATURE AND QUANTITY OF MATTER IN THE UNIVERSE.

Apart from discussion of Link (2), and of the state of things up to the split of gravity from the other three forces (that is, preceding (3), and the processes involved in (3) itself), which are to be discussed in Part IV, below, the final link to be treated in the present Part is (8). This concerns the origin of galaxies and larger groupings of galaxies.[52] However, I will consider this problem in the context of two other problems from which it has become inseparable. These are the problem of whether the universe is open or closed, and that of the quantity and nature of matter (mass-energy) in the universe.

The *basic* problem is this. The universe is observed to be extremely uniform ("smooth"), at least on large scales, and the predominant cosmological theories - Friedmann models with Robertson-Walker metric - hold space-time to be homogeneous and isotropic. But in such a universe, how could density differences - the inhomogeneities we observe on smaller scales - have arisen, ultimately to become galaxies or clusters of galaxies? Attempts to locate the source of inhomogeneity in quantum fluctuations in the very early universe have not been particularly successful, though hope remains for that approach. Such early slight density fluctuations would have to yield appropriately-sized magnifications by the time of galaxy formation, but there remain serious ambiguities about when that epoch occurred, and about what the appropriate sizes of the inhomogeneities should have been. Should they have been galaxy-sized, clusters forming only later, through the mutual gravitational attraction of individual galaxies? Or did cluster-sized inhomogeneities form first, galaxies arising in the process of collapse of such clouds? Theories of the formation of galaxies disagree on which came first, though much evidence has been accumulated in favor of the former process, where galaxies (or perhaps even smaller units) come first. On the other hand, however, evidence is also accumulating that there are still larger organizations of matter than even clusters of galaxies in the universe: not only superclusters, but long lines of superclusters, arranged in cell-like form with vast voids in between. With every extension of the limits of observational capability, the universe has turned out to be clumpy. At this writing, evidence has been offered that very large-scale clustering of galaxies exists - on scales of 200 to 400 megaparsecs, depending on the value of the Hubble constant. If that evidence is borne out, there will be serious troubles for all existing theories of galaxy formation; and if such theories must be based on the physics of the very early universe, as they are now and seem to have to be, there may be serious repercussions for those theories also. What would account for the presence of such structure?

The most widely-considered view is that galaxies - or clusters - formed by density fluctuations reaching a point where the gas underwent gravitational contraction. (Alternatives will be discussed in Part IV.) Thus, theories of the origin of structures in the universe are dependent on knowledge about the masses of those structures. This problem in turn is intimately related to the more general question of the amount and distribution of matter in the universe. To see more deeply into these relations of issues, let us examine the cosmological issues and the questions about the quantity and nature of mass.

It would take a certain "critical density" of mass-energy in the universe to ensure that the universe is exactly balanced between collapsing and expanding forever. The ratio of actual to critical mass-density (strictly, mass-energy density) is referred to as Ω (omega).[53] If Ω is equal to one - the actual density being precisely equal to the critical density - the universe is "flat." (See the discussion in Part II, Section 3, above.) In that case, its expansion, while continuing forever, will slow asymptotically toward a rate of zero. If Ω is greater than one (and the cosmological constant is zero), the expansion will ultimately come to a halt, and the universe will contract; it is said to be "closed."[54] If Ω is less than one (and the cosmological constant is zero), the expansion will continue forever: the universe is "open." The value of the critical density (and therefore of Ω) depends crucially on the rate of expansion (the Hubble constant), whose value still remains uncertain by a factor of about two. Within those limits, the critical density is found to be on the order of 10^{-29} gm/cm^3, which comes out to about ten hydrogen atoms per cubic meter, or one brick per cubic light-year.

Thus the quantity (or density) of mass in the universe determines its future closure or collapse. How can this quantity or density be determined? One way is to determine the actual density of all matter which is detectable through electromagnetic radiation. The result is found to be on the order of less than two percent of the critical density (in the neighborhood of 10^{-31} gm/cm), and perhaps considerably less than that - perhaps less than one percent. It would thus appear that, unless there is other matter besides this, the amount of matter-energy in the universe is far too small to close it, and that it will therefore go on expanding forever. However, there are two pieces of evidence indicating that there is more matter in the universe than is found in the electromagnetically-observable - that there is some sort of "dark matter" which does not reveal itself through any electromagnetic radiation detectable so far, and, indeed, perhaps not through any electromagnetic radiation, observed or not.[55] These two arguments are based on gravitational considerations (the way the bodies move under the force of gravity) rather than on electromagnetic ones. (We shall see yet a third argument, a theoretical one, for the existence of dark matter in Part IV, below.)

First: There is another method of ascertaining the mass of a system, and that is by consideration of its gravitational behavior. The rotation of galaxies has been found to be not that of Keplerian rotation, in which the velocity of rotation decreases with increasing distance from the central mass, but rather to be analogous to that of a solid wheel. Such a rotation curve is possible only if the mass-density of the galaxies is constant, rather than decreasing with distance. Evidence indicates that there must be five to ten times the amount of matter in a galaxy, of whatever size or type, as shows up in its electromagnetic radiation. The matter extends well beyond the visible limits of its disk, and is arranged in a spherical halo.

Second: As indicated by studies of the dynamics of clusters of galaxies (using the virial theorem, with the well-founded assumption that at least some of the clusters are in dynamical equilibrium), the same problem occurs with respect to those objects. Even more matter is called for: the quantity of "dark" matter required is an order of magnitude greater than can be accounted for on the basis of the masses of the constituent galaxies, even when *their own* dark matter, as discussed above, is added in.

There thus seems to be a great deal of additional unseen matter in clusters of galaxies. Even with the addition of these two types of dark matter, however, the total mass-density is still well short of sufficient to close the universe, Ω remaining at a value estimated in the range 0.2-0.3. However, these arguments are based on examination only of the neighborhoods of galaxies - at most, within a few tens of millions of light-years of galaxies. It remains an open question as to whether there is still further dark matter in the deeper regions of space, where there are few or no galaxies, the visible objects forming only in a "biased" way, on the "crests" of the density fluctuations of a far more prevalent kind of matter than the baryonic type with which we have been familiar. Is there any positive reason to think that there may indeed be large amounts of matter there, enough even to close the universe? This question will be explored further in Part IV.

The problem of galaxy formation (or the formation of other types of large-scale structure) has thus become inseparable from that of the closure of the universe as a whole, which in turn is dependent on the density of matter-energy in the universe. However, we see also that the problem of the formation of large-scale structure is now inseparable from the issue of the *nature* of that matter. Calculations of the amount of *baryonic* matter called for by calculations of helium production in the very early universe may be up to about 0.15 of critical density. So there is more dark matter than is revealed by electromagnetic radiation, but probably still more than can be accounted for by baryons - "ordinary" matter - alone. This problem, too, will be examined further in Part IV.

Thus the problem of the origin of galaxies ultimately rests on a far deeper problem, that of the origin, nature, and quantity (or rather, the density) of mass in the universe. This problem, as we have seen, affects much else in our understanding of the universe, and its complex of subproblems constitutes the weakest link in our chain.

Indeed, a host of subproblems surrounds the central issue of the nature of mass: clearly, the whole question of the nature of mass is not yet understood. Some of the major ones, besides the above, are the following:

(a) *Origin and explanation of the spectrum of particle masses.* Why do the fundamental particles have the masses they do? Why are the ratios of their masses what they are? Why, for example, is the electron so much lighter than other particles? (The masslessness of the neutrino in the Standard Model is a consequence of its chirality.) These questions remain unanswered in any accepted theory today. Presumably the mass spectrum of elementary particles should be explainable in the same fundamental terms as other quantum numbers and their spectra.

(b) *Generation problem.* Fermions fall into three "generations": (i) up and down quarks and electrons and electron neutrinos; (ii) charmed and strange quarks and muons and muon neutrinos; (iii) truth (or top) and beauty (or bottom) quarks and tauons and tauon neutrinos. (The top quark and tau neutrinos have not yet been observed, but a number of independent lines of reasoning indicate strongly that they do exist, and that they will probably be found in the foreseeable future.) Only the first generation seems necessary for nature. Why, then, are there more than one such generation; as Rabi said when told of the discovery of the muon, "Who ordered that?" (He was in a Chinese restaurant at the time.)

A long-outstanding problem concerned whether there are further generations beyond three, and if so, how many. That problem now appears to be solved: evidence has recently been reported that strongly suggests that there are only three.

(c) *Cosmological constant.*[56] The curvature of the universe, and the question of whether it is closed or not, depends on the mass-energy density. The contribution of ordinary and exotic forms of matter and radiation have already been considered. A further contribution comes from the energy of the vacuum, and this energy behaves like the cosmological constant which Einstein had originally introduced to stabilize the universe against expansion or contraction (an introduction he later considered to be the greatest error he had ever made). However, when the various factors that, theoretically, could be expected to contribute to the energy of the vacuum are estimated (no mean feat, since some of them depend on theories or knowledge that we do not have in any direct way), the sum total of that energy is, very roughly, some 120 orders of magnitude larger than the observational limits (which are so small as to be consistent with a value of zero). A cosmological constant of that size would produce a curvature of space that would be noticeable on an everyday scale. But such curvature is not noticed, and indeed would not be noticed, according to observational evidence, on a scale of the order of 10^{10} light-years.[57] However, the factor of 10^{120} is arrived at by adding up the various contributions as if they were independent of one another; presumably in a truly unified theory, those factors would be related, and the discrepancy would be reduced accordingly - even perhaps making the cosmological constant equal to zero, or, alternatively, just enough to make $\Omega=1$. From this perspective, the cosmological constant problem is simply one more source *telling us* of the need for further unification. We must wait and see. Other solutions have been suggested, none of which appears to be satisfactory.

The cosmological constant problem is obviously more general than the problem of the quantity and nature of matter, and could be considered independently of the latter. The same, indeed, could be said of the two preceding issues. However, as in those two cases, I place it here in my discussion because of its relations to that question, and because these problems will be discussed together in the later parts of this paper. The relevant point is that a cosmological constant of a reasonable magnitude could furnish enough energy to close the universe, and, if properly sized, to make it "flat." Other problems may arise from this, however, and, despite the considerations given in the last paragraph, it is usually assumed to have a value of zero.

Of the problems discussed so far, these interrelated three - the problem of galaxy (and other large-scale structure) formation, the problem of whether the universe will continue to expand forever, and the problem of the quantity and nature of mass in the universe - pose the most serious difficulties concerning our present understanding of the universe. In the terms used earlier, we must ask about the extent to which they threaten to force major revisions in our present picture of the universe: are they problems which we can justifiably consider to be *problems of incompleteness*? or is there reason to believe that they may in fact force us to revise our present picture in drastic ways? These questions will be taken up in Parts IV and V.

IV THE SEARCH FOR HIGHER UNIFICATION

SECTION 1. GRAND UNIFICATION THEORIES AND THEIR COSMOLOGICAL IMPLICATIONS.

It must not be forgotten that, beyond the electroweak theory, unity is promised but not achieved. Even the electroweak theory is not a full unification of its two constituent forces, as we have seen. We must now ask this question: *What are the prospects for higher unification? In particular, what are the prospects that our best candidates for higher unification may resolve the problems considered in Part III, particularly those having to do with the formation of structure in the universe, the quantity and nature of mass-energy, and the closure of the universe?*

Despite the immense successes of the electroweak unification and its juxtaposition with quantum chromodynamics to form the Standard Model, the need for a deeper theory is made clear by a number of considerations. For example, even in the electroweak theory, the electromagnetic and weak forces are described in terms of different group structures, U(1) and SU(2) respectively, with different coupling strengths. Although those couplings are related in the theory by a "mixing angle" (Weinberg angle), the value of that angle is not specified by the theory. The juxtaposition of the electroweak and quantum chromodynamic theories was, of course, based largely (though as we shall soon see, not entirely) on similarity of mathematical structure. A number of parameters, including the origins and values of the masses of quarks and leptons, were left undetermined, to be "put in by hand." The existence of three distinct "generations" of Standard-Model particles, and the question of whether there were still more generations (or families), was not explained. These last two issues, as we have seen, have to do with the lack of understanding which our present theories give of the ultimate nature of matter.

In addition to such reasons suggesting the necessity of higher unification, there were also clues that such a theory was possible, and that gave directions in which to search for it. Earlier, one of the clues that had suggested the possibility of a unified theory of the electromagnetic and weak forces had been the fact that those two forces, so different in strength on ordinary scales, converge to a common value at higher energies. Asymptotic freedom brings out an analogous relation between these two unified forces and the strong one[58]: if the strong force gets weaker with increasing energy, then at some energy - much higher than the unification energy of the electroweak force - the latter might be expected to become equal in strength to the strong force. Thus, not only did the theories of the electroweak and strong forces resemble each other in mathematical form; there were also deeper reasons for suspecting a possible unification of them.

Hence the search for a gauge unification of the electroweak and strong forces - the search for what has come to be called a Grand Unified Theory (GUT) - was embarked upon. Such theories were already being proposed by the early 1970's, and were soon being subjected to experimental test.

The most straightforward way to construct a "GUT" unifying the theories of the strong force, quantum chromodynamics, with the electroweak theory, was to attempt to show that the respective group-theoretic structures of those theories, SU(3) for quantum chromodynamics and SU(2) x (U(1) for the electroweak, and so SU(3) x SU(2) x U(1) for the Standard Model, are subgroups of a higher group. There were several possibilities, the simplest and in many ways most appealing being the group SU(5), employed in the theory proposed by Georgi and Glashow.[59] In this theory, the SU(5) symmetry exists at energies of 10^{14} GeV and above, and at distances of 10^{-28} cm. and below. At such energies, eight GUT exchange particles (SU(5) gauge bosons), the X particles, act on baryons to convert them into leptons. The theory successfully accounted for all known particles and their interactions. Further, the Weinberg angle could be deduced, as could the exact equality of the electric charges on the proton and the electron.

However, the theory, at least in its simplest form, faces serious difficulties. Chief among these is the following. Since this theory proposes a symmetry between baryons and leptons, it implies that baryons can decay. A number of experiments have been set up around the world to test this prediction with regard to protons, but no unambiguous cases of proton decay have been found.[60] The experiments appear to set a lower limit of 10^{32} years for the lifetime of the proton. Since the simplest ("minimal") version of the SU(5) theory predicts a lifetime of 10^{29} years, it is widely agreed that the minimal SU(5) theory is eliminated by these results. However, more complicated unifications of this group-theoretical sort give a proton lifetime that exceeds the experimental lower bound and in fact can be made to impart a lifetime that is beyond any possible experimental test. Perhaps its successes indicate at least that SU(5) is a subgroup of some higher unifying group.

Similar moves are open in the case of the as-yet unfound Higgs particle (if that is indeed the, or a, mechanism of symmetry breaking). While there are some reasons to hope that the next generation of accelerators (especially the Superconducting Supercollider, if it is built) may be able to find that particle, its mass is an undetermined parameter of GUTs, and failure to find it even then may be answered by arguing that its mass is yet larger than experiments have been able to reach.

Despite the lack of a definitive choice of GUT, most versions have some basic features in common. When applied to Big Bang cosmology, the GUT symmetry - the unification of the strong and electroweak forces - is broken (by the Higgs mechanism or some reasonable alternative) at 10^{-35} seconds after the Big Bang, when the temperature is of the order of 10^{28} °K (that is, at the unification energy of 10^{14} GeV). It was only much later, when the universe was older by a factor of 10^{25} and the energy lower by a factor of 10^{12}, that a further symmetry breaking separated the electromagnetic and weak forces.

GUTs in general offer certain benefits in addition to those mentioned in connection with the SU(5) theory. The Big-Bang origins of other sorts of particles than the exchange-particles of the four forces can (with limitations to be discussed later) also be understood: when those particles arose, the processes by which they gained mass, and when and how they began interacting with one another. For example, I have noted on several occasions that powerful support for the Big Bang theory comes from calculations of the amount of helium, deuterium, and (with less certainty) lithium that would have been produced in the early Big-Bang scenario. Those calculations could now be put into the new and larger context of particle physics and cosmology. The synthesis of those elements took place at approximately 100 seconds after the Big Bang.

GUTs are also appealing in that they may provide a way of explaining the presumed predominance of matter over antimatter in the universe.[61] Quantum field theories of elementary particles require that every particle have its antiparticle (though in some cases the particle is its own antiparticle), and that in the conversion of energy into matter, both a particle and its antiparticle must be produced. However, antimatter is actually very rare in the universe, being produced (as far is known) only under rather exotic conditions, as decay products in cosmic ray showers, or in the physicists' accelerators. Although the spectra of antimatter galaxies would be indistinguishable from those of galaxies consisting of ordinary matter, any contact between such galaxies and intergalactic matter of the ordinary sort would, through the annihilation of the matter and antimatter, produce gamma rays. This radiation should be observable in galaxies in our own Virgo cluster, but are not. Though this argument extends only to nearby galaxies, no mechanism is known which could have separated matter from antimatter on the grander scales of clusters and superclusters of galaxies. Thus the available evidence indicates that the universe consists almost completely of ordinary matter. Where, then, is the antimatter that would have been produced in pair production at the beginning of the universe, in quantities equal to that of ordinary matter? The proposed answer is this. In the present-day world, a certain conservation law, that of CP conservation, is violated in certain weak interactions.[62] (C is charge and P is parity, the two being conserved in combination except in these weak interactions. A further aspect of CP violation, having to do with the strong interactions, will be mentioned later.) If this violation also occurred in the case of the original GUT X particles, then, the explanation asserts, there would have been slightly more of the particles we call matter than there were of antimatter. Calculations suggest that for every billion or so annihilations of antimatter and antimatter X-particles, there would have been one surplus matter particle. When the universe cooled to the point where there was not enough energy to create new X and anti-X pairs, the one-per-billion remnant matter particles would have been left. The ratio of the number of cosmic background-radiation photons (the ultimate left-overs of those early matter-antimatter annihilations) to the number of (baryonic) particles of ordinary matter is in fact found to be on the order of 10^9 to 1. However, these results of the calculations which seem to agree so remarkably with those of observation are by no means conclusive, since they depend on details of the particular GUT used in the calculations.

SECTION 2. INFLATIONARY COSMOLOGY.

Another possibility opened by GUTs lies in the inflationary theory. Our inability to determine the correct GUT (if there is one) leaves open two possibilities. Either the phase transition (symmetry breaking) at 10^{-35} seconds, which separated the strong from the electroweak force, occurred immediately when the universe (or a region of it) reached the critical GUT temperature of 10^{28} °K, or else it was delayed by the phenomenon of supercooling. (Supercooling is illustrated by the possibility of cooling pure water as much as 20 °K below its freezing point - the temperature at which the "phase transition" from liquid to ice usually occurs - before it undergoes that transition.) Intermediate situations are improbable, and the first alternative, immediate phase transition, would have produced large numbers of massive magnetic monopoles and other topological defects, enough to have caused the universe to stop expanding and collapse to its initial singularity in only a few thousand years. Thus the supercooling alternative seems quite plausible, even though its necessity has not been established. But as we shall see, the most important argument in favor of the "inflationary theory" is its ability to provide answers to very serious questions which are not addressed by standard Big Bang theory, questions that appear to be difficult to deal with in any other way.[63] Among these questions are the following.

According to the inflationary theory, as supercooling proceeded below the GUT phase transition temperature, the universe (or the region of it in question) would have been in what is known as a "false vacuum" state. (The possibility and general features of false vacuum states are predicted by quantum theory.) A field is ordinarily considered to be in a vacuum state when it is in a state of least energy. This occurs when there are no real particles in the field. But the state of lowest possible energy density is not that in which the Higgs fields have a value of zero: that state, in other words, is not one of a "true" vacuum, even though it resembles such a vacuum in that it contains no real Higgs particles. Energy is required to keep the Higgs fields at zero, and thus the false vacuum contains energy. Thus, the energy density will be greater than that of the "true" vacuum. Due to the gravitational force, a pressure would exist in such a vacuum, and general relativity implies that that pressure - bizarrely for an effect of gravitation -

is repulsive. The repulsive pressure causes an extremely rapid (exponential) expansion of the false-vacuum region. In such a scenario, the region in question doubles in radius every 10^{-34} seconds, and this "inflation" would have continued for a period of 10^{-32} seconds or more.

Further, the peculiar state of energy density of the false vacuum requires that the energy density remain *constant* as the volume expands, and thus that the *total* energy *increase* as the volume expands.[64] At the end of this inflationary episode, having expanded by a factor of 10^{50} or more, the region has its symmetry broken, the phase transition occurs, and the enormous energy (analogous to the "latent heat" released in the melting of ice) available from the vacuum state is released.[65] The released energy reheats the region to close to (though somewhat below) the unification temperature, providing the energy for the region's particles to arise. That is the source of all the matter and energy in the universe that we can observe, and indeed in the entire region of the universe with which the inflationary expansion began. After that symmetry-breaking event, the universe (or, more accurately, our region thereof) expands according to the standard Big-Bang scenario. Our present observable universe is a very small part of the region that underwent this inflationary and subsequent Big-Bang expansion. Other such regions of the universe must be unobservable, even though their existence is called for by the theory.[66]

The inflationary theory offers answers to a number of serious problems in standard Big Bang cosmology. Three examples of such problems will be summarized here: the horizon problem, the flatness problem, and the monopole problem. The *horizon problem* arises from the fact that in the early universe, expansion (even that in the standard Big Bang) was so rapid that regions separated sufficiently could not have influenced one another causally.[67] (The sizes of causally-independent volumes existing prior to the GUT phase transition were such that 10^{90} such regions would fit into the present horizon volume.) And yet, in every direction, the cosmic background radiation is today at the same temperature to an accuracy of one part in ten thousand.[68]

How could regions in, say, opposite directions in the sky - 10^{90} such regions, all told - be so similar, that is, have reached such a degree, or even any degree, of thermodynamic equilibrium? The inflationary answer to this problem is that our present (observable) universe was originally only a small region of the total universe, less than a billionth the radius of the proton. Such a region was small enough for communication to have taken place, thus making it possible, for example, for all parts of it to have achieved the same temperature. Inflation, followed by the standard Big Bang expansion, thus ultimately resulted in our homogeneous and isotropic "universe." (It is implied, of course, that there are other regions of the universe, well beyond our horizon, in which conditions are quite different.)

The *flatness problem* has to do with the fact that the present universe is so close to the critical value of mass (energy) density, on the order of 10^{-29} gm/cm^3. As we have seen, if the density is higher than that, the universe will be "closed" - that is, at some point (depending on how much higher) it will stop in its expansion, after which it will contract back to a "Big Crunch." If the density is lower than the critical, the universe is "open," expanding forever. If the value is exactly equal to the critical, the universe is said to be "flat"; it will then go on expanding forever, but will gradually slow down, asymptotically approaching an expansion rate of zero. (The terms "closed," "open," and "flat" are derived from the property of curvature, and the associated geometry, that would hold in the universe in the three cases.) *A priori*, we would expect that the density could have *any* value whatever; and yet the value it does in fact have is very close to the critical value, being estimated to lie somewhere between 0.1 and 2 of the critical value. Why, then, does it happen to be so close to the critical mass-density? The inflationary response rests on the fact that, if Ω, the ratio of the actual to the critical density, originally departed from one, the degree of departure would have increased rapidly over time. In order for it to be so close to one today, then, it would have had to be much, much closer to one originally - indeed, differing from one by a factor of less than one part in 10^{15} at a second after the Big Bang.[69] The inflationary theory provides an explanation: the equation governing the way the ratio of actual to critical mass-density evolves is different from its correlate in the standard Big Bang theory. The equations show that instead of being driven *away* from unity as in the latter theory, the ratio is driven *toward* unity in the inflationary epoch. The theory further predicts that the value of the ratio should be extremely close to one, and perhaps equal to it, at present. I will return to this question of the mass-density of the universe several times in what follows.

The *monopole problem* results from the unavoidable implication of GUTs that large numbers of magnetic monopoles - particles which, unlike ordinary magnets, have only one magnetic pole - would have been formed, with enormous masses. In an ordinary phase transition, for example the freezing of water, the formation of crystals, or the magnetization of a bar of iron, not all regions need pass into the same final state (in different small regions of the formed crystal, the axes of the crystal may differ in direction; orientations of magnetization may differ from one part of the iron bar to another.) Defects will form at the junctions between such regions or "domains." Such "topological discontinuities" would also arise in the GUT phase transition: one-, two-, and three-dimensional defects (magnetic monopoles, cosmic strings, and domain walls, respectively) would generally form at the boundaries of the 10^{90} early-universe "domains" which could not have causally influenced the course of one another's phase change. The energy, and therefore the masses, of such objects would have been of the order of the enormous energy of the GUT phase transition. (Thus the mass of a monopole would have been 10^{16} times that of a proton - comparable to that of a small bacterium.) Such huge numbers of massive defects would have been sufficient to close the universe quite quickly; and even if they did not, if such monopoles were even reasonably plentiful in the present universe, they would have neutralized the magnetic fields that are found to exist in galaxies. Inflation takes care of this problem: the boundaries of domains, where the defects would have formed, were expanded far out of sight in the inflationary epoch. Few if any such defects would have remained within our horizon. (However, we shall see later that it might be convenient to have a comfortable supply of cosmic strings around.)

By "smoothing out" the universe, or at least our region of it, the inflationary epoch bypasses any question of the arbitrariness of initial conditions in the universe. If there were any particles or other structure prior to inflation, that structure would have been diluted to insignificance in the period of exponential expansion, and an entirely new set of initial conditions would have emerged naturally as a consequence of the inflation. Thus, for example, instead of, as in the standard Big Bang theory, having to *assume* homogeneity and isotropy, those conditions are consequences of the enormous expansion, as is the existence of the mass-energy - both the particles and the background radiation - of the universe. (The asymmetric formation of matter over antimatter would presumably have taken place after the inflationary epoch.) This in itself makes the inflationary theory enormously appealing.

However, the inflationary theory is not without its problems. In particular, so far it has not been able to illuminate the problem of how galaxies (or clusters of galaxies, whichever came first) form. Fluctuations which would be expected to arise would be magnified immensely by the expansion during the inflationary era. At least some of those fluctuations should form differences in density which would ultimately lead to the gravitational collapse of the denser regions of matter to form galaxies or galaxy clusters. Unfortunately, the predicted magnitude of the density differences varies greatly depending on the GUT used in the calculations. And in GUTs which do give fluctuations of more or less the right "size," it seems necessary (again, so far in the development of this young theory) to resort to rather artificial and uncomfortable expedients, such as abandonment of the idea that Higgs fields drive the inflation, in order to arrive at those values.

SECTION 3. DARK MATTER AND GALAXY FORMATION.

The problem of galaxy formation is compounded by addition of the hypothesis of inflation. For, according to that hypothesis, local inhomogeneities would be smoothed out in the enormous expansion supposed to have taken place in that epoch. On the other hand, inflationary cosmology has something relevant to say about the problem of the existence of unseen matter: in suggests that the value of Ω must be very close to unity, despite the observational evidence that it is well below this.

This argument is theoretical, being based on the answer provided by the inflationary theory to the "flatness problem." Recall that the flatness problem asks not why the observed mass-density deviates so much from the critical value, but rather why it is so close to it: why, that is, Ω is so close to unity when, in principle, it could have any value whatever. Because the inflationary theory (in some version) is so enormously successful in meeting a number of other problems, its answer to the flatness problem is taken very seriously by many astronomers and physicists, so that the huge amount of matter still required to make Ω very close or exactly equal to one must be present in some form or other. Thus, as much as 99% of the total amount of matter (the 1% being the observed electromagnetically-radiating matter) in the universe may not yet have been accounted for.

As we have seen, the amount of *baryonic* matter - particles, like protons and neutrons, consisting of quarks - is tightly constrained by the arguments which have accounted so very well for the amount of helium and deuterium produced in the Big Bang. On the basis of those arguments, the contribution of baryonic matter to Ω can *at best* be about 15%. (Remember that this raises a further problem that is more than incidental: if electromagnetic radiation gives less than .02 of Ω, where is the rest of the baryonic matter?) Thus it is possible that *some* of the missing matter - in galaxies and clusters, perhaps - is the ordinary matter with which we are familiar as the baryonic particles which make up atomic nuclei. But, especially if the inflationary theory is correct, there must still be an enormous amount of matter which is not of an ordinary type at all, even if the full allowable 15% is baryonic.

Many candidates are available for dark matter. As regards the missing *baryonic* matter, "brown dwarfs," stars whose masses are too small to support nuclear fusion, and thus to shine, or "Jupiters," still smaller planetary-sized objects, are a possibility. One might also suspect a large amount of neutral or extremely hot intercluster hydrogen, existing in the vast voids between clusters and superclusters of galaxies. Again, perhaps a good bit of baryonic matter is locked up in black holes. There are, however, rather strict limits on the amount of matter that could exist in these forms. We have seen that the baryonic total is tightly restricted. Further, there cannot be any significant amount of neutral hydrogen in intercluster space, or it would absorb light from distant quasars. Mechanisms for maintaining an intergalactic gas (ionized hydrogen) at very high temperatures are extremely hard to imagine. If there are black holes, they must be large enough so that they would not have evaporated by now (by Hawking radiation) and small enough so that we would not observe their effects on visible bodies. Still, there are plenty of places to look for dark baryonic matter.

The situation is not hopeless, either, with regard to non-baryonic candidates. For a while, neutrinos were suspected of constituting the non-baryonic matter. If at least one type of neutrino had a small mass - between 10 and 50 electron volts - then the large number of neutrinos left over from early calculable process of the Big Bang (about 100 per cubic centimeter) might contribute sufficient total mass to close the universe. However, while a few experiments have suggested that neutrinos have masses in the range required, such results have not been found when the experiments were tried elsewhere. Upper limits on the possible mass of the neutrino, for instance calculated from the characteristics of neutrinos received from Supernova 1987a, are now very close to the minimum required to close the universe. (Although the neutrinos from that supernova are electron neutrinos, other considerations cast doubt on the other two types of neutrinos having a mass sufficient to close the universe.)

But more is involved in the rejection of the neutrino dark-matter hypothesis than this. There is also its failure to account for galaxy formation. They were once favored in this regard. After all, neutrinos do interact only weakly with ordinary matter, and thus would have been missed in estimates of matter-density based on searches of electromagnetic radiation. Furthermore, the density fluctuations which would be allowed to persist in a universe dominated by neutrinos with a mass of, say, 30 eV, could be calculated to be approximately the size of superclusters of galaxies. This suggested a "top-down" scenario in which superclusters formed first and galaxies formed only later. However, numerical simulations performed on the basis of reasonable assumptions about such a universe disagree strongly with observations of the actual distribution of both galaxies and clusters in the universe. Also, the density fluctuations required would be large enough (by a factor of several) to show up in the cosmic background radiation as observed at the present time; the fluctuations are not present. Galaxies are formed late in the top-down scenario - perhaps earlier than a red shift of 3, yet galaxies are observed at greater red shifts than this. The neutrino dark-matter hypothesis, and the associated "top-down" scenario for galaxy/cluster formation, are not yet ruled out: the objections depend on simplifying assumptions; other factors not yet taken into account might change the results; the relevant observations are difficult; and the observational limits do not completely exclude the theory. Nevertheless, those views are in enough difficulty for most astronomers to have turned to other ideas.

Currently, the favorite approach is that of *cold dark matter.* Many of the difficulties of the neutrino dark-matter hypothesis stem from the fact that massive neutrinos would have relativistic velocities (be "hot") for too long in the early history of the universe. What is needed seems to be some type of weakly interacting particle (or several types of such particles) that becomes nonrelativistic at times much earlier than massive neutrinos. In such a scenario, galaxies are formed first and clusters later ("bottom-up" or "hierarchical" scenario). The particles could not travel far enough to destroy density fluctuations of galactic size before they could collapse gravitationally. Galaxies could have formed earlier than in the neutrino view, and the fluctuations from which they arose would have been below (but uncomfortably near) the limits imposed by current observations of the microwave background radiation. Galaxies would form where the cold dark matter was denser, and thus their electromagnetic radiation would indicate the presence of such density fluctuations ("biased" galaxy formation). Such a picture agrees well with numerical simulations of galaxy formation. There are many difficulties of detail in such a picture, but there are also fundamental objections. It has a great deal of difficulty accounting for the largest observed structures in the universe: the vast voids, and sheets or lines of clusters of galaxies which have recently been found. (If the existence of such structures continues to be borne out, how can they be accounted for? It seems not by gravitational collapse - the mode of galaxy or cluster formation considered earlier, - for the universe is too young for them to have formed that way. Perhaps by explosions of some sort, for instance of large numbers of supernovae from a generation of pre-galactic stars? Yet another possibility will be noted at the end of this section.) It also requires the universe to be fairly young, *i.e.*, it requires a large value of the Hubble constant. And while it agrees with many observations for a value $\Omega=0.2$, it does not fare as well if, with inflationary theory, we have $\Omega=1$. Finally, if the cosmic background radiation does not show anisotropies soon - at only relatively slight improvements in observational discrimination, say an order of magnitude better than present levels - cold dark matter theories will face still further trouble.

In any case, there remains the question of what the cold dark non-baryonic matter could be. In fact, whether the non-baryonic mass is sufficient to close the universe or not, there are several candidates for that mass. And it is important to recognize that *all these candidates are serious ones, in the sense that they were introduced for physical reasons that were independent of the dark matter problem*. That is, they have other justification than merely that they would solve that particular problem. One current favorite is the axion, a particle proposed in the attempt to preclude CP violation in strong interactions.[70] In its modern form, the mass of the axion should be on the order of 10^{-5} eV, much less than that of the electron. Axions and supersymmetric particles can be expected to interact weakly with ordinary matter, which would explain why they have not been observed. They are predicted by their theories to have been created in profusion in the Big Bang. Some light particle called for by supersymmetric theories (to be discussed shortly), such as the gravitino or photino, is another possibility. Of course neither axions nor photinos nor gravitinos have been observed to exist, and supersymmetry, as we shall see, is as yet only a good possibility. For all the objections raised above, at least neutrinos are known to exist!

Even if it ultimately becomes necessary to conceive of galaxy formation by some sort of "seeds" other than primeval fluctuations, all is not lost for the modern scientific picture, since candidates for being such seeds are available within current theories - for example, cosmic strings, the relic two-dimensional topological defects spoken of in Section 2 of this Part, in connection with the inflationary theory (not to be confused with superstrings).

SECTION 4. BEYOND GUTS.

Going well beyond GUTs, still grander unification is being sought, to include the fourth force, gravitation, in a synthesis of all four forces. The pattern of reasoning upon which unification of the electromagnetic and weak forces had been constructed, and the reasonable expectation that the electroweak and strong forces might be unified in a similar pattern, offered fresh grounds for hoping that there should be some still higher energy at which the unified electroweak-strong force would be equal to the gravitational force, and that a complete symmetry might be found between these two types of force. Perhaps, again, some process of symmetry-breaking could be envisioned whereby, in successive stages of lowered energy, first the gravitational force might be broken from the electroweak-strong (GUT), then the electroweak broken from the strong, and, finally, the electromagnetic separated from the weak. A single all-embracing scientific theory might thus account for all the diversity of forces and particles of nature. Such a theory, with a unification energy of 10^{19} GeV, and a corresponding temperature of 10^{32} °K, would press still earlier in the history of the universe. It would go back to when the universe was a hundred-millionth its age at the GUT transition (10^{-35} seconds), to the "Planck time" of 10^{-43} seconds after the Big Bang, when quantum effects could be expected to arise in gravitation. It would thus be a theory unifying quantum theory and general relativity: a theory of quantum gravity, a full unification of the domains of the very small and the very large.[71] Perhaps, too, it would be only through such a unification that problems of lesser theories could be resolved - for example, the problem of choosing the proper unifying group for a grand unified theory of the electroweak and strong forces, and the problem of infinities which has plagued the lesser theories. Three major programs of such unification have been the objects of wide attention at various points in recent years.

(1) *Kaluza-Klein theories.*[72] In 1921, Theodor Kaluza rewrote Einstein's equations of general relativity in five dimensions (instead of the four Einstein had used), thereby developing a theory in which electromagnetism was unified with gravitation. The fifth dimension incorporated classical electromagnetism. In 1926, Oskar Klein extended Kaluza's theory, applying it to strong fields as well as the weak ones to which Kaluza's theory had been applied, introducing quantum mechanical considerations into it, and, most importantly from the viewpoint of later development, giving an explanation of why the fifth dimension is not observed. That dimension is collapsed or "rolled up," each point of four-dimensional space-time being associated with a cylinder which constitutes the fifth dimension at that point; mathematically speaking, the fifth dimension is compactified. Klein's quantum-mechanical elaboration of Kaluza's theory enabled him to calculate the radius of the cylinder as 10^{-32} cm., so that it is unobservable. In this theory, electrodynamics falls naturally out of the compactification of the fifth dimension - clearly an attractive feature. Though long ignored, this compactification-from-higher-dimensions approach was revived in the late 1970's, in the hope that, with a sufficient number of higher dimensions, one could get a theory in which some of the dimensions would be compactified - by some presumably dynamical process - to produce the GUT

regime, four dimensions remaining to form the space-time of general relativity and the gravitational force which it describes. However, despite their appeal, Kaluza-Klein theories soon ran into difficulties. Eleven-dimensional theories seemed to show promise. However, they yielded bosons, but had trouble producing the observed characteristics of fermions (specifically, their chirality); fermion fields had to be put in "by hand." However, a new approach, able to incorporate the Kaluza-Klein approach within its framework, resolved this problem.

(2) *Supersymmetry and supergravity.*[73] It was Heisenberg's idea that the strong force does not recognize the distinction between the proton and the neutron: that as far as the strong force is concerned, those two particles are symmetrical (indistinguishable), the symmetry being that of "isotopic spin." In isotopic spin space, the proton and neutron are partners; if the isotopic spin arrow points "up" in this space, the particle is a proton, and if "down," a neutron. This idea of an "internal (as opposed to a space-time) symmetry was incorporated into later theories of the strong force, and provided a model for subsequent unification theories. GUTs postulate a symmetry between baryons and leptons, which is broken in a phase transition similar to, but at a higher energy than, that of the electroweak symmetry breaking. But can one not go farther with such an approach? Is it possible to construct a theory in which there is a symmetry between fermions and bosons? Since fermions are the descendants of classical concepts of matter, and bosons are descendants of classical concepts of force, such a theory could be seen as removing the fundamentality of the distinction postulated in classical physics between matter and force.

Supersymmetry is such a theory. In the "superspace' of the theory, the particle is a fermion if the supersymmetric arrow points "up," and a boson if it points "down." In one version, for example, rotating a boson (more precisely, performing a transformation) from the subspace of four regular dimensions to the additional fermionic dimensions would convert it into a fermion, and vice versa. Thus every fermionic particle has its bosonic partner, and vice versa. Supersymmetry theory can also be extended to show a deep relation between space-time and internal symmetries.

Supersymmetric theories did not succeed in showing that the partner of each of the *known* fermions is a *known* boson, and vice versa. Rather, the supersymmetric partners are hypothesized particles, none of which has been found in experiment. The postulated supersymmetric bosonic partners of ordinary fermions are given names beginning with s- (the electron's partner is the "selectron," and there are also squarks and sleptons), while names of fermionic partners of ordinary bosons end in "-ino" (so that, persisting in the long history of particle-theoretic whimsy, the W's partner is the "wino"). In general, these partner particles differ not only in spin but also in mass, presumably as a result of the breaking of the original symmetry.

When - after some difficulty - supersymmetry theory was formulated as a local gauge theory, it happened, in the usual way, that a particle fell out of the theory. However, that particle has the remarkable characteristic of having spin 2 - the spin of the hypothesized graviton that would have to be present in a quantized theory of gravity. Thus, in addition to calling for a symmetry between fermions and bosons, the theory automatically incorporated a theory of gravity. The resulting supersymmetric theory, supergravity, held out the promise - if it could be worked out successfully - of providing the long-sought unification of quantum field theory and general relativity. Supergravity is indeed a generalization of Einstein's general relativity, reducing to the latter when gravitational interactions of the gravitino are ignored.

The simplest supergravity theory (N=1 supergravity) contains only the graviton and its partner, the gravitino. However, supergravity theory can be formulated at eight different levels, with more particles being required at each successive level. The highest-level theory, N=8 supergravity, contains 256 particles, including the one graviton and eight gravitinos (hence the '8' in N=8). A space of eleven dimensions (ten of space and one of time) is the largest in which it is possible to formulate supergravity. N=8 supergravity is the result of compactifying such a space (an N=1 supergravity in 11 dimensions) to four dimensions.[74]

To be successful, some such higher-N theories should fulfill their promise of providing a symmetry between fermions and bosons, while at the same time yielding a theory of gravity. Further, the higher-dimensional theory should be broken in some understandable way to yield the world we know, the low-energy world of the Standard Model and of general relativity. Four of the dimensions would provide the latter, while the remaining ones would compactify in such a way as to account for the particles and forces, ultimately furnishing those of the Standard Model. All known particles should, of course, be among the particles predicted by the theory, and as to those which do not so correspond, either they would have to be found in further experiments, or the failure to find them explained.

Unfortunately, the initial promise of supergravity, and more generally of supersymmetry, has not been borne out, at least as yet. Early arguments to the effect that infinities would be cancelled in supergravity, making the theory finite, turned out to be shaky. Also, as we have seen, none of the supersymmetric partners of known particles have been observed in experiment. It is of course possible to explain this by arguing that supersymmetry has been broken so badly in the very early stages of the universe that those partners obtained masses too great to be produced by present accelerators. But even so, obtaining the Standard Model from the exalted level of supersymmetry and supergravity has proved very difficult, and indeed success in doing this has not been achieved. In the first place, physical mechanisms for breaking supersymmetry are hard to come by, and those which have been proposed lead to serious difficulties, such as producing a very large cosmological constant (see below). In the second place, as we have seen, the Standard Model has not been shown to be the ultimate result of supersymmetry breaking, and in some cases, such as the following, it appears difficult to see how it could be. In 1956, Yang and Lee found that the weak interactions recognize a distinction between "left-handedness" and "right-handedness": the weak force violates the (three-dimensional analogue of) mirror-image symmetry called parity.[75] Where the weak force is concerned, nature shows a definite chirality, or handedness. But it was found that chirality could be obtained only when space-time has an even number of dimensions. Eleven-dimensional space-time supergravity was thus placed in peril. Still a further difficulty of supersymmetry theories is that it proves difficult to prevent the lifetime of the proton from being shorter than the lower limits imposed by the results of proton decay experiments.

One motive for investigating supersymmetry theories is that they hold promise of answering a problem that otherwise might prove difficult. This is the so-called "hierarchy problem." The problem is this. GUTs in general go on the supposition that the energy-region between the GUT and electroweak levels is, physically, a "desert" (to use Glashow's skeptical characterization). In all the thirteen or so orders of magnitude between our attainable energies and those of the GUT level and beyond, there is supposed to be no new physics: no intermediate symmetry breaking, no new levels of forces and particles, no new and as-yet unknown processes. At first glance, this might seem an appalling assumption to make, particularly when most of that energy region cannot be investigated by experiment. However, it can be shown that the radiative corrections of field theory, which modify the coupling constants of the forces, scale as the logarithm of the energy, not as the energy, so that the difference of thirteen orders of magnitude between the strong and electroweak energies does not have the alarming consequences one might expect. Unfortunately, the same is not true for the masses; the same radiative corrections should give the electroweak gauge bosons masses more like those appropriate to the level of the strong force. In order to preclude the assignment of such enormous masses, SU(5) must assume very unnatural cancellations of those corrections. This state of affairs is called the "hierarchy problem."

In supersymmetric theories (including superstring theories, discussed next), the corrections which threaten to endow the W bosons with an enormous mass (on a GUTs or even Planck scale) cancel automatically. The ability of supersymmetry theories to deal so neatly with the hierarchy problem is indeed a major motivation for taking them seriously. Unfortunately, as we have seen, other problems arise for such theories. All we can say is that *if* supersymmetry works out in some form, we can expect with confidence that the hierarchy problem will be a problem no more.

(3) *Superstring theory.*[76] When the theory of the strong force was still intractable, models of quarks were developed in which those particles were the ends of strings, whose lengths were conceived as being on the order of the range of the strong force, 10^{-13} cm. The models - they were never more than that - involved serious difficulties, among them being that the model apparently called for a spin-2 particle, which seemed to have no place in a theory of the strong force. When quantum chromodynamics was introduced, such string models were abandoned by all but a small number of people. These people - John Schwarz and Joel Scherk, most significantly - reasoned that spin 2 was that attributed to the graviton, and hence that a string theory might be more appropriate as a basis for a theory of gravity than of the strong force. They therefore reconceived the strings as having lengths appropriate to the Planck scale, where unification of gravitational and quantum theories was expected to be found. The extremely small length of such strings, on the order of the Planck length, 10^{-33} cm., means that they can be approximated by points on the scales that physics had dealt with previously. Applying supersymmetry to the string transformations provided a superstring theory, which did indeed call for a spin-2 particle. As supersymmetry already incorporated supergravity, the latter could be expected, in some sense, to be an approximate result of the superstring theory.

In 1984, Green and Schwarz showed that the resulting 10-dimensional superstring theory had consequences of major importance.[77] Infinities cancelled automatically, at least for lower-order calculations, and as for higher-order ones, the case for finitude, though not absolutely proven, was much stronger for superstrings than were the corresponding arguments for supergravity theories. Chirality, the bane of supergravity, was a natural feature of superstring theory. Furthermore, another sort of mathematical disaster afflicting many previously-popular theories, a sort referred to as "anomalies,"[78] and which would involve the violation of fundamental conservation laws, could be avoided by appropriate choice from among the possible superstring theories. Still other afflictions of earlier theories - "ghosts" and "tachyons" - were also not present in superstring theory. And finally, Green and Schwarz showed that the constraints on the construction of a superstring theory were so tight that only two group structures, SO(32) and E(8) x E(8), were possible. Although they considered both closed-loop and open strings, the specific version they advocated was an open-string one, in which interactions were described in terms of the joining and separation of the ends of strings. The group transformations were those of SO(32). The closed-loop alternative, with E(8) x E(8) as group, was developed in a quite different approach by four Princeton physicists, who together were duly christened the "Princeton String Quartet."[79] In that "heterotic superstring theory," particles are viewed as modes of vibrations of the loop, fermions going around the string in one direction, bosons in the other. Heterotic string theory combines a ten-dimensional theory, in which fermions are dealt with successfully, with a twenty-six dimensional treatment of bosons. The extra sixteen dimensions make possible the E(8) x E(8) symmetry. An advantage of the heterotic theory is that it eases the problem of getting chiral fermions, which we have seen to be a trouble for other sorts of theories.

Even with two groups as possibilities, the powerful restrictions on choice of fundamental group were welcome after the wide range of possibilities open to GUTs. Coupling this boon with the other remarkable advantages of the superstring approach, it is no wonder that a host of physicists turned their attention to trying to develop the theory.

The formulation of superstring theory is, however, still in its infancy. To date, the theory has not been put into the form of a field theory, but only employs the "first quantization" of quantum mechanics itself, and not the "second quantization" that marks the transition to a field theory. Indeed, even within its first-quantization limits, superstring theory calculations are crude as compared to its professed aim as a unifying theory. The ideal of superstring theory is to treat the strings not - as material particles had been conceived in earlier theories - as objects distinct from but occupying space, but as identical with it. For superstring theory, attempting to be the ultimate unifying theory, fuses the concepts of space (space-time) and matter (matter-force), so that the strings are themselves not *in* ordinary space-time. Yet in practice, calculations with superstrings thus far have employed methods in which space-time is treated as a background within which the strings exist and interact.

Two especially pressing difficulties confront superstring theory at present. The first has to do with how the compactification is supposed to take place, so the four spatiotemporal dimensions of our universe will remain uncompactified, while the remaining dimensions compactify in such a way as to result in the particles and forces of the Standard Model. It appears, however, that there are an enormous number of ways in which compactification could occur, and no clear way of choosing from among them. Why the familiar four dimensions and the Standard Model result is unexplained; they are far from being the only possibility. There are different four-dimensional-plus-six-compactified-dimensional possibilities. There are still other possibilities in which the result is something other than four-plus-six. Perhaps it might have been four dimensions that compactified, leaving the other six to form an "everyday" spacetime world. Thus one of the most attractive features of superstring theory, its satisfaction of a large number of very tight constraints governing the formulation of an adequate theory, appears to dissipate, as lack of constraint comes in through the back door.

The second major problem for the theory concerns the stability of the compactified dimensions. For they would be expected to undergo large quantum fluctuations. And if so, one or more might "uncurl," suddenly making its presence known. Also, what keeps them from expanding along with the four spatiotemporal dimensions? This is not to say that these problems will not be taken care of with a fuller formulation of the theory. It is only to point out that such questions still remain before the theory, and that much more work must be done before it will be possible to determine whether the great initial promise of the theory will be realized.

As of this writing, superstring theory remains an area of intensive investigation. A variety of approaches are being taken. For example, some workers maintain that the talk of higher dimensions is not to be taken literally. Also, deep relations have been revealed between string theory, and more generally quantum field theory, and knots, links, and braids. Knot theory itself, and the associated braids and links (also related to conformal field theory), and its relations to physics, is an area of mathematics which is being developed in new ways at present by mathematicians and physicists alike.[80] It may yet illuminate these fundamental physical theories in important ways.

Whatever the unifying theory may be, if there is one - whether it be superstring theory or something else - it should throw new light on the nature of matter, space, and time. Superstrings, for example, do not exist *in* space and time; space and time would emerge from those 2-dimensional entities. This would also be true for theories in which space-time and matter emerge from compactification or analogous process from some higher-dimensional predecessor. And after all, such a higher unifying theory would at last join the theory of material processes, quantum field theory, with the theory of space-time, general relativity. Perhaps the difficulties about space and time with which philosophers have concerned themselves will be answered then, by physical understanding rather than by mere reflection. Perhaps also difficulties in our understanding of quantum mechanics will also be resolved then - again, rather than by considering those difficulties to be only ones of "interpretation," to be solved by more careful reflection or examination of the content of present physics.

SECTION 5. OTHER CURRENT APPROACHES.

The issues and approaches in physics and cosmology which I have surveyed are of course not the only ones that have been considered, nor have I mentioned by any means all of the directions being taken within those examined here. The total range of scientific activity in these fields is much more wide-ranging, covering many possibilities, some more speculative than others. Although some of these alternative approaches are potentially important, I will only briefly mention three of them here.

(1) The history of science has revealed one level after another of the structure of matter, and considerable effort has been devoted to the question of whether there is a level of composition still deeper than that of quarks and leptons.[81] Despite the fact that experiments at the limit of attainable energies indicate that quarks and leptons behave like point-particles without internal structure to a level of 10^{-16} cm, this possibility remains an open one which could conceivably solve a number of problems. Variations on this theme include proposals of the existence of "preons" and of "rishons" as constituents of quarks and leptons.

(2) The Higgs particle, which provides the mechanism of symmetry breaking in the major unification theories described earlier, has not been observed. Attempts have been made to detail an alternative mechanism in terms of "technicolor theory." Technicolor is also aimed at solving the hierarchy problem. To date, however, the theory has not proved satisfactory: for one thing, its own particle, the technipion, has no more been observed than has the Higgs particle.

(3) The perennially-arising suggestion that the law of gravitation should be modified is also present.[82] The suggestion is that the dark matter problem is an artifact of an incorrect law of gravitation.

Two other possibilities have been widely publicized in recent years, but have now largely faded from the scene: that the red shifts of galaxies are not indicative of distance, or at least that they cannot be uncritically taken to be so;[83] and that there may be other forces than the four discussed here. It should be said that, even if new forces were found, they could probably be absorbed into the scope of present theories without any trouble. Suggestions that there may be a fifth - short-range - force have now been almost universally rejected on the basis of a large number of experiments.

Science should not be seen as monolithic, as consisting, even in specific areas, of one approach, guided by one theoretical approach only, ignoring or suppressing alternatives.[84] A multitude of approaches are always being taken. Human variety ensures that open alternatives will in fact be explored, even though sometimes such explorations are treated far more harshly than is justified. On the other hand, it is not true that "anything goes,"[85] that any and all avenues are equally deserving of exploration. Scientific research needs to be guided by a clear understanding that there are alternatives, and by a clear understanding of what the range of legitimate possibilities is. (Here is one place, among many, where *genuine* philosophy of science, based on close examination of science and its history, and on an understanding of historical and contemporary misunderstandings of science and its methods, is a necessity.)

V THE FUTURE OF THE CONTEMPORARY SCIENTIFIC PICTURE

SECTION 1. A COHERENT PICTURE: INCOMPLETE OR POSSIBLY INCORRECT?

The question of the exact nature of a deeper unifying theory is still open. However, a history of the universe back to the time of electroweak-strong symmetry breaking has already been outlined in terms of the available and highly successful Standard-Model-Big Bang combination. The deeper unifying theory being sought can be expected to carry that history still further back, to the Planck time, and can be expected to provide a unification of *all* fields and *all* forces, and therefore of *all* particles, in the universe. Further, the theory would form the framework within which we can seek and place an understanding of the formation and evolution of galaxies, stars, chemical elements, and life. (It would also ultimately have to provide the framework for explaining the possibility of rational thought, and therefore the possibility of scientific understanding itself.) To the extent discussed in this paper, such understanding is, in many cases though to varying degrees, already present. We have, already, an understanding of much, and well-founded reason to believe that we can get more.

In his inaugural lecture as Lucasian Professor at Cambridge University, Stephen Hawking declared that

> In this lecture, I want to discuss the possibility that the goal of theoretical physics might be achieved in the not too distant future, say, by the end of the century. By this I mean that we might have a complete, consistent, and unified theory of the physical interactions which would describe all possible observations. Of course, one has to be very cautious about making such predictions ... Nevertheless, we have made a lot of progress in recent years and, as I shall describe, there are some grounds for cautious optimism that we may see a complete theory within the lifetime of some of those present here.[86]

Hawking's view is not an isolated one. Many physicists have spoken, with similar optimism, of the prospect of attaining what they unabashedly call a "Theory of Everything."

It is impossible to read such words without reflecting that, at various stages of the history of science, at least some scientists have proclaimed that all important theoretical principles in some field had been found, or that all important empirical discoveries in some area had been made, and that the end of exciting research in that area had been achieved or was in sight. We remind ourselves of William Dampier's account of the situation in physics at the end of the nineteenth century:

> It seemed as though the main framework had been put together once for all, and that little remained to be done but to measure physical constants to the increased accuracy represented by another decimal place, and to carry further those investigations which had seemed at intervals to be on the point of solving

> the problem of the structure of the luminiferous aether.[87]

Although there were many who held this attitude in the late nineteenth century - the sentiment about seeking only the next decimal place goes back to Maxwell in 1871, - it is an exaggeration to look on scientists of that period as uncritical, and by no means all of them believed that physics was in or near such a final state.[88] But when we do find such expectations in the history of science, as we do, we tend to view them as complacent, presumptuous, inflated, arrogant, fully deserving of the rejection which was the fate of the theories they exalted in their overweening pride.

How does the situation at the end of the twentieth century, with so many scientists expressing grand (or grandiose) expectations of attaining a "Theory of Everything," stand up to the charge that it is ignoring the lessons of the history of science? Are today's scientists only repeating the folly of their ill-fated predecessors? Or are they on surer ground, the expectations of today based more firmly on a sober assessment of the total situation in physics?

In this connection, three specific questions arise.

1. Are there "dark clouds on the horizon" of the contemporary scientific picture - *specific* problems that may require us to revise our present theories as profoundly as nineteenth-century physics had to be revised?

2. Is it possible that the present directions of research will not achieve unification, but that ultimate unification, if it is achieved, will be in a form very different from anything we have now, as twentieth-century physics differed from that of the nineteenth?

3. Is it possible that we are "on the wrong track" more generally, and that perhaps scientific claims cannot be considered "knowledge" at all?

Answers to the first two of these questions must be based on our detailed assessment of the present scientific picture: on a close examination of whether there exist, in the science itself (rather than merely in the wishful mind of the physicist), adequate reasons for thinking that such a goal might be achieved. We must ask whether there are solid grounds, good reasons, indicating that such a final unification is in the offing, and we must soberly assess the strength of those indications and the problems that might make the promise go awry. The third question, in most of its various ramifications, is of a more general kind, and will be discussed in the final two sections of this paper. In this section I will consider the first two questions.

1. ARE THERE SPECIFIC "DARK CLOUDS ON THE HORIZON" OF THE CONTEMPORARY SCIENTIFIC PICTURE?

In a famous lecture delivered in 1900, Lord Kelvin (William Thomson) gave this assessment of the status of one crucial piece of the physics to whose development he had contributed:

> The beauty and clearness of the dynamical theory, which asserts heat and light to be modes of motion, is at present obscured by two clouds. The first involved the question, How could the earth move through an elastic solid, such as essentially is the luminiferous ether? The second is the Maxwell-Boltzmann doctrine of partition of energy.[89]

Kelvin's statement was both prescient and ironic; for his two "dark clouds" were among the central problems which led to the development of theories whose very character as explanations were different in fundamental ways from those of nineteenth-century physics. We must ask whether there are any such "dark clouds" in the present scientific picture. In order to consider this question, let us review the account given above of the problems existing with regard to the major parts of the present scientific picture.

Our survey of problems existing in the present scientific picture has led us to conclude that the Big Bang theory, as deepened by the Standard Model, and the Standard Model itself, are well-established: there is *no specific reason* to suppose that any of them will be overthrown, or that they will not be incorporated into a higher synthesis without fundamental alteration. The "standard models" of stellar formation and evolution, the origin of chemical elements, and the origin of planetary systems are all in good order: though incomplete, those theories are quite good, and where there are alternatives, they are constructed in conformity with the background framework of the Standard Model-Big Bang theory. Even the problem of the origin of galaxies and larger structures in the universe, and the associated questions about the nature of mass and the closure of the universe, are not as yet threats to the overall picture. Although recent discoveries have complicated the problem of the origin of structure, and do constitute a very real danger to cold dark matter theories, that threat cannot yet be viewed as a "problem of incorrectness": cold dark matter theories still provide a good way of making galaxies, even if those theories have to be supplemented by other hypotheses (*e.g.*, cosmic strings, massive neutrinos, positive cosmological constant) for making the large-scale structures. Finally, even if we are forced to adopt a very different picture of galaxy formation than we have today, one which, for example, might be the same for galaxies as for larger structures, we can still expect it formulated within the larger background framework of present physical cosmology: the homogeneity and isotropy of the universe, and the Big Bang theory itself, are not - as yet - threatened by the new discoveries of large-scale structure. In short, the weight of evidence is that the problems associated with galaxy formation, like those in the cases of star formation, the origin of elements, the origin of planetary systems, and the origin of life, constitute *gaps in a unified picture which we can, with reasonable confidence, expect to see filled.* Even if theories of that particular domain - galaxy and large-structure origins - have to be replaced, there is no specific reason at present to suppose that the broader background framework will have to be replaced. That is not, of course, to say that such reasons will never appear, but only that there possibility is a *mere* possibility at present.

We have seen that present theories provide an abundance of possibilities for the dark matter. Indeed, dark matter is almost necessary for the formation of galaxies as well as being called for if we are to retain the benefits of the inflationary theory or anything like it. As to the dark matter itself, many candidates, both for the baryonic and non-baryonic cases, exist within current research and current lines of promising research - candidates which, in the case of non-baryonic candidates, have not been introduced merely to answer the dark-matter problem, but to answer other, independent problems. Finally the questions of how much dark matter there is, and of whether the universe is open or closed, are clearly problems of the incompleteness of our present knowledge.

All in all, we may reasonably draw the following conclusion: that *In the light of careful and conscientious appraisal of the totality of the best evidence available today, the problems in the picture of the universe as portrayed by science at the present time are ones of incompleteness.*[90] Though the details may change - even the standard models in some areas - the overall picture, the Standard Model of elementary processes as applied to Big Bang cosmology, are, *in the light of a close, detailed assessment of present evidence*, here to stay. As in the case of many past theories (see Section 2, below), future modifications will preserve much of them, placing them in a still larger context of deeper and perhaps unifying theories.

Of course, one might claim that the same might have been said at the end of the nineteenth century. Nevertheless, the situation at the end of the twentieth century differs greatly from that at the end of the nineteenth. Today, as we have seen, we have a remarkably coherent picture. It gives a very well-supported account of elementary particles and forces. It provides the broad outlines, and much in the way of detail, about the origin, evolution, and destiny of the universe and its contents. The portions of it which are widely accepted - especially the Standard Model of elementary particles and forces and General Relativity, as applied to Big

Bang cosmology - are remarkably successful in accounting for the facts. The "standard models" in other areas - the origins of stars, chemical elements, and planets - are framed within the context of the physically-enriched Big Bang theory. Alternatives to those standard models are possible, but nevertheless, both the standard models and the more serious alternatives are compatible with the larger picture given by physical cosmology. Although there is not a similar "standard model" for the origin of life with the specificity of those in the preceding areas, all serious possibilities are today seen as being within the context of antecedent chemical and astronomical conditions. Together, all these present theories give, in the best cases, a superb, and in others a very promising or at least not unpromising, account of the phenomena of nature from 10^{-17} centimeters to the largest observed scales of the universe, and from 10^{-35} seconds after the Big Bang to the farthest reaches of the future. In addition, although occasional suggestions have been made of the existence of phenomena that required a further force or forces, such possibilities are by now largely abandoned, at least as of the present, and in any case, no further force has been suggested that could not be easily taken into account within the present framework. Finally, the present picture gives unambiguous directions as to how to proceed in a search for higher unification.

This state of affairs is very far from that which existed at the end of the nineteenth century. Although there were "standard models" in various areas, *most* of them were incompatible with one another in fundamental ways - the time-scales of biological evolution and physics, for example.[91] Although there were a variety of speculations, there was no good theory of the origin of the solar system, and theories of stellar evolution and the origin of chemical elements were highly primitive and almost (but not quite) non-existent. Even within physics itself, it was known that there was an unconformity between theories of matter and light. (How, for instance, do they interact?); there was also one between electrodynamics and mechanics - there to be seen, whether seen or not. Further, theories for the removal of those unconformities varied from scientist to scientist, and there were conflicting "philosophical" approaches, for instance the opposing "mechanical," the "electromagnetic," and the "energeticist" views of how to go about proceeding.[92] In short, there were reasons *within the total subject-matter* of nineteenth-century scientist for believing that parts of the total science were in conflict with one another, and that there was much that was not known. This is not a retrospective assessment: the situation was realized by a great many scientists at the time.

This situation contrasts with the present picture, in which the various "standard models" not only are successful with respect to their own particular domains - more so, and with respect to far wider domains and in far greater detail, than any of the nineteenth-century theories - but also are *coherent* with one another, giving a unified, consistent picture to an extent that was at the very best only a hope in the nineteenth century. Further, present theories and directions of research give clear possibilities for and constraints on answers to outstanding questions to an extent and specificity that was lacking in the nineteenth century. (In the latter connection, recall that potential answers to the dark matter problem are given in terms of possibilities already introduced in physics for independent reasons.) Though there is still a major bifurcation - between the Standard Model (quantum theories) and General Relativity - specific ways exist, in the prevailing lines of research, for potentially bridging the gap: in particular, supersymmetric theories (and therefore also superstring theories) include General Relativity as a subgroup.[93] There was no analogue to this situation in the nineteenth century.

All this is not to minimize existing problems, but only to say that we must undertake a sober, careful, and serious assessment of the actual science and its problems, and their degree of seriousness, on the basis of the best and most complete evidence we can assemble. Many of the problems, as has been indicated, are very worrisome, and it is possible that new discoveries may make them still more serious. The problem of the age of globular clusters and that of the timing of galaxy formation, and that the large-scale structures in the universe, are three examples of problems that are on the *verge* of placing current "standard models" in jeopardy, and must be fully recognized as such. Nevertheless, a thorough, objective assessment of these and other problems must still lead to the conclusion that the major problems of the contemporary scientific picture are as yet problems of incompleteness. And as we have seen, even if they ultimately reach a more troublesome stage, it is not clear that they will require some fundamental change in that picture.

Contrariwise, as will be argued more fully in the following Sections, we must not base our assessments on facile generalities, such as that science always has been wrong, so it must be wrong now (even if the evidence does not warrant such a conclusion); or that we should take a "humble" attitude toward what we claim to be knowledge; or that it is (merely) possible that we may be wrong, therefore we cannot legitimately conclude that any of our theories can be trusted. Nevertheless, thorough and repeated reviews of the picture at any given time must be made, to ascertain whether we have missed something critical.

2. IS IT POSSIBLE THAT PRESENT DIRECTIONS OF RESEARCH WILL NOT RESULT IN UNIFICATION?

The theoretical shortcomings of present physics and cosmology - undetermined parameters in the case of the Standard Model, breakdown at the quantum level because of inevitable singularities in the case of general relativity - strongly suggest the need for a deeper theory. What form such a theory will take is still uncertain. However, as we have seen, that form is increasingly constrained as to its general mathematical structure (it should be a gauge field theory) and the specific group structure imposed. What *specific* doubts can be raised, on the basis of what we now believe, that these directions of research will succeed in leading to such a theory, and that that theory will have the form of theories that are now under consideration? After all, it can be said, we have been disillusioned in turn by minimal SU(5), supergravity, and a host of more specific proposals.

We must be careful to understand what this question asks. There can be no idea here of trying to get a *guarantee* that some particular direction of present research will lead to a successful unifying theory, or even that such a theory will be found among the directions now being sought.[94] The only question is whether there are good reasons to think that they may. Indeed, there are such reasons, despite the fact that difficulties remain which may ultimately cause us to abandon those particular lines.

In the case of the search for a GUT, the promise that a unifying gauge group could be found of which the Standard Model [SU(3) x SU (2) x U(1)] and General Relativity (Poincaré) groups are subgroups is supported by the existence of many candidates. At the same time, that plethora of alternatives poses a problem: how to close in on one of the possibilities as the one that would provide the successful unification. The range of such candidates is severely restricted in superstring theory, to either SO(32) or E(8) x E(8). This constraint is one of the advantages of superstring theory - one of the reasons leading many scientists to take it seriously as a direction for present research. Superstring theory is also anomaly-free, and there is strong reason to believe that all infinities cancel in it. Being supersymmetric, superstring theory is connected with the group-theoretic tradition in which the gauge theories are also located. Being supersymmetric means that superstring theories include (though not without some attendant difficulties) supergravity, and therefore general relativity. Thus the promise of superstring theory rests on the following grounds: (1) its further restriction of alternative possible theories; (2) its character as removing problems that have plagued earlier theories; and (3) its comprehensiveness in including gravity within its scope. *These are powerful considerations which, while by no means guaranteeing success, make the investigation of superstring theory a promising direction of research. It is one kind of possibility whose investigation is justified in the light of what we have learned.* Of course we must remember, in the light of the problems discussed in connection with superstring theory earlier (Part IV, Section 4), that its claims are promises, and are far from achievements of success, at least yet. As Wilczek notes, "The theory is still promising ... and promising ... ."[95] In particular, it has yet to make a specific prediction that can be tested by observation. It may turn out not be the all-encompassing theory called for by present well-established theories; after all, its vaunted uniqueness is already damaged by the host of possible compactifications that it apparently leaves open. Nevertheless, it does deserve investigation, perhaps alongside many other research programs, all with the same kind of justification in the light of what we have learned.

Thus, it is *possible* that current lines of research will not result in the sought-after unification. Nevertheless, there are reasons for devoting investigative attention to some lines of research and not others, at least at present. In order to determine whether they can succeed, the theories must be worked out in detail. Perhaps new mathematical or physical techniques will have to be developed in order to do that (to an unparalleled extent, fields as complex and diverse as topology, algebraic geometry, and Riemann surface theory must be brought to bear; and as we have seen, Witten has turned to developing the mathematical subject of knot theory in the hope that it will be useful in clarifying superstring theory and quantum field theory in general). But until we have done that, we must look thoroughly into the various directions indicated by current theories as possible and promising.

The investigation of unifying theories, from the GUT level up, is still in its infancy. We have seen, for example, that superstring theories are still formulated in a quite primitive way, given the way we expect them to be formulated ultimately. It is quite plausible that further development will in fact turn up experimentally-accessible predictions and further constraints (perhaps on allowable compactifications, one might dare to hope), and that the experiments and/or constraints may yet narrow the possible theories, even to the point of selecting one as the only possible one. Given all the successes and well-founded promise of the theories, it is far too early to abandon them because of either this sort of problem or the first one.

Suppose we do succeed in getting the hoped-for Theory of Everything: would that be the end of physics or even of science in general? Surely not: understanding does not necessarily include calculability and prediction, and, as we have seen, treatment of fields which are understood in deeper terms may still require explanation in terms appropriate to that field.

SECTION 2. GENERAL REASONS FOR DOUBT: THE NATURE OF SCIENTIFIC INQUIRY.

Our third question is:

3. IS IT POSSIBLE THAT WE ARE "ON THE WRONG TRACK" MORE GENERALLY?

In Section 1, we asked whether there were any *specific* indications in present science that our present theories and directions may be incorrect. The present question *can* also be directed specifically at the present physical picture. Then it takes such forms as, Might (in the sense that there are specific reasons for supposing that) the whole present approach be rejected as false - not just superstrings, but the Standard Model, gauge theory in general, and perhaps even quantum theory? Might our whole present approach have to be abandoned, as that of classical physics was earlier this century, in favor of some radical new approach? And in such cases, one can assert the necessity of looking closely at the present scientific situation, to see whether there are any concrete, specific indications that this may be so.

However, the present question is frequently asked - or its positive answer asserted - in a far more general spirit. That is, the kinds of doubts that are raised have nothing to do with the *specific* considerations relevant to the first two questions. Such considerations are even treated as irrelevant to the question, which is taken to be one concerning the nature of human belief, or the lessons of history, or some such broader issue. Asked in this spirit, this third question could be raised *independently of what specific theories or specific evidence there may be*, and thus asked *with respect to any scientific conclusion at any stage of scientific history*. In this sense, the question is an "in-principle" one, and thus it is no wonder that it has lain in the domain of philosophy.

The answer to it must always be, Yes, it is always possible that we may be on the wrong track in this general sense. This is true *no matter how strong the evidence is that we are not*, and *no matter how weak the evidence that we are*. It is *always* possible, in principle, that considerations may arise that will require rejection of our most fundamental theories. Too, it is always possible that the directions of our current research will not turn out to be successful, *no matter how strong the indications are that they may be*. Indeed, and in the same sense of the word "possible," it is always possible that *all* our theories are wrong - not only our present directions of research, but, despite the evidence, the Standard Model, gauge theory in general, quantum theory, the Big Bang theory, all our theories about the origins of stars, elements, and so on for all our beliefs.

But the possibility involved, insofar as it is divorced from any considerations specific to the scientific theory or theories concerned, is a *mere* possibility, applicable equally to any belief whatsoever. What are we to make of such mere possibility? Are we to say that there is no reason to believe them, because they are not certain? Are we to abandon the searches indicated by them as promising, because they always might fail, or might be wrong even if found? But if we are to do these things, what else will serve as our guide? Is there some source of belief and action other than investigation of nature, which will give surer guarantees? There is no other way. This is shown not only by the failure of traditional ideas about "necessities of thought and inquiry," but still more by the kinds of considerations adduced in Part I of this essay about the human condition. The human mind is a product of evolution, and as such has no access to eternal, necessary truths. We should have learned by now that, when an idea *seems* to be a necessary truth, it should be distrusted all the more. We have no ironclad guarantees; science does not deal in certainties. But we do have reasons for believing some things rather than others, even though everything about those beliefs - those conclusions of investigation of nature - is not clear. Science builds on what it has learned, using its best past conclusions as a framework within which to form new possibilities to investigate, always with the possibility of criticizing or rejecting any part of that framework itself.

The point of this general lack of certainty is not that we must be skeptical of science, but rather that we must always be cautious. We must be prepared for the appearance of problems which cannot have been specified in terms of present beliefs, which cannot have been anticipated. Nevertheless, this caution does not preclude the possibility of reasoned judgment: this caution is the true humility in the face of the universe and our circumstances within it, so far as is known; it is not a false humility based on vague and facile generalities. Our judgments must be, as they have been in this essay, based on the evidence we do have available, not arrived at in terms of *mere* possibilities of what might turn up. For *such* doubts are not reasons for complete skepticism with regard to science - with regard either to its conclusions (as telling us, perhaps in an incomplete way, how things are and are to be understood, and therefore what to believe is true) or its directives (as actions to take in the search for further knowledge). They are only reasons for a general caution because our knowledge is never absolutely certain. But to say that our present theories and expectations are not certainties, or even, in the case of some expectations and hopes, strongly justified, is not necessarily by itself sufficient reason to reject them or to consider them "myths," having no deeper status than any other myths. They are reasonable beliefs, the bases of *the best-justified lines of research available, given what we have learned so far - ones which we have solid, if not conclusive, reason to expect will succeed.*

For similar reasons, doubt should not be cast on the present scientific picture, or aspects of it, *merely* on the basis of generalities about the history of science, such as that every time such claims have been made in the past, they have been dashed to pieces by subsequent developments.[96] Certainly - again - that is reason for caution; but rejection or doubt must based on specific and known problems, not on mere general possibilities that our present views and approaches might turn out to be wrong. Our only recourse is to review, continually and carefully, the actual, specific problems in our current views. We must approach these questions by looking at the science, not by generalities.

Even if it turns out that our present theories and approaches must be rejected or radically altered, it is still a highly misleading oversimplification to say that those present views would *necessarily* turn out to be simply false, and much less that they are simply "myths," on a par with other myths of origin, nature, and development. Consider past scientific theories which have had to be rejected or altered in radical ways. Were they, equally in all cases, simply rejected as false? Even where the fundamental concepts are changed quite drastically, as in the transition from Newtonian to quantum mechanics, or from the former to special and general relativity, *the new theory often explains why things appear as they do - why nature is as it is - at a certain level of experience (size, velocity, energy, etc.) according to the older theory*. When - and of course to the extent that - such an explanation is given by a new theory, it is often (though not always) still correct to say that the old theory *is true at that level of experience*: things really do appear as the theory says they do at that level, and we can, in terms of the new theory, understand why they do, and furthermore why its description does not hold at a deeper level. The older theory is thus not simply "false." Indeed, in the case of at least some theories, they were true in a precisely specifiable sense.

We can, with considerable confidence based on present detailed evidence, expect that any new theory which replaces our present ones will maintain this status for at least some of them. For our best-established current theories - those embodied in the Standard Model and its application to the history of the universe - continue to be accepted at least in this way: that "model" does too good a job. Even if we eventually find a fundamental theory which is radically new in the sense that it is not a gauge theory, or even a quantum theory, that new theory will still have to explain why gauge field theory, and the Standard Model in particular, was so successful: why things appear in a Standard-Model way at a certain level of experience; why things are as they are at that level. The same considerations hold for general relativity. And though the more speculative current approaches - superstring theory, for example - might conceivably not be preserved to the same extent as the Standard Model, a radically new theory would still have to explain why superstring theory appeared so promising - why it was anomaly-free, for example. (In ways that can be specified, Aristotelian physics and cosmology is not "true on its level' to the same extent as Newtonian physics; but we can nevertheless understand, in terms of later theories, why it was true to the extent that it was, on its level (*i.e.*, for its physics, on the "level" of being located in a gravitational field with friction. With appropriate modifications, similar considerations apply to phlogiston, caloric, and ether theories.)

In treating reasons for caution as reasons for doubt, many philosophers of science have supported a view that the conclusions of science are not to be taken seriously as rivals of other "perspectives": that those conclusions cannot be considered as true or promising. Such a view, as we have seen, misconceives the role of doubt in science, and thus misinterprets the knowledge-seeking enterprise itself; indeed, it rests largely on ignorance of the achievements of that enterprise. In supporting such views, philosophers have contributed to the relativism and antiscientism that is so widespread in our present culture. In thus giving comfort to ideas of the special place and capacities of humanity in the universe, philosophy remains a last bastion of superstition. But that is not the only error of philosophy.

SECTION 3. THE AIMS AND NATURE OF THE PHILOSOPHY OF SCIENCE.

In addition to misconceiving the nature and role of doubt in science, philosophical theories of science and its method have tended to focus on aspects supposed to be unchanging, independent of the results of inquiry. Inquiry, it is said, presupposes a method of inquiry; and what counts as an explanation, a theory, or a piece of evidence or observation remains the same, independent of the actual results of science. This approach is diametrically opposed to that of the present essay. Here, what counts as scientific method, explanation, theory, evidence, observation, *etc.*, can change with new results of inquiry.[97] Indeed, a major theme of Part I of this essay was that we have to *learn* what the goals of science are, and that we have to *learn* what understanding, explanation, consists in. That is only to be expected, given the fact that the human mind is a product of evolution. In this part, I will focus on two further aspects of scientific innovation: how what counts as a theory can change, and how the relation between theory and observation can alter as a result of new findings. These two aspects are well illustrated in the current scientific picture, which, in addition to its factual conclusions about the universe, also brings about changes in what we are to count as a good theory, and in the role of experiment or observation in science.

(1) *Learning what to count as a theory, and in particular as a good theory.*

The field-theoretic approach to understanding fundamental particles and forces had been abandoned as hopeless by many physicists by the 1960's, due to the failure to find adequate, and in particular, renormalizable, theories of the weak and strong forces. The electroweak unification, employing the gauge-theoretic (Yang-Mills) approach, however, was an enormous success both in its mathematical structure and its experimental test. Gauge theories were shown to be renormalizable, and the electroweak theory was confirmed by a remarkable series of experimental results. Field theory - in its new gauge form - was restored to pre-eminence as a strategy for theory-formulation. A theory of the strong force, quantum chromodynamics, was constructed, modelled on the electroweak theory. It has provided, even though only for high energies, the first calculable theory of the strong interactions. Asymptotic freedom, the source of the possibility of such calculation, has been shown to arise only in a non-Abelian gauge theory. The success of quantum electrodynamics, attained by the end of the 1940's, was understood more deeply as due to its being a gauge theory (in its case, Abelian). The theories of the electroweak and strong interactions, so similar in mathematical form, and so universally successful that there is no experimental result for which they do not account, could now be juxtaposed in the Standard Model. The Standard Model found a natural application in the by-then (1970's) highly successful Big Bang theory, where the energies available at the beginning of the universe would have been sufficient, before a calculable time of 10^{-10} seconds, for the electroweak unity to have been realized, and then, at the energy prevailing at that time, broken into the separate weak and electromagnetic forces. Since the Big Bang theory was itself a product of general relativistic cosmology, a link was indicated between the quantum field theories and general relativity. To this story must be added the change in character of fundamental theory brought about through Noether's Theorem (that every conservation principle follows from a symetry principle): a shift from conservation laws to symmetry principles as the fundamental laws of nature. That shift placed group theory, and specifically Lie groups, at the focus of physical theory.

Something had been learned about how to construct theories, and about what a good scientific theory should be like. It was a lesson that had been built, after fifty years of intensive work, on the shoulders of quantum mechanics and its field-theoretic extensions, and on the long line of theories on which those theories had in turn been built. Even if there were abstract, "formal" conditions for being a theory (such as its being an axiomatic system with correspondence rules), such conditions would not capture the dynamic process of seeking knowledge, which includes alterations not only of our factual beliefs, but also our beliefs about how to explain.

(2) *The possibility of experimental test, and the changing character of theory-construction and establishment.*

The energies at which GUT unification is predicted are far beyond any accelerators we could ever hope to build. This is even more the case with theories aiming at the unification of the standard-model forces with gravity. In the face of this impossibility of direct test of all those theories, we are left with only two hopes.

On the one hand, the high-energy theories might have unambiguous testable low-energy consequences. Proton decay (as a test of GUTs) is an example; but what we would like would be predictions which would help establish some candidate theory rather than, as with proton decay, being the basis of their rejection. On the other hand, we can hope that the constraints on unification theories will be found to be so tight that the correct choice, given compatibility with the low-energy standard model, will be determinable on purely mathematical grounds. In this latter case, if it is realized, the ways of attempting to establish physical theories will have undergone a profound departure from the previous experimental approach. The new approach might nevertheless be justified should the desired constraints appear. As so often in the history of science, we would then have learned something new about scientific method, about how to learn. (The contemporary scientific view thus gives us not only a coherent picture of the facts of the universe, but also a lesson in the epistemology of science.[98]) As in the case of the hope of uncovering predictions at experimentally-attainable energies, this hope is not without basis. Many constraints have in fact been found, successively narrowing the possible kinds of theories that can be candidates. The very accumulation of such constraints as new and promising theories have been developed is reason enough to continue to hope for even more. Perhaps the most attractive feature of superstring theory is that it embodies so many such restrictions. However, as we have seen, one of its great disappointments, to date, has been that even it (or at least certain forms of it) leaves open an indefinitely large number of alternative compactifications. The body of constraints is still not tight enough.

These two examples once again illustrate a major theme of this essay: that it is the business of the philosopher of science to investigate the emergence of mind, of how we have departed from the expectations which have been imposed on the human mind, ultimately by the very limited experience of everyday life. In fulfilling this task, the philosopher must survey, in depth, the science of his day, reviewing it and comparing it with the past views upon which it has built. The picture which the science of his age gives of the universe in which we live, and of our place in it, including the problems - the inadequacies and the degree of coherence in it - and directions of research which that picture provides, must constitute the object of his investigation, but also, in its best aspects, the framework within which it is conducted. The task must be repeated by every generation. Philosophy of science cannot confine itself to investigation of the eternal and unchanging, even if there is any such thing.

The two examples considered in this final section once again illustrate this task, and, more generally, what has been a major theme of this essay: that the expectations we have of how nature must behave and be investigated and understood, which have their origins ultimately in the very limited everyday experience of the human mind, may or may not be fulfilled when we go beyond those everyday limits. It is one of the wonders of the human mind that it is able to transcend the limits which its evolutionary circumstances have imposed on it, and that it can arrive, through investigation of nature, at novel ways of understanding, and of understanding what it is to understand.

Wake Forest University, Winston, Salem, USA

NOTES

1. Conference presentation of this paper was aided by National Science Foundation Grant Int-8905411

2. A representative sampling from various fields, illustrating not only the sorts of conclusions I have in mind, but also the methods by which those conclusions are arrived at: articles by Lieberman, Dibble, and Alexander in P. Mellars and C. Stringer (eds.), *The Human Revolution: Behavioural and Biological Perspectives on the Origins of Modern Humans*, Princeton, Princeton University Press, 1989; R. Byrne and A. Whiten (eds.), *Machiavellian Intelligence: Social Expertise and the Evolution of Intellect in Monkeys, Apes, and Humans*, Oxford, Clarendon,

1988; R.G. Klein, *The Human Career: Human Biological and Cultural Origins*, Chicago, University of Chicago Press, 1989; D.B. Dickson, *The Dawn of Belief: Religion in the Upper Paleolithic of Southwestern Europe*, Tucson, University of Arizona Press, 1990; J. Puhvel, *Comparative Mythology*, Baltimore, Johns Hopkins University Press, 1987; J. Crook, *The Evolution of Human Consciousness*, Oxford, Clarendon, 1985; T. Wynn, *The Evolution of Spatial Competence*, Urbana, University of Illinois Press, 1989; E. A. Hovelock, «The Linguistic .Task of the Presocratics», in K. Rehb (ed.), *Language and Thought in Early Greek Philosophy*, La Salle, Monist Library of Philosophy, 1983; G. S. Kirk, J. E. Raven, and M. Schefield, *The Presocratic Philosophers*, Second Edition, Cambrige University Press, 11983.

3. For studies of ways in which traditional ideas of space, time, matter, and force - descendants of more primitive concepts arising in everyday life - shaped the forms of explanation which were imposed in seventeenth-century mechanics, see D. Shapere, "The Mechanical Philosophy of Nature," forthcoming. For further discussion of the relations between classical philosophical concepts, classical mechanics, and contemporary physical theories, see D. Shapere, "Newton's Place in History," *i*n G. Brush, A, Seeff and P. Theerman (eds.), *Action and Reaction,Newark, Del., Universsity of Delaware Press, forthecoming;* "Modern Physics and the Philosophy of Science," in A. Fine and J. Leplin (eds.), *PSA 1988*, Vol. 2, East Lansing, Philosophy of Science Association, 1989, pp. 201-210; and "The Origin and Nature of Metaphysics," forthcoming in *Philosophical Topics*.

4. D. Shapere, "The Origin and Nature of Metaphysics," *loc. cit.*

5. The necessity of using background knowledge in the sense described here does not inevitably destroy the objectivity of inquiry. For the background that is brought to inquiry is not just any arbitrary belief; there are constraints on what can count as background knowledge, and those constraints themselves have been learned through the long process of inquiry. Indeed, true objectivity does not consist (as many traditional views supposed) in approaching an inquiry with no presuppositions at all: that would be an impossible undertaking. Rather, it consists in bringing to the inquiry the best prior relevant information available, despite the fact that information *may* later become subject to doubt. Further details on the concept of "objectivity" are given in D. Shapere, "Objectivity,

Rationality, and Scientific Change," in P. Kitcher and P. Asquith (eds.), *PSA 1984*, East Lansing, Philosophy of Science Association, 1985, Vol. 2, 637-662.

6. L. Wittgenstein, *Philosophical Investigations*, Oxford, Blackwell, 1953, #437, p. 129e.

7. A number of the articles referred to (but by no means all) are included in one or another of the following collections: E.W. Kolb and M.S. Turner (eds.), *The Early Universe: Reprints*, Reading, Mass., Addison-Wesley, 1988; J. Bernstein and G. Feinberg (eds.), *Cosmological Constants: Papers in Modern Cosmology*, New York, Columbia University Press, 1986; A. Zee (ed.), *Unity of Forces in the Universe*, Singapore, World Scientific, 1982; K.R. Lang and O. Gingerich (eds.), *A Source Book in Astronomy and Astrophysics, 1900-1975*, Cambridge, Harvard University Press, 1979. Many of the topics covered here are examined in articles in E.W. Kolb, M.S. Turner, D. Lindley, K. Olive, and D. Seckel (eds.), *Inner Space/Outer Space: The Interface Between Cosmology and Particle Physics*, Chicago, University of Chicago Press, 1986. Less technical surveys are given in P. Davies (ed.), *The New Physics*, Cambridge, Cambridge University Press, 1989.

8. Fermi's original (1934) theory of the weak interaction did not involve exchange particles. For the early development of weak interaction theory, see C. Strachan (ed.), *The Theory of Beta-Decay*, New York, Pergamon, 1969; P.K. Kabir (ed.), *The Development of Weak Interaction Theory*, New York, Gordon and Breach, 1963.

9. Some surveys of the development of particle physics are: A. Pais, *Inward Bound*, Oxford, Clarendon, 1986; L.M. Brown and L. Hoddeson (eds.), *The Birth of Particle Physics*, Cambridge, Cambridge University Press, 1983; R.P. Crease and C.C. Mann, *The Second Creation: Makers of the Revolution in 20th-Century Physics*, New York, MacMillan, 1986.

10. S. Weinberg, "A Model of Leptons," *Physical Review Letters*, *19* (1967), 1264-1266; A. Salam, "Weak and Electromagnetic Interactions," in N. Svartholm (ed.), *Proceedings of the Eighth Nobel Symposium on Elementary Particle Theory*, New York, Interscience, 1968, 367-377.

11. C.N. Yang and R. Mills, "Conservation of Isotopic Spin and Isotopic Gauge Invariance," *Physical Review*, *96* (1954), 191-195.

12. G. 't Hooft, "Renormalizable Lagrangians for Massive Yang-Mills Fields," *Nuclear Physics, B35* (1971), 167-188.

13. The story of the experiments has been surveyed in many works, among them P. Galison, *How Experiments End*, Chicago, University of Chicago Press, 1987; C. Sutton, *The Particle Connection*, New York, Simon and Schuster, 1984; P. Watkins, *Story of the W and Z*, Cambridge, Cambridge University Press, 1986.

14. For a survey, see W. Marciano and J. Pagels, "Quantum Chromodynamics," Nature, 279 (1979), 479-483.

15. D.J. Gross and F.A. Wilczek, "Ultraviolet Behavior of Non-Abelian Gauge Theories," *Physical Review Letters, 30* (1973), 1343-1346; H.D. Politzer, "Reliable Perturbative Results for Strong Interactions," *Physical Review Letters, 30* (1973), 1346-1349.

16. A. Einstein, "Kosmologische Betrachtungen zur allgemeinen Relativitätstheorie," *Sitzungsberichte der Preussischen Akad. d. Wissenschaften*, 1917, 831-839; English translation, "Cosmological Considerations on the General Theory of Relativity," in Einstein *et al.*, *The Principle of Relativity*, New York, Dover, 1923, 177-188. Einstein's work and its relation to other work in physics and cosmology are examined in penetrating depth in A. Pais, *Subtle is the Lord ...*, Oxford, Clarendon, 1982. For surveys of the development of modern cosmology, see J. North, *The Measure of the Universe*, Oxford, Clarendon, 1965; E.H. Harrison, *Cosmology*, Cambridge, Cambridge University Press, 1981; R.W. Smith, *The Expanding Universe: Astronomy's 'Great Debate' 1900-1931*, Cambridge, Cambridge University Press, 1982; G.F.R. Ellis, "The Expanding Universe: A History of Cosmology from 1917 to 1960," in D. Howard and J. Stachel (eds.), *Einstein and the History of General Relativity*, Boston, Birkhauser, 1989.

17. G. Lemaitre, "A Homogeneous Universe of Constant Mass and Increasing Radius Accounting for the Radial Velocity of Extragalactic Nebulae," *Ann. Soc. Sci. Bruxelles, A47* (1927), 49-59 [English translation in *Monthly Notices of the Royal Astronomical Society, 91* (1931), 483-490]; *The Primeval Atom: An Essay on Cosmogony*, New York, Van Nostrand, 1950. Lemaitre developed his views in a number of significant papers between these works.

18. G. Gamow, "Expanding Universe and the Origin of the Elements," *Physical Review, 70* (1946), 572-573; *The Creation of the Universe*, New York, Viking, 1952.

19. R.A. Alpher and R.C. Herman, "Theory of the Origin and Relative Abundance Distribution of the Elements," *Reviews of Modern Physics*, *22* (1950), 153-212.

20. H. Bondi and T. Gold, "The Steady State Theory of the Expanding Universe," *Monthly Notices of the Royal Astronomical Society*, *108* (1948), 252-270; F. Hoyle, "A New Model for the Expanding Universe," *Monthly Notices of the Royal Astronomical Society*, *108* (1948), 372-382.

21. W. Baade, "Extragalactic Nebulae: Report to IAU Commission 28," *International Astronomical Union Transactions*, *8* (1952), 397-399. Means of obtaining distances of extragalactic objects are surveyed critically in M. Rowan-Robinson, *The Cosmological Distance Ladder: Distance and Time in the Universe*, New York, W.H. Freeman, 1985.

A Cepheid variable is a star which varies in brightness with an extremely regular period. The length of the period is directly related to its intrinsic luminosity. With this knowledge of the intrinsic brightness and the apparent luminosity, the distance is easily calculated. Long-period Cepheids are quite bright, and so are easily distinguished in nearby galaxies. It was the confusion of such regularly-varying stars with periods of two to forty or so days with equally regularly varying stars with periods less than two days (now called RR Lyrae stars) that led to the age problem. The relation of period and luminosity is different for the two classes, but the distance-scale for both groups had been based on the short-period variables.

22. A. Sandage, "Current Problems in the Extra-Galactic Distance Scale," *Astrophysical Journal*, *127* (1958), 513-526.

23. The age problem has not been entirely eliminated even yet. The most reasonable age for the universe in a Big-Bang scenario lie somewhere between about 9 and 20 billion years. The ages of some globular clusters have been estimated at 14 to 20 billion years. Since galaxies, along with their globular clusters, probably did not begin to form for some time after the Big Bang, the compatibility of these ages remains marginal at best.

24. M. Ryle and P. Scheuer, "The Spatial Distribution and the Nature of Radio Stars," *Proceedings of the Royal Society, A230* (1955), 448-462.

25. E.g., R.A. Alpher and R.C. Herman, "Remarks on the Evolution of the Expanding Universe," *Physical Review, 75* (1949), 1089-1095; G. Gamow, "The Physics of the Expanding Universe," *Vistas in Astronomy, 2* (1956), 1726-1732. Strictly speaking, as will come out later in this paper, the radiation is from (carries information about) a time on the order of 10^5 years after the Big Bang itself.

26. A.A. Penzias and R.W. Wilson, "A Measurement of Excess Antenna Temperature at 4080 Mc/s," *Astrophysical Journal, 142* (1965), 419-421; R.H. Dicke, P.J.E. Peebles, P.G. Roll, and D.T. Wilkinson, "Cosmic Black Body Radiation," *Astrophysical Journal, 142* (1965) 414-419. Several years of study were required to establish the black-body nature of the radiation.

27. S. Weinberg, *Gravitation and Cosmology: Principles and Applications of the General Theory of Relativity*, New York, Wiley, 1972. The publication of this ground-breaking book came only twelve years after Bondi's *Cosmology*, in which a four-page chapter on "Microphysics and Cosmology" (concerned primarily with "a few numerical 'coincidences'") opened with the admission that "Unfortunately present-day understanding of atomic and nuclear physics is not deep enough to indicate in any direct way which results are of particular significance for cosmology." (H. Bondi, *Cosmology*, Cambridge, Cambridge University Press, 1960, p. 59.)

28. F.A. Wilczek, "Foundations and Working Pictures in Microphysical Cosmology," in G.W. Gibbons, S.W. Hawking, and S.T.C. Siklos (eds.), *The Very Early Universe*, Cambridge, Cambridge University Press, 1983, p. 19. For a discussion of the problem as it was understood before the discovery of asymptotic freedom, see S. Weinberg, *Gravitation and Cosmology: Principles and Applications of the General Theory of Relativity*, New York, Wiley, 1972, pp. 588-597; see also Weinberg's post-asymptotic-freedom comment - still quite cautious, limiting the "simple picture" to "the first hundredth of a second" and after - in his *The First Three Minutes*, New York, Basic Books, 1977, p. 141.

29. C.F. von Weizsäcker, "Element Transformation Inside Stars," I, *Physikalische Zeitschrift*, *38* (1937), 176; II, *ibid.*, *39* (1938), 633-646; H.A. Bethe, "Energy Production in Stars," *Physical Review*, *55* (1939), 434-456.

30. Of particular importance was E.M. Burbidge, G.R. Burbidge, W.A. Fowler, and F. Hoyle, "Synthesis of the Elements in Stars," *Reviews of Modern Physics*, *29* (1957), 547-650.

31. For detailed discussions of philosophical analyses of "reduction," see E. Nagel, *The Structure of Science: Problems in the Logic of Scientific Explanation*, New York, Harcourt, Brace, and World, Inc., 1961, Ch. 11; M. Spector, *Concepts of Reduction in Physical Science*, Philadelphia, Temple University Press, 1978.

32. This difference could remain even if explanation in terms of contexts were to prove necessary in physics - for example, because of quantum non-locality.

For the distinction between compositional and other forms of explanation, see D. Shapere, "Scientific Theories and Their Domains," in F. Suppe (ed.), *The Structure of Scientific Theories*, Urbana, University of Illinois Press, 1974, pp. 518-565, and "On the Relations between Compositional and Evolutionary Theories," in F. Ayala and T. Dobzhansky (eds.), *Studies in the Philosophy of Biology*, London, Macmillan, 1974, pp. 187-202. What is referred to in those writings as "evolutionary" explanation is a special form of explanation in terms of context.

33. This view also makes non-mystical sense of the much maligned concept of "emergence." Further, as can be seen from the above discussion, the problems of reduction and explanation are closely related; explanation will be discussed again later. A fuller treatment of the problem of reduction (and of explanation) would include a critique and fundamental alteration of the two conditions, mentioned above in the text, of the usual philosophical doctrines of reduction and their variations.

The view of reduction sketched here depends on the claim that what counts as an explanation can change in the light of inquiry, and can differ from one field of inquiry to another. Positivist philosophers tended to focus only on what is *essential and unchanging* about explanation. Being immune to revision in the light of scientific results, such a concept of explanation was to be "analyzed" in terms independent of science: "explanation," and the related concept of "reduction," were made transcendental concepts, independent of experience. As

Part I of this paper makes clear, such reification is an error: like all our ideas, our explanatory expectations - what we suppose understanding must be like - arose in the context of human evolutionary history, in the attempt to come to terms with the everyday world of human environment. And those expectations are subject to modification and even rejection in the face of further inquiry. We must learn what understanding, explanation, consists in.

There is a close parallel between the errors committed with regard to explanation and those made with regard to ethical concepts. There, too, concepts like "good," "right," and "ought" have often been transformed into ones transcending, and therefore divorced from, anything empirical: "is," we are told, does not imply "ought," and there is a "naturalistic fallacy" involved in trying to make any such connection. The diagnosis and cure of this kind of error is the same as in the case of the concept of explanation: to understand the problems of ethics as having arisen from the existence of contexts, the relation between the individual and his context, and the emergence of new sorts of contexts which altered the problems and the concepts in specifiable ways, for specifiable reasons.

34. J.N. Islam, "Possible Ultimate Fate of the Universe," *Quarterly Journal of the Royal Astronomical Society, 18* (1977), 3; F.J. Dyson, "Time Without End: Physics and Biology in an Open Universe," *Reviews of Modern Physics, 51* (1979), 447.

35. Strictly speaking, this is a misnomer, since nuclei and electrons had not been combined before.

36. *E.g.*, J. Maddox, "Down with the Big Bang," *Nature, 340* (1989), 425; H.C. Arp, G. Burbidge, F. Hoyle, J.V. Narlikar, and N.C. Wickramasinghe, "The Extragalactic Universe: An Alternative View," *Nature, 346* (1990), 807-812.

37. For experimental evidence concerning General Relativity and its rivals, see the following works by C.M. Will: *Theory and Experiment in Gravitational Physics*, Cambridge, Cambridge University Press, 1981; *Was Einstein Right? Putting General Relativity to the Test*, New York, Basic Books, 1986; "General Relativity at 75: How Right Was Einstein?" *Science, 250* (1990), 770-776.

38. See, for example, D. Sugimoto, D.Q. Lamb, and D.N. Schramm (eds.), *Fundamental Problems in the Theory of Stellar Evolution*, Dordrecht, Reidel, 1981; R.S. Roger and P.E. Dewdney (eds.), *Regions of Recent Star Formation*,

Dordrecht, Reidel, 1982; M. Peimbert and J. Jugaku (eds.), *Star Forming Regions*, Dordrecht, Reidel, 1987; F.H. Shu, F.C. Adams, and S. Lizano, "Star Formation in Molecular Clouds: Observation and Theory," in G. Burbidge, D. Layzer, and J.G. Phillips (eds.), *Annual Review of Astronomy and Astrophysics, 25* (1987), Palo Alto, Annual Reviews, Inc., 23-81; M. Cohen, *In Darkness Born: The Story of Star Formation*, Cambridge, Cambridge University Press, 1988; H.A. Weaver and L. Danly (eds.), *The Formation and Evolution of Planetary Systems*, Cambridge, Cambridge University Press, 1989; G.W. Wetherill, "Formation of the Earth," *Annual Review of Earth and Planetary Sciences, 18* (1990), Palo Alto, Annual Reviews, Inc., 205-256 (this article also surveys the formation of the solar system).

Although there is a "standard model" for planetary formation, there remain dissenters; for example, *cf.* M.M. Woolfson, "The Capture Theory and the Origin of the Solar System," and "The Evolution of the Solar System," in S.F. Dermott (ed.), *The Origin of the Solar System*, New York, Wiley, 1978, pp. 179-198 and 198-217 respectively; J.R. Dormand and M.M. Woolfson, *The Origin of the Solar System: The Capture Theory*, New York, Wiley, 1989. A summary of Woolfson's view is given in I.P. Williams, *The Origin of the Planets*, Bristol, Adam Hilger, 1978, pp. 46-49.

39. See, for example, K.A. Kvenvolden (ed.), *Geochemistry and the Origin of Life* [Benchmark Papers in Geology], Stroudsburg, Pa., Dowden, Hutchinson and Ross, Inc., 1974; C. Ponnamperuma (ed.), *Cosmochemistry and the Origin of Life*, Dordrecht, Reidel, 1983; S. Chang, D. DesMarais, R. Mack, S.L. Miller, and G.E. Strathearn, "Prebiotic Organic Syntheses and the Origin of Life," in J.W. Schopf (ed.), *Earth's Earliest Biosphere: Its Origin and Evolution*, New Haven, Yale University Press, 1984; R. Shapiro, *Origins: A Skeptic's Guide to the Creation of Life on Earth*, New York, Summit Books, 1986; J. Oró, S.L. Miller, and A. Lazcano, "The Origin and Early Evolution of Life on Earth," in *Annual Review of Earth and Planetary Science, 18* (1990), Palo Alto, Annual Reviews, Inc., 317-356.

40. T.R. Cech, "The Chemistry of Self-splicing RNA and RNA Enzymes," *Science, 236* (1987), 1532. Important related work by Altman and his group is described in S. Altman *et al.*, "Enzymatic Cleavage of RNA by RNA," *Trends in the Biochemical Sciences, 11*, 515.

41. G.F. Joyce, "RNA Evolution and the Origins of Life," *Nature*, *338* (1989), 217-224.

42. See J.W. Truran, ""Nucleosynthesis," *Annual Review of Nuclear and Particle Science*, *34* (1984), 53-97; C.E. Rolfs and W.S. Rodney, *Cauldrons in the Cosmos*, Chicago, University of Chicago Press, 1988; M. Lozano, M.I. Gallardo, and J.M. Arias (eds.), *Nuclear Astrophysics*, New York, Springer-Verlag, 1989.

43. A.G. Petschek (ed.), *Supernovae*, New York, Springer-Verlag, 1990.

44. Further details are found in D. Shapere, "Objectivity, Rationality, and Scientific Change," in P. Kitcher and P. Asquith (eds.), *PSA 1984*, East Lansing, Philosophy of Science Association, 1985, Vol. 2, pp. 637-662.

45. P. Duhem, *The Aim and Structure of Physical Theory*, Princeton, Princeton University Press, 1954; W.V. Quine, *Word and Object*, Cambridge, M.I.T. Press, 1960 (and also later writings); P.K. Feyerabend, *Against Method*, London, Verso, 1978

46. G.W. Wetherill, "Formation of the Earth," *Annual Review of Earth and Planetary Sciences*, *18* (1990), 209-210. The expression "standard model" in this connection is Wetherill's own, and I will also speak in this essay of "standard models" in various fields, reserving the capitalized "Standard Model" for the one prevailing in physics.

47. For an excellent survey of uncertainties and alternatives regarding Big-Bang synthesis of the light elements, see A.M. Boesgaard and G. Steigman, "Big Bang Nucleosynthesis: Theories and Observations," *Annual Review of Astronomy and Astrophysics*, *23* (1985), Palo Alto, Annual Reviews, Inc., 319-378.

48. K.R. Popper, *The Logic of Scientific Discovery*, New York, Basic Books, 1959; *Conjectures and Refutations*, New York, Basic Books, 1962. Popper's view is that science proceeds by making "conjectures," for which there is no rationale; the only "logic" of science comes after the conjecture or hypothesis is presented, and then it consists only in attempts to falsify the conjecture. This viewpoint is directly opposed to that of this paper, for which science builds new conjectures or hypotheses on what it has learned in past investigations.

49. Contrast the views expressed in P.K. Feyerabend, *Against Method: Outline of an Anarchistic Theory of Knowledge*, London, Verso, 1978, with those presented in D. Shapere, "On the Introduction of New Ideas in Science," forthcoming in Leplin (ed.), *The Creation of Ideas in Science.* According to Feyerabend, "anything goes" in science: it is irrational to ignore any possible hypothesis, even those, like witchcraft, which have long been rejected. This view is similar to that of Popper, who claims that hypotheses are introduced in science on the basis of no rational considerations whatever.

50. See, for example, B.J.T. Jones and J.E. Jones (eds.), *The Origin and Evolution of Galaxies*, Dordrecht, Reidel, 1983; J. Audouze and J. Tran Thanh Van (eds.), *Formation and Evolution of Galaxies and Large Structures in the Universe*, Dordrecht, Reidel, 1984; J. Bergeron, D. Kunth, B. Rocca-Volmerange, and J. Tran Thanh Van (eds.), *High Redshift and Primeval Galaxies*, Editions Frontières, 1987; C.S. Frenk, R.S. Ellis, T. Shanks, A.F. Heavens, and J.A. Peacock (eds.), *The Epoch of Galaxy Formation*, Dordrecht, Kluwer, 1988.

51. The value of the Hubble constant H is presently known only within a factor of two: some estimates give 50 km/sec/megaparsec, others give 100. Because of this uncertainty, many cosmological parameters must be multiplied by some function of a "fudge factor," h, which is generally taken to be between 1/2 and 1. In the case of Ω, for instance, Trimble gives $0.015<\Omega h^2<0.15$. (V. Trimble, "Existence and Nature of Dark Matter in the Universe," *Annual Review of Astronomy and Astrophysics*, *25* (1987), Palo Alto, Annual Reviews, Inc., p. 453. The presence of this fudge factor will be implicit in the discussions in this paper.

52. These statements depend on the assumption that the value of the cosmological constant is zero. Further complications must be introduced if it is not. These complications will be passed over in this paper. *Cf.*, G. Börner, *The Early Universe: Facts and Fiction*, New York, Springer-Verlag, 1988, p. 5.

53. J. Bahcall, T. Piran, and S. Weinberg (eds.), *Dark Matter in the Universe*, Singapore, World Scientific, 1987.

54. S. Weinberg, "The Cosmological Constant Problem," *Reviews of Modern Physics*, *61* (1989), 1-23.

55. E.D. Loh and E.J. Spillar, "Photometric Redshifts of Galaxies," *Astrophysical Journal*, *303* (1986), 154.

56. H. Georgi, H. Quinn, and S. Weinberg, "Hierarchy of Interactions in Unified Gauge Theories," *Physical Review Letters*, *33* (1974), 451-454.

57. H. Georgi and S.L. Glashow, "Unity of All Elementary-Particle Forces," *Physical Review*, *32* (1974), 236.

58. D.H. Perkins, "Proton Decay Experiments," in J.D. Jackson, H.E. Gove, and R.F. Schwitters (eds.), *Annual Review of Nuclear and Particle Science*, *34* (1984), Palo Alto, Annual Reviews, Inc., 1-53; P. Langacker, "The Present Status of Grand Unification and Proton Decay," in E.W. Kolb, M.S. Turner, D. Lindley, K. Olive, and D. Seckel (eds.), *Inner Space/Outer Space: The Interface Between Cosmology and Particle Physics*, Chicago, University of Chicago Press, 1986, 3-24.

59. A.D. Sakharov, "Violation of CP Invariance, C Asymmetry, and Baryon Asymmetry of the Universe," *Zh.Ek. Teor. Fiz.* *5* (1967), 32; English translation, *JETP Letters*, *5* (1967), 24; E.W. Kolb and M.S. Turner, "Grand Unified Theories and the Origin of the Baryon Asymmetry," in J.D. Jackson, G.E. Gove, and R.F. Schwitters (eds.), *Annual Review of Nuclear and Particle Science*, *33* (1983), Palo Alto, Annual Reviews, Inc., 645-696.

60. L. Wolfenstein, "Present Status of CP Violation," *Annual Review of Nuclear and Particle Science*, *36* (1986), Palo Alto, Annual Reviews, Inc., 137-170.

61. A. Guth, "Inflationary Universe: A Possible Solution to the Horizon and Flatness Problems," *Physical Review*, *S23* (1981), 347-356; A. Albrecht and P. Steinhardt, "Cosmology for Grand Unified Theories with Radiatively Induced Symmetry Breaking," *Physical Review Letters*, *48* (1982), 1220-1223; and other papers contained in L.F. Abbott and S. Pi (eds.), *Inflationary Cosmology*, Singapore, World Scientific, 1986.

62. In more prosaic circumstances, the total energy in an expanding isolated volume would remain conserved, and would thus be diluted in density as it spread through a larger volume. It can be shown that no contradiction of the law of conservation of energy is involved in the contention that the false-vacuum

energy increases with expansion. Indeed, the sum of the negative gravitational and positive false-vacuum energies involved may add up to exactly zero. (The negative gravitational energy increases along with that of the false vacuum.)

Thus the total energy of the universe, or this region of it, may be nil. As Guth is fond of saying, the universe may be the ultimate free lunch. This possibility (and the equally arguable possibility that the sum of all other conserved quantities is also zero) leads to a further intriguing one: "If it is true that our Universe has a zero net value for all conserved quantities, than it may simply be a [quantum] fluctuation of the vacuum, the vacuum of some larger space in which our Universe is imbedded." (E.P. Tryon, "Is the Universe a Vacuum Fluctuation?" *Nature*, *246* (1973), 396-397.)

63. The end of the inflationary epoch, together with some other important details, differs in the original (Guth, *op. cit.*) version and the "new inflationary theory" offered by Albrecht and Steinhardt, *op. cit.* These differences are passed over here, my discussion being in terms of the new inflationary theory.

64. The inadequacy of naïve verificationist theories of scientifically-admissible concepts is shown clearly here. Though inherently unobservable (because signals from other regions - beyond our own horizon - could never reach us), those other regions nevertheless are required to exist by theory. Although the inflationary theory cannot be said to be established, still it is a legitimate theory, constructed in the light of a background acceptable on the basis of good reasons. The very existence of a horizon, into which previously-inaccessible events are continually entering, points to the same conclusion in the light of a theory that is accepted. Science cannot allow itself to be limited by *a priori* epistemological constraints: we have to learn what is admissible in science, and what is admissible can change in the light of new knowledge.

65. That is, the expansion exceeded the velocity of light. Since this was an expansion of *space* rather than a movement of matter in space, the requirement of special relativity that nothing can move faster than light is not violated.

The "horizon" at any given epoch is determined by the distance light could have travelled in the time since the beginning of the universe. (Thus the horizon grows with time.) Because of the finite velocity of light and the above-mentioned requirement of special relativity with regard to that velocity, it is impossible *in principle* to get information beyond the horizon distance. The portion of the universe accessible to our observing instruments is considerably smaller than this. The total region of the universe that expands in an exponential way according to the inflationary theory is much larger than the universe as defined by the horizon, either at the beginning of the inflationary episode or today.

66. Preliminary results from the COBE (Cosmic Background Explorer) satellite are already indicating that the isotropy is even greater. The satellite is also rapidly removing any residual hesitancy about its black-body character, and thus the cosmological significance, of the radiation. It also has disturbing implications for the problem of galaxy formation, however, as we shall see. ("The good news is that the three-degree cosmic microwave background has the exact spectrum of black-body radiation, and is spatially featureless. The bad news is the same." [D. Lindley, "An Excess of Perfection," *Nature*, *343* (18 Jan. 1990), p. 207])

67. At the Planck time, 10^{-43} seconds after the Big Bang, the departure of Ω from unity would have been on the order of one part in 10^{59} These arguments assume a zero value of the cosmological constant.

68. The solution to the problem, which had been uncovered by 't Hooft, was offered by R. Peccei and H. Quinn, "Constraints Imposed by CP Conservation in the Presence of Instantons," *Physical Review D*, *16* (1977), 1791. The particle associated with the Peccei-Quinn solution was described by S. Weinberg, "A New Light Boson?" *Physical Review Letters*, *40* (1978), 223-226, and by F. Wilczek, "Problem of Strong P and T Invariance in the Presence of Instantons," *Physical Review Letters*, *40* (1978), 279-282. Wilczek is responsible for the name "axion," and in honor of the particle he has mounted a box of the detergent booster on his office wall.

69. As I have formulated the issue, it is one of approaching gravitation in terms of the ways in which particle physics has been so successfully dealt with. However, this is not the only approach, and there are serious problems with it. The attempt to develop a theory of "quantum gravity" is approached very differently (and in many different ways) by many theorists working from a background of general relativity. For an excellent survey of that approach, its differences from particle approaches, and some fundamental problems of the latter, see. C. Isham, "Quantum Gravity," in P. Davies (ed.), *The New Physics*, Cambridge, Cambridge University Press, 1989, pp. 70-93. However, Isham, like

many particle theorists, sees great promise in superstring theories, so that there may be a merger of these approaches. The history of the general-relativistic attempts to quantize gravity is important, and deserves to be written.

70. T. Applequist, A. Chodos, and P.G.O. Freund (eds.), *Modern Kaluza-Klein Theories*, Reading, Mass., Addison-Wesley, 1987. The Introduction to this book gives a historical sketch. In addition to fundamental modern papers, the volume includes both Kaluza's original paper, "On the Unity Problem of Physics," is included in that book (in the original German and in English translation) and Klein's 1926 paper, "Quantum Theory and Five Dimensional Theory of Relativity."

71. P. West (ed.), *Supersymmetry: A Decade of Development*, Boston, Adam Hilger, 1986; J. Wess and J. Bagger, *Supersymmetry and Supergravity*, New York, Springer-Verlag, 1983.

72. For discussion of complexities passed over here, see P.G.O. Freund, *Introduction to Supersymmetry*, Cambridge, Cambridge University Press, 1986.

73. T.D. Lee and C.N. Yang, "Question of Parity Conservation in Weak Interactions," *Physical Review*, *104* (1956), 254-258. Shortly after this paper appeared, confirmation of the predicted parity violation in weak interactions was reported in C.S. Wu, E. Ambler, R.W. Hayward, D.D. Hoppes, and R.P. Hudson, "Experimental Test of Parity Conservation in Beta-decay," *Physical Review*, *105* (1957), 1413-1415.

74. M.B. Green, J.H. Schwarz, and E. Witten, *Superstring Theory*, Cambridge, Cambridge University Press, 1988, 2 vols.

75. M.B. Green and J.H. Schwarz, "Anomaly Cancellations in Supersymmetric D=10 Gauge Theory and Superstring Theory," *Physics Letters*, *149B* (1984), 117-122.

76. L. Alvarez-Gaumé and E. Witten, "Gravitational Anomalies," *Nuclear Physics B*, *234* (1983), 269-330.

77. D.J. Gross, J.A. Harvey, E. Martinec, and R. Rohm, "Heterotic String Theories (I). The Free Heterotic String," *Nuclear Physics B*, *256* (1985), 253-284.

78. E. Witten, "Quantum Field Theory and the Jones Polynomial," Institute for Advanced Study preprint, 1988; J.S. Birman, "Recent Developments in Braid and Link Theory," *The Mathematical Intelligencer*, Vol. 13, #1 (Winter 1991), 52-60.

79. A survey is given in O.W. Greenberg, "A New Level of Structure," *Physics Today*, *38* (1985), 22-30.

80. *E.g.*, M. Milgrom, "A Departure from Newtonian Dynamics at Low Accelerations as an Explanation of the Mass Discrepancy in Galactic Systems," in J. Bahcall, T. Piran, and S. Weinberg (eds.), *Dark Matter in the Universe*, Singapore, World Scientific, 1987.

81. *Cf.*, G.B. Field, H. Arp, and J. Bahcall (eds.), *The Redshift Controversy*, Reading, Mass., W.A. Benjamin, Inc., 1973.

82. A view made popular by T.S. Kuhn, *The Structure of Scientific Revolutions*, Chicago, University of Chicago Press, 1970; for a critical review, see D. Shapere, "The Structure of Scientific Revolutions," *Philosophical Review*, LXXIII (1964), 383-394.

83. P. Feyerabend, *Against Method: Outline of an Anarchistic Theory of Knowledge*, London, Verso, 1978. The theme of Feyerabend's view is that "The only principle that does not inhibit progress is: *anything goes*." (p. 10)

84. S. Hawking, "Is the End in Sight for Theoretical Physics?" Inaugural Lecture as Lucasian Professor of Mathematics (delivered 29 April 1980), Cambridge University, *CERN Courier*, *21* (1981), 3 (Part I), 71 (Part II); published as a separate reprint by Cambridge University Press, 1980.

85. W.C. Dampier, *A History of Science and Its Relations with Philosophy and Religion*, Cambridge, Cambridge University Press, 1949, p. 369.

86. S.G. Brush, "Thermodynamics and History," *The Graduate Journal*, *7* (1967), 522-523, and defended by L. Badash, "The Completeness of Nineteenth-Century

Science," *Isis*, *63* (1972), 48-58. Many quotations, including Maxwell's, are given in Badash's article illustrating nineteenth-century attitudes. It is interesting that Badash takes it for granted that predictions of the completion of physics are to be described as "pessimistic." Whether that word accurately describes the attitudes of all who made such predictions, contemporary physicists tend to view the predicted end as an achievement to be hailed.

87. The lecture was published in Lord Kelvin [William Thomson], "Nineteenth Century Clouds Over the Dynamical Theory of Heat and Light," *Philosophical Magazine*, *2* (1901), 1-40; reprinted in Kelvin, *Baltimore Lectures on Molecular Dynamics and the Wave Theory of Light*, London, Baltimore, 1904, pp. 486-527. The lecture is not included in the recent (1987) M.I.T. press reprint of the lectures.

88. A systematic analysis of "problems of incompleteness" and "problems of incorrectness, and indeed of types of scientific problems in general, is given in D. Shapere, "Objectivity, Rationality, and Scientific Change," in P. Kitcher and P. Asquith (eds.), *PSA 1984*, East Lansing, Philosophy of Science Association, 1985, Vol. 2, 637-662.

89. J.D. Burchfield, *Lord Kelvin and the Age of the Earth*, Chicago, University of Chicago Press, 1975.

90. See, for instance, B.G. Doran, "Origins and Consolidation of Field Theory in Nineteenth Century Britain: From the Mechanical to the Electromagnetic View of Nature," *Historical Studies in the Physical Sciences*, Princeton, Princeton University Press, Vol. 6 (1975), 133-260.

91. There are, however, some important dissenters who doubt that a quantum-field-theoretic approach is the way to go. See, for example, R. Penrose, *The Emperor's New Mind*, New York, Oxford, 1989; C. Isham, "Quantum Gravity," in P. Davies (ed.), *The New Physics*, Cambridge, Cambridge University Press, 1989, 70-93.

92. Unification is not necessarily achievable, even in principle: there is no *a priori* guarantee that it is even possible, much less that it is a necessary condition of explanatory success. For there is no absolute necessity that there *must* be a fundamental unity to the laws of nature, or that scientific understanding *requires*

unification. Why, from an *a priori* point of view, should there not be independent and irreducible entities and agencies in nature? Could nature not be fundamentally pluralistic in its fundamental features, and still be intelligible? Wolfgang Pauli warned of that possibility when he said - only half-jokingly - that "What God hath put asunder let no man join." And surely we must admit that our searches for unification could, in principle, fail, not because of our lack of ability to find the unity, but simply because it is not there to be found; and we might come to understand that disunity itself, why it exists. Indeed, a major theme of this essay is that we have to *learn* what the goals of science are, and that we have to *learn* what understanding, explanation, consists in. The question of unification and its achievability must be answered in terms of evidence taken from the content of the best scientific results available, not on the basis of *a priori* arguments concerning the "nature" of science or of explanation.

93. F. Wilczek, "Perspectives on Particle Physics and Cosmology," summary talk presented at the 1990 Nobel Symposium on the Birth and Evolution of Our Universe, Gräftåavallen, Östersund, Sweden, June 11-16, 1990, p. 9; preprint IASSNS-HEP-90/64.

94. That sort of argument is very popular among philosophers today, and is known and widely accepted under the name of "the inductive skeptical argument."

95. For an account of how the concept of observation can change due to new scientific beliefs, see D. Shapere, "The Concept of Observation in Science and Philosophy," *Philosophy of Science*, *49* (1982), 485-525.

96. Further elaboration of this point is given in D. Shapere, "Modern Physics and the Philosophy of Science," in A. Fine and J. Leplin (eds.), *PSA 1988*, Vol. 2, East Lansing, Philosophy of Science Association, 1989, pp. 210-210.

97. M. Gell-Mann, "A Schematic Model of Baryons and Mesons," *Physics Letters*, *8* (1964), 214-215; G. Zweig, "An SU(3) Model for Strong Interaction Symmetry and Its Breaking," *CERN Preprint 8182/TH401*, 1964.

98. A beautiful and profound elementary study is given in H. Weyl, *Symmetry*, Princeton, Princeton University Press, 1952.

99. J.F. Gunion, H.E. Haber, G. Kane, and S. Dawson, *The Higgs Hunter's Guide*, New York, Addison-Wesley, 1990.

BIBLIOGRAPHY

Abbott, L.F., and S. Pi (eds.), *Inflationary Cosmology*, Singapore, World Scientific, 1986.

Albrecht, A., and P. Steinhardt, "Cosmology for Grand Unified Theories with Radiatively Induced Symmetry Breaking," *Physical Review Letters*, *48* (1982), 1220-1223.

Alpher, R.A, and R.C. Herman, "Remarks on the Evolution of the Expanding Universe," *Physical Review*, *75* (1949), 1089-1095.

Alpher, R.A., and R.C. Herman, "Theory of the Origin and Relative Abundance Distribution of the Elements," *Revues of Modern Physics*, *22* (1950), 153-212.

Altman, S., *et al*, "Enzymatic Cleavage of RNA by RNA," *Trends in Biochemical Sciences*, *11*, 515.

Alvarez-Gaumé, L., and E. Witten, "Gravitational Anomalies," *Nuclear Physics B*, *B234* (1983), 269-330.

Applequist, T., A. Chodos, and P.G.O. Freund (eds.), *Modern Kaluza-Klein Theories*, Reading, Mass., Addison-Wesley, 1987.

Arp, H.C., G. Burbidge, F. Hoyle, J.V. Narlikar, and N.C. Wickramasinghe, "The Extragalactic Universe: An Alternative View," *Nature*, *346* (1990), 807-812.

Audouze, J., and J. Tran Thanh Van (eds.), *Formation and Evolution of Galaxies and Large Structures in the Universe*, Dordrecht, Reidel, 1984.

Baade, W., "Extragalactic Nebulae: Report to IAU Commission 28," *International Astronomical Union Transactions*, *8* (1952), 397-399.

Badash, L., "The Completeness of Nineteenth-Century Science," *Isis*, *63* (1972), 48-58.

Bahcall, J., T. Piran, and S. Weinberg (eds.), *Dark Matter in the Universe*, Singapore, World Scientific, 1987.

Bergeron,, J., D. Kunth, B. Rocca-Volmerange, and J. Tran Thanh Van (eds.), *High Redshift and Primeval Galaxies*, Editions Frontieres, 1987.

Bernstein, J., and G. Feinberg (eds.), *Cosmological Constants: Papers in Modern Cosmology*, New York, Columbia University Press, 1986.

Bethe, H.A., "Energy Production in Stars," *Physical Review*, *55* (1939), 434-456.

Birman, J.S., "Recent Developments in Braid and Link Theory," *The Mathematical Intelligencer*, Vol. 13, #1 (Winter 1991), 52-60.

Boesgaard, A.M., and G. Steigman, "Big Bang Nucleosynthesis: Theories and Observations," *Annual Review of Astronomy and Astrophysics*, *23* (1985), Palo Alto, Annual Reviews, Inc., 319-378.

Bondi, H., *Cosmology*, Cambridge, Cambridge University Press, 1960.

Bondi, H., and T. Gold, "The Steady State Theory of the Expanding Universe," *Monthly Notices of the Royal Astronomical Society*, *108* (1948), 252-270.

Börner, G., *The Early Universe: Facts and Fiction*, New York, Springer-Verlag, 1988.

Brown, L.M., and L.Hoddeson (eds.), *The Birth of Particle Physics*, Cambridge, Cambridge University Press, 1983.

Brush, S.G., "Thermodynamics and History," *The Graduate Journal*, *7* (1967), 477-565.

Burbidge, E.M., G.R. Burbidge, W.A. Fowler, and F. Hoyle, "Synthesis of the Elements in Stars," *Reviews of Modern Physics*, *29* (1957), 547-650.

Burchfield, J.D., *Lord Kelvin and the Age of the Earth*, Chicago, University of Chicago Press, 1975.

Byrne, R., and A. Whiten (eds. *Machiavellian Intelligence: Social Expertise and the Evoution of Intellect in Monkeys, Apes, and Humans*, Oxford, Clarendon, 1988.

Chang, S., D. DesMarais, R. Mack, S.L. Miller, and G.E. Strathearn, "Prebiotic Organic Syntheses and the Origin of Life," in J.W. Schopf (ed.), *Earth's Earliest Biosphere: Its Origin and Evolution*, Princeton, Princeton University Press, 1983, 53-92.

Cech, T.R., "The Chemistry of Self-splicing RNA and RNA Enzymes," *Science*, *236* (1987), 1532.

Cohen, M., *In Darkness Born: The Story of Star Formation*, Cambridge, Cambridge University Press, 1988.

Crease, R.P., and C.C. Mann, *The Second Creation: Makers of the Revolution in 20th-Century Physics*, New York, MacMillan, 1986.

Crook, J., *The Evolution of Human Consciousness*, Oxford, Clarendon, 1985.

Dampier, W.C., *A History of Science and Its Relations with Philosophy and Religion*, Cambridge, Cambridge University Press, 1949.

Davies, P. (ed.), *The New Physics*, Cambridge, Cambridge University Press, 1989.

Day, W., *Genesis on Planet Earth*, New Haven, Yale University Press, 1984.

Dicke, R.H., P.J.E. Peebles, P.G. Roll, and D.T. Wilkinson, "Cosmic Black Body Radiation," *Astrophysical Journal*, *142* (1965), 414-419.

Dickson, D.B., *The Dawn of Belief: Religion in the Upper Paleolithic of Southwestern Europe*, Tucson, University of Arizona Press, 1990.

Doran, B.G., "Origins and Consolidation of Field Theory in Nineteenth Century Britain: From the Mechanical to the Electromagnetic View of Nature," *Historical*

Studies in the Physical Sciences, Princeton, Princeton University Press, Vol. 6 (1975), 133-260.

Dormand, J.R., and M.M. Woolfson, *The Origin of the Solar System: The Capture Theory*, New York, Wiley, 1989.

Duhem, P., *The Aim and Structure of Physical Theory*, Princeton, Princeton University Press, 1954.

Dyson, F.J., "Time Without End: Physics and Biology in an Open Universe," *Reviews of Modern Physics*, *51*, 447.

Einstein, A., "Kosmologische Betrachtungen zur allgemeinen Relativitätstheorie," *Preuss. Akad. Wiss. Sitzb.*, 1915, 831-839; English translation, "Cosmological Considerations on the General Theory of Relativity," in Einstein *et al.*, *The Principle of Relativity*, New York, Dover, 1923, 177-188.

Ellis, G.F.R., "The Expanding Universe: A History of Cosmology from 1917 to 1960," in Howard, D., and J. Stachel (eds.), *Einstein and the History of General Relativity*, Boston, Birkhauser, 1989.

Feyerabend, P., *Against Method: Outline of an Anarchistic Theory of Knowledge*, London, Verso, 1978.

Field, G.B., H. Arp, and J. Bahcall (eds.), *The Redshift Controversy*, Reading, Mass., W.A. Benjamin, Inc., 1973.

Frenk, C.S., R.S. Ellis, T. Shanks, A.F. Heavens, and J.A. Peacock (eds.), *The Epoch of Galaxy Formation*, Dordrecht, Kluwer, 1988.

Freund, P.G.O., *Introduction to Supersymmetry*, Cambridge, Cambridge University Press, 1986.

Galison, P., *How Experiments End*, Chicago, University of Chicago Press, 1987.

Gamow, G., "Expanding Universe and the Origin of the Elements," *Physical Review*, *70* (1946), 572-573.

Gamow, G., *The Creation of the Universe*, New York, Viking, 1952.

Gamow, G., "The Physics of the Expanding Universe," *Vistas in Astronomy, 2* (1956), 1726-1732.

Gell-Mann, M., "A Schematic Model of Baryons and Mesons," *Physics Letters, 8* (1964), 214-215.

Georgi, H., and S.L. Glashow, "Unity of All Elementary-Particle Forces," *Physical Review, 32* (1974) *Letters*, 36.

Georgi, H., H. Quinn, and S. Weinberg, "Hierarchy of Interactions in Unified Gauge Theories," *Physical Review Letters, 33* (1974), 451.

Gibbons, G.W., S.W. Hawking, and S.T.C. Siklos (eds.), *The Very Early Universe*, Cambridge, Cambridge University Press, 1983.

Green, M.B., and J.H. Schwarz, "Anomaly Cancellations in Supersymmetric D=10 Gauge Theory and Superstring Theory," *Physics Letters, 149B* (1984), 117-122.

Green, M.B., J.H. Schwarz, and E. Witten, *Superstring Theory*, Cambridge, Cambridge University Press, 1988, 2 vols.

Greenberg, O.W., "A New Level of Structure," *Physics Today, 38* (1985), 22-30.

Gross, D.J., J.A. Harvey, E. Martinec, and R. Rohm, "Heterotic String Theories (I). The Free Heterotic String," *Nuclear Physics B, 256* (1985), 253-284.

Gross, D.J., and F.A. Wilczek, "Ultraviolet Behavior of Non-Abelian Gauge Theories," *Physical Review Letters, 30* (1973), 1343-1346.

Gunion, J.F., H.E. Haber, G. Kane, and S. Dawson, *The Higgs Hunter's Guide*, New York, Addison-Wesley, 1990.

Guth, A., "Inflationary Universe: A Possible Solution to the Horizon and Flatness Problems," *Physical Review, D23* (1981), 347-356.

Harrison, E.H., *Cosmology*, Cambridge, Cambridge University Press, 1981.

Havelock, E.A., "The Linguistic Task of the Presocratics," in K. Robb (ed.), *Language and Thought in Early Greek Philosophy*, La Salle, Ill., Monist Library of Philosophy, 1983.

Hawking, S., "Is the End in Sight for Theoretical Physics?" Inaugural Lecture as Lucasian Professor of Mathematics (delivered 29 April 1980), Cambridge University, *CERN Courier*, *21* (1981), 3 (Part I), 71 (Part II); reprinted separately: Cambridge, Cambridge University Press, 1980.

Hoyle, F., "A New Model for the Expanding Universe," *Monthly Notices of the Royal Astronomical Society*, *108* (1948), 372-382.

Isham, C., "Quantum Gravity," in P. Davies (ed.), *The New Physics*, Cambridge, Cambridge University Press, 1989, 70-93.

Islam, J.N., "Possible Ultimate Fate of the Universe," *Quarterly Journal of the Royal Astronomical Society*, *18* (1977), 3.

Jones, B.J.T., and J.E, Jones (eds.), *The Origin and Evolution of Galaxies*, Dordrecht, Reidel, 1983.

Joyce, G.F., "RNA Evolution and the Origins of Life," *Nature*, *338* (1989), 217-224.

Kabir, P.K. (ed.), *The Development of Weak Interaction Theory*, New York, Gordon and Breach, 1963.

Kaku, M., *Introduction to Superstrings*, New York, Springer-Verlag, 1988.

Kelvin, Lord [William Thomson], *Baltimore Lectures on Molecular Dynamics and the Wave Theory of Light*, London, Baltimore, 1904.

Kelvin, Lord [William Thomson], "Nineteenth Century Clouds Over the Dynamical Theory of Heat and Light," *Philosophical Magazine*, *2* (1901), 1-40.

Kirk, G.S., J.E. Raven, and M. Schofield, *The Presocratic Philosophers*, Second Edition, Cambridge, Cambridge University Press, 1983.

Klein, R.G., *The Human Career: Human Biological and Cultural Origins*, Chicago, University of Chicago Press, 1989.

Kolb, E.W., and M.S. Turner, "Grand Unified Theories and the Origin of the Baryon Asymmetry," in J.D. Jackson, G.E. Gove, and R.F. Schwitters (eds.), *Annual Review of Nuclear and Particle Science, 33* (1983), Palo Alto, Annual Reviews, Inc., 645-696.

Kolb, E.W., and M.S. Turner (eds.), *The Early Universe: Reprints*, Reading, Mass., Addison-Wesley, 1988.

Kolb,, E.W., M.S. Turner, D. Lindley, K. Olive, and D. Seckel (eds.), *Inner Space/Outer Space: The Interface Between Cosmology and Particle Physics*, Chicago, University of Chicago Press, 1986.

Kuhn, T.S., *The Structure of Scientific Revolutions*, Chicago, Universiy of Chicago Press, 1970.

Kvenvolden, K.A. (ed.), *Geochemistry and the Origin of Life* [Benchmark Papers in Geology], Stroudsburg, Pa., Dowden, Hutcinson and Ross, Inc., 1974.

Lang, K.R., and O. Gingerich (eds.), *A Source Book in Astronomy and Astrophysics, 1900-1975*, Cambridge, Harvard University Press, 1979.

Langacker, P., "The Present Status of Grand Unification and Proton Decay," in E.W. Kolb, M.S. Turner, D. Lindley, K. Olive, and D. Seckel (eds.), *Inner Space/Outer Space: The Interface Between Cosmology and Particle Physics*, Chicago, University of Chicago Press, 1986, 3-24.

Lee, T.D., and C.N. Yang, "Question of Parity Conservation in Weak Interactions," *Physical Review, 104* (1956), 254-258.

Lemaitre, G., "A Homogeneous Universe of Constant Mass and Increasing Radius Accounting for the Radial Velocity of Extragalactic Nebulae," *Ann. Soc. Sci. Bruxelles*, A47 (1927), 49-59 [English translation in *Monthly Notices of the Royal Astronomical Society, 91* (1931), 483-490].

Lemaitre, G., *The Primeval Atom: An Essay on Cosmogony*, New York, Van Nostrand, 1950.

Lindley, D., "An Excess of Perfection," *Nature, 343* (18 Jan. 1990), 207.

Loh, E.D., and E.J. Spillar, "Photometric Redshifts of Galaxies," *Astrophysical Journal, 303* (1986), 154-161.

Lozano, M., M.I. Gallardo, and J.M. Arias (eds.), *Nuclear Astrophysics*, New York, Springer-Verlag, 1989.

Maddox, J., "Down with the Big Bang," *Nature, 340* (1989), 425.

Marciano, W., and H. Pagels, "Quantum Chromodynamics," *Nature, 279* (1979 479-483.

Mellars, P., and C. Stringer (eds.), *The Human Revolution: Behavioural and Biological Pespectives on the Origins of Modern Humans*, Princeton, Princeton University Press, 1989.

Milgrom, M., "A Departure from Newtonian Dynamics at Low Accelerations as an Explanation of the Mass Discrepancy in Galactic Systems," in J. Bahcall, T. Piran, and S. Weinberg (eds.), *Dark Matter in the Universe*, Singapore, World Scientific, 1987.

Mohapatra, R.N., *Unification and Supersymmetry*, New York, Springer-Verlag, 1986.

Nagel, E., *The Structure of Science: Problems in the Logic of Scientific Explanation*, New York, Harcourt, Brace and World, Inc., 1961.

North, J., *The Measure of the Universe*, Oxford, Clarendon, 1965.

Or6, J., S.L. Miller, and A. Lazcano, "The Origin and Early Evolution of Life on Earth," in *Annual Review of Earth and Planetary Science, 18* (1990), Palo Alto, Annual Reviews, Inc., 317-356.

Pagels, H., *The Cosmic Code: Quantum Physics as the Language of Nature*, New York, Simon and Schuster, 1982.

Pais, A., *Inward Bound*, Oxford, Clarendon, 1986.

Pais, A., *Subtle is the Lord ...*, Oxford, Clarendon, 1982.

Peccei, R., and H. Quinn, "Constraints Imposed by CP Conservation in the Presence of Instantons," *Physical Review D*, *16* (1977), 1791.

Peimbert, M., and J. Jugaku (eds.), *Star Forming Regions*, Dordrecht, Reidel, 1987.

Penrose, R., *The Emperor's New Mind*, New York, Oxford, 1989.

Penzias, A.A., and R.W. Wilson, "A Measurement of Excess Antenna Temperature at 4080 Mc/s," *Astrophysical Journal*, *142* (1965), 419-421.

Perkins, D.H., "Proton Decay Experiments," in J.D. Jackson, H.E. Gove, and R.F. Schwitters (eds.), *Annual Review of Nuclear and Particle Science*, *34* (1984), Palo Alto, Annual Reviews, Inc., 1-53.

Petschek, A.G., (ed.), *Supernovae*, New York, Springer-Verlag, 1990.

Politzer, H.D., "Reliable Perturbative Results for Strong Interactions," *Physical Review Letters*, *30* (1973), 1346-1349.

Ponnamperuma, C. (ed.), *Cosmochemistry and the Origin of Life*, Dordrecht, Reidel, 1983.

Popper, K.R., *Conjectures and Refutations*, New York, Basic Books, 1962.

Popper, K.R., *The Logic of Scientific Discovery*, New York, Basic Books, 1959.

Puhvel, J., *Comparative Mythology*, Baltimore, Johns Hopkins University Press, 1987.

Quine, W.V., *Word and Object*, Cambridge, M.I.T. Press, 1960.

Roger, R.S., and P.E. Dewdney (eds.), *Regions of Recent Star Formation*, Dordrecht, Reidel, 1982.

Rolfs, C.E., and W.S. Rodney, *Cauldrons in the Cosmos*, Chicago, University of Chicago Press, 1988.

Rowan-Robinson, M., *The Cosmological Distance Ladder: Distance and Time in the Universe*, New York, W.H. Freeman, 1985.

Ryle, M., and P. Scheuer, "The Spatial Distribution and the Nature of Radio Stars," *Proceedings of the Royal Society*, *A230* (1955), 448-462.

Sakharov, A.D., "Violation of CP Invariance, C Asymmetry, and Baryon Asymmetry of the Universe, *Zh. Ek. Teor. Fiz.*, *5* (1967), 32; English translation, *JETP Letters*, *5* (1967), 24.

Salam, A., "Weak and Electromagnetic Interactions," in N. Svartholm (ed.), *Proceedings of the Eighth Nobel Symposium on Elementary Particle Theory*, New York, Interscience, 1968, 367-377.

Sandage, A., "Current Problems in the Extra-Galactic Distance Scale," *Astrophysical Journal*, *127* (1958), 513-526.

Shapere, D., "Modern Physics and the Philosophy of Science," in A. Fine and J. Leplin (eds.), *PSA 1988*, Vol. 2, East Lansing, Philosophy of Science Association, 1989, 201-210.

Shapere, D., "Newton's Place in History," in S. Brush, A. Sieff, and P. Theerman (eds.), *Action and Reaction: Proceedings of a Conference on the 300th Anniversary of Publication of Newton's Principia*, Newarh, Del., University of Deleware Press, 1991.

Shapere, D., "Objectivity, Rationality, and Scientific Change," in P. Kitcher and P. Asquith (eds.), *PSA 1984*, East Lansing, Philosophy of Science Association, 1985, Vol. 2, 637-662.

Shapere, D., "On the Introduction of New Ideas in Science," forthcoming in J. Leplin (ed.), *The Creation of Ideas in Science*.

Shapere, D., "On the Relations between Compositional and Evolutionary Theories," in F. Ayala and T. Dobzhansky (eds.), *Studies in the Philosophy of Biology*, London, Macmillan, 1974, 187-202.

Shapere, D., "Scientific Theories and Their Domains," in F. Suppe (ed.), *The Structure of Scientific Theories*, Urbana, University of Illinois Press, 1974, 518-565.

Shapere, D., "The Concept of Observation in Science and Philosophy," *Philosophy of Science*, *49* (1982), 485-525.

Shapere, D., "The Mechanical Philosophy of Nature," forthcoming.

Shapere, D., "The Origin and Nature of Metaphysics," forthcoming in *Philosophical Topics*.

Shapere, D., "The Structure of Scientific Revolutions," *Philosophical Review*, LXXIII (1964), 383-394.

Shapiro, R., *Origins: A Skeptic's Guide to the Creation of Life on Earth*, New York, Summit Books, 1986.

Shu, F.H., F.C. Adams, and S. Lizano, "Star Formation in Molecular Clouds: Observation and Theory," in G. Burbidge, D. Layzer, and J.G. Phillips (eds.), *Annual Review of Astronomy and Astrophysics*, *25* (1987), Palo Alto, Annual Reviews, Inc., 23-81.

Smith, R.W., *The Expanding Universe: Astronomy's 'Great Debate' 1900-1931*, Cambridge, Cambridge University Press, 1982.

Spector, M., *Concepts of Reduction in Physical Science*, Philadelphia, Temple University Press, 1978.

Strachan, C. (ed.), *The Theory of Beta Decay*, New York, Pergamon, 1969.

Sugimoto,, D., D.Q. Lamb, and D.N. Schramm (eds.), *Fundamental Problems in the Theory of Stellar Evolution*, Dordrecht, Reidel, 1981.

Sutton, C., *The Particle Connection*, New York, Simon and Schuster, 1984.

't Hooft, G., "Renormalizable Langrangians for Massive Yang-Mills Fields," *Nuclear Physics, B35* (1971), 167-188.

Trimble, V., "Existence and Nature of Dark Matter in the Universe," *Annual Review of Astronomy and Astrophysics, 25* (1987), Palo Alto, Annual Reviews, Inc., 425-472.

Truran, J.W., "Nucleosynthesis", *Annual Review of Nuclear and Particle Science, 34* (1984), Palo Alto, Annual Reviews, Inc., 53-97.

Tryon, E.P., "Is the Universe a Vacuum Fluctuation?" *Nature, 246* (1973), 396-397.

von Weizsacker, C.F., "Element Transformation Inside Stars," I, *Physikalische Zeitschrift, 38* (1937), 176; II, *ibid, 39* (1938), 633-646.

Watkins, P., *Story of the W and Z*, Cambridge, Cambridge University Press, 1986.

Weaver, H.A., and L. Danly (eds.), *The Formation and Evolution of Planetary Systems*, Cambridge, Cambridge University Press, 1989.

Weinberg, S., "A Model of Leptons," *Physical Review Letters, 19* (1967), 1264-1266.

Weinberg, S., "A New Light Boson?" *Physical Review Letters, 40* (1978), 223-226.

Weinberg, S., *Gravitation and Cosmology: Principles and Applications of the General Theory of Relativity*, New York, Wiley, 1972.

Weinberg, S., "The Cosmological Constant Problem," *Reviews of Modern Physics, 61* (1989), 1-23.

Weinberg, S., *The First Three Minutes: A Modern View of the Origin of the Universe*, New York, Basic Books, 1977.

Wess, J., and J. Bagger, *Supersymmetry and Supergravity*, Princeton, Princeton University Press, 1983.

West, P. (ed.), *Supersymmetry: A Decade of Development*, Boston, Adam Hilger, 1986.

Wetherill, G.W., "Formation of the Earth," *Annual Review of Earth and Planetary Sciences*, *18* (1990), Palo Alto, Annual Reviews, Inc., 205-256.

Weyl, H., *Symmetry*, Princeton, Princeton University Press, 1952.

Wilczek, F.A., "Foundations and Working Pictures in Microphysical Cosmology," in G.W. Gibbons, S.W. Hawking, and S.T.C. Siklos (eds.), *The Very Early Universe*, Cambridge, Cambridge University Press, 1983.

Wilczek, F., "Perspectives on Particle Physics and Cosmology," summary talk presented at the 1990 Nobel Symposium on the Birth and Evolution of Our Universe, Gräftåavallen, Östersund, Sweden, June 11-16, 1990; preprint IASSNS-HEP-90/64.

Wilczek, F., "Problem of Strong P and T Invariance in the Presence of Instantons," *Physical Review Letters*, *40* (1978), 279-282.

Will, C.M., "General Relativity at 75: How Right Was Einstein?" *Science*, *250* (1990), 770-776.

Will, C.M., *Theory and Experiment in Gravitational Physics*, Cambridge, Cambridge University Press, 1981.

Will, C.M., *Was Einstein Right? Putting General Relativity to the Test*, New York, Basic Books, 1986.

Williams, I.P., *The Origin of the Planets*, Bristol, Adam Hilger, 1978.

Witten, E., "Quantum Field Theory and the Jones Polynomial," Institute for Advanced Study preprint, 1988.

Wittgenstein, L., *Philosophical Investigations*, Oxford, Blackwell, 1953.

Wolfenstein, L., "Present Status of CP Violation," *Annual Review of Nuclear and Particle Science*, *36* (1986), Palo Alto, Annual Reviews, Inc., 137-170.

Woolfson, M.M., "The Capture Theory and the Origin of the Solar System," in S.F. Dermott (ed.), *The Origin of the Solar System*, New York, Wiley, 1978, 179-198.

Woolfson, M.M., "The Evolution of the Solar System," in S.F. Dermott (ed.), *The Origin of the Solar System*, New York, Wiley, 1978, 198-217.

Wu, C.S., E. Ambler, R.W. Hayward, D.D. Hoppes, and R.P. Hudson, "Experimental Test of Parity Conservation in Beta-decay," *Physical Review*, *105* (1957), 1413-1415.

Wynn, T., *The Evolution of Spatial Competence*, Urbana, University of Illinois Press, 1989.

Yang, C.N., and R. Mills, "Conservation of Isotopic Spin and Isotopic Gauge Invariance," *Physical Review*, *96* (1954), 191-195.

Zee, A., (ed.), *Unity of Forces in the Universe*, Singapore, World Scientific, 1982.

Zweig, G., "An SU(3) Model for Strong Interaction Symmetry and Its Breaking," *CERN Preprint 8182/TH401* 1964.

Robert Engel

FROM MOLECULES TO LIFE

A chemist will tend to look at the phenomenon of "life" from a rather different perspective than either a physicist or a biologist, practitioners of scientific disciplines most closely related to that of a chemist. Chemists view even complex phenomena in terms of a structural-functional relationship at a molecular level. Biologists, on the other hand, tend to view complex structures in a much more phenomenological manner, and physicists are often concerned not at all with the structure of matter, but simply with the relationship of that matter to energy in a given phenomenon. The molecular view taken by chemists (and biochemists) provides a different perpective of the nature of "living systems" and of the phenomenon of "life".

For example, with regard to the system described by Professor Mosterin of the Spanish fox tracking the Spanish rabbit across the Spanish field, chemists and biologists take different views. The biologist has a particular concern for the balance of the two animal organisms in the surrounding world, a concern for the relative well being and the continuation of each of the two species. A chemist or biochemist has a different concern with this system. A particular interest might lie in the nature of the molecules left behind by the rabbit which allow the fox to trail it. Further, there is an interest in not only the structure of these molecules, but in the mechanism of their perception by the fox, and their source and purpose in the rabbit. Possibly, the chemist is also concerned

E. Agazzi and A. Cordero (eds.), Philosophy and the Origin and Evolution of the Universe, 203–218.

with ways in which fox could be misled by the substitution of a different molecule for that left behind by the rabbit. (The concern of the physicist with this system is even yet another matter.)

To look at the concern of the chemist in more detail, consider first the molecule shown in Fig.l, glycylglycylglycine.

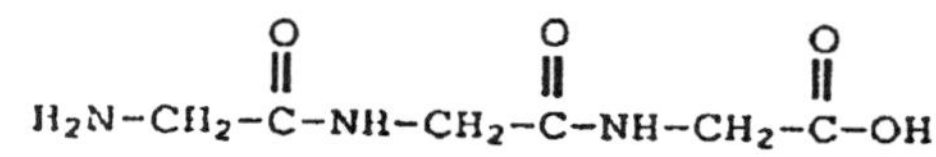

Fig.1 -Molecular structure of glycyglycyglycyne

This is a simple tripeptide, composed of three amino acid (glycine) units joined by amide linkages. This is one type of molecule which has been of interest to us in our work on the systematic regulation of metabolic processes. A phenomenon observed with this molecule (and other tripeptides) is that it is taken up by strains of Escherichia coli and other bacteria (as well as yeast strains and other organisms) from the surrounding medium and used as a nutrient. There exists with these species a "tripeptide transport system" facilitating its uptake against a concentration gradient. This phenomenon has been well documented by others.[1]

Considering a bacterium, a biologist might look at such a material and the uptake phenomenon and have a concern with their effect on the continued growth of the bacterium. A biophysicist might have a particular concern for the energy changes involved in the uptake process, without really caring about the molecular nature of the compound or its ultimate fate.

A chemist, while having some interest in both the energetics of the process and the growth phenomenon of the bacterium, is most concerned with the molecular structural features of the system. Knowing something of the fundamental chemistry of the particular structural components of the molecule, how do these structural components serve in the overall uptake process, and to what end? A chemist will look for the effect of structural variation on the efficiency of the process, i.e. the structure-function relationship. What are the structural features which allow for the uptake process to occur? (Specifically, the structural requirements are the presence of the two peptide linkages and the free carboxyl group. There is little or no effect of substitution of larger structural components for hydrogen on the internal carbons, or with the presence of electrical charges, although there is some decrease in uptake efficiency with substitution of larger structural components on the free amino group.) Our own work has been concerned with using this type of molecule as a delivery system for other compounds which would have specific interactions within the organism, but have no "normal" way of entering the cell.[2]

Chemists have a concern for the molecular structural relationships of interacting components in any process Venturing in either direction, i.e. understanding the energetics, or the larger effect of a particular process, chemists look at the dependency of the process on molecular structure. From a chemist's point of view, any observable phenomenon of interacting matter has a dependence on the molecular structure of the interacting components. It is their job to understand the molecular structural factors in an chemical change, including those of living systems.

Given this view we make predictions and modify chemical reactions, hopefully in a systematic manner to search for results, even those occurring as part of the operation of a living system. Consider an example of this approach we have used in investigations within our laboratory. Consider the two molecular structures shown in Fig. 2.

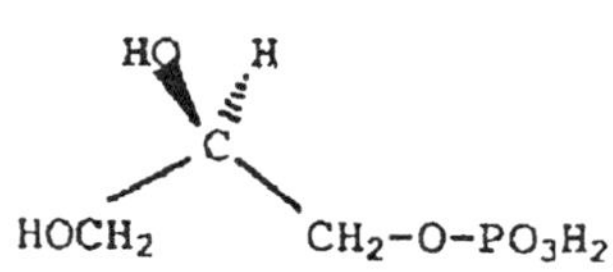

sn-glycerol 3-phosphate

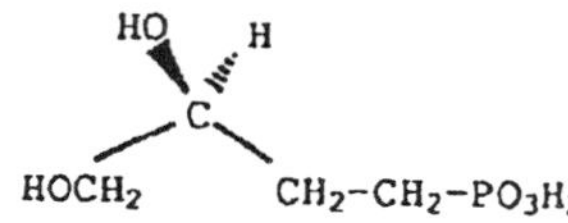

(S) -3,4-dihydroxybutyl-1-phosphonic acid

Fig. 2 -Structure of *sn*-glycerol 3-phosphate and its isosteric analogue

At the top is shown the structure of sn-glycerol 3-phosphate, a material which is required in the generation of components of cell walls of living organisms. In the biochemical reactions of this material leading to phosphatidyl glycerol, a major component of cell wall in a variety of organisms, the phosphorus-oxygen bond need be broken. The structure shown at the bottom is a molecule designed to be like sn-glycerol 3-phosphate in all ways except for the replacement of the critical esteric oxygen by a methylene group. We say that this molecule is "isosteric" with the natural compound. This analogue undergoes virtually all of the same biochemical reactions as the natural material, except for the (now impossible) breakage of the phosphorus-oxygen bond. This material thus is designed and serves as a specific inhibitor of cell wall synthesis for particular organisms. A molecular structural consideration of the chemical process of a living organism allows us to modify that life process in a highly systematic manner. Specifically, by introducing this analogue to an organism we induce it to use it in the initial stages of cell wall synthesis, but the total process is interrupted by the inability of the analogue to take part in the final reaction.

Now, let us move from these specific examples and consider the fundamental concept of a living system as it is viewed by a chemist or biochemist. We will spend some time looking at living systems in contrast to non-living systems, and consider both with a view of their molecular structure-function nature. We will consider individual functions of living systems and compare them with functions of non-living systems.

To a chemist or biochemist, the fundamental characteristic of a living system is its dynamic metabolism. By dynamic metabolism we mean the continuous performance of critical functions involving the processing of matter. The uptake of tripeptides or the processing of sn-glycerol 3-phosphate for the generation of cell wall are specific parts of the dynamic metabolism of an organism. These critical functions in the processing of matter are for two purposes: 1) the regeneration and maintenance of the living system, and 2) the reproduction of the living system. A living system performs both of these functions, or has available the basic mechanism for performing both of these functions of dynamic metabolism.

By regeneration and reproduction we mean two related, albeit different actions. Regeneration is the metabolic turnover of components of the system. When we say that there is metabolic turnover we mean the continual replacement of molecules or parts of molecules in a "steady-state" synthesis and breakdown of structural components. These are chemical processes which maintain the parent species as an entity. A bacterial cell, even in a phase in which it is not undergoing subdivision and reproduction, will have a dynamic metabolism which maintains it as a system with a potential purpose, reproduction.

As an example of metabolic turnover, phospholipids of exterior cell walls and cell component walls undergo constant replacement with identical molecules, newly synthesized by chemical (biochemical) processes occuring within the cell (intermediary metabolism). This does not involve a change in the macrostructure of the system, except as could be noted by isotopic labelling. A new brick replaces a brick previously in place. But the new brick is identical to the one replaced, and has been generated inside the unit, by the unit. Individual phospholipid molecules in structural components of Escherichia coli, not undergoing growth and subdivision, have metabolic turnover half-lives measured in hours. Components of living systems all undergo this metabolic turnover at some rate, in some instances quite rapidly while in others with long half-lives. Living systems in "dormant" states have their rates of metabolic turnover greatly decreased, but still in existence.

We can see an example of "turnover" in a relatively simple "non-living" chemical system as shown in Fig. 3. This is part of a reaction system investigated by Wilke, et al.,[3] in which a cyclododecatriene-nickel complex undergoes continuous regeneration of its carbon-hydrogen periphery, synthesizing new cyclododecatriene molecules as a by-product. Cyclododecatriene by itself does not perform this type of process, nor are any "new" cyclododecatriene-nickel complex molecules made. Upon continuous supply of the raw material 1,3-butadiene, a continuous replacement of the periphery component occurs with "molting" of the old. This is simple regeneration and not reproduction. There is no mechanism available for the molecule to replicate its entire self, only its peripheral portion.

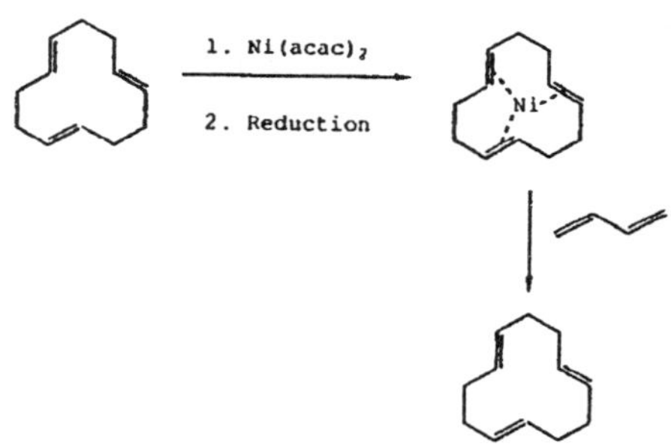

Fig. 3 -Catalytic synthesis of cyclododecatriene by reaction of butadiene with cyclododecatrienenickel(0)

Turnover of another type also occurs in other simple chemical systems at equilibrium. For example, a single crystal of copper sulfate resting in a saturated aqueous solution is a dynamic situation with regard to replacement. Ionic components of surface sites of the crystal lattice are constantly moving off into solution, being replaced in the lattice by ionic components from the saturated solution. However, this is not a metabolic turnover. The crystal is not synthesizing new components for its lattice, but is simply receiving them from the surrounding medium.

Niether of these examples represents a "living" system, although they have some type of dynamism. Niether is capable of true reproduction, the formation of a "new" unit of its own type, separate from itself. Moreover, the example of the crystal dynamism is not a metabolic turnover from within, but rather simply a replacement from without. There is no production of the replacing unit from within; there is no metabolism.

We can look at other chemical species which in a sense undergo reproduction, but are not truly "living". They are static with regard to regeneration, and reproduce only under particular circumstances. Viruses might be thought to be at the border of "living" and "non-living" species. Virus species by themselves act as chemical macromolecules having no dynamic metabolism. They express reproductive capacities only when in association with another species which (according to our definition) is living, i.e. capable of undergoing both replacement and reproduction. A wide variety of viruses are now known to us. Each particular virus has its own peculiar characteristics, but all consist fundamentally of two portions with defined chemical structures. These two portions are: 1) a nucleic acid portion involved in the reproduction of the virus, and 2) a protective covering for the nucleic acid portion made up of protein or glycoprotein. In some instances a third portion may be defined as a receptor region for interaction with host species, although at times this is indistinct from the protective covering region.

An intact, complete virus acts like an "ordinary" organic molecule, albeit a large one. They can aggregate to form crystals. For example, poliomyelitis virus and coxsackie virus species form crystal lattice structures in the same way as do "ordinary" organic molecules. Crystals as aggregates of individual virus particles can exhibit the dynamism of the type noted for a copper sulfate crystal, individual virus particles can be replaced at the surface. However, portions of individual virus particles do not undergo regeneration. As with the copper sulfate crystal, there is no dynamic metabolism. New replacement parts for the virus crystal are not being created, but rather simply taken up from the external environment. New parts of the virus molecule are not being synthesized.

In Fig. 4 is shown an electron micrograph of an individual T4 bacteriophage, a virus which undergoes a specific interaction with bacteria. We can identify for the virus specific regions with particular purposes. The "head" region and the "sheath" connecting it to the "tail" region serve as a surrounding protection for the reproductive portion contained therein. The "tail" and "tail fibers" serve as receptor interaction regions for its association with a bacterium. By itself the T4 bacteriophage has no reproduction and no metabolic turnover. In fact, it has no metabolism; it niether uses any external molecular species to replace portions of its structure, synthesizes any molecular species, nor undergoes any change in components or structure. Isolated, it lacks fundamental characteristics of "living" species.

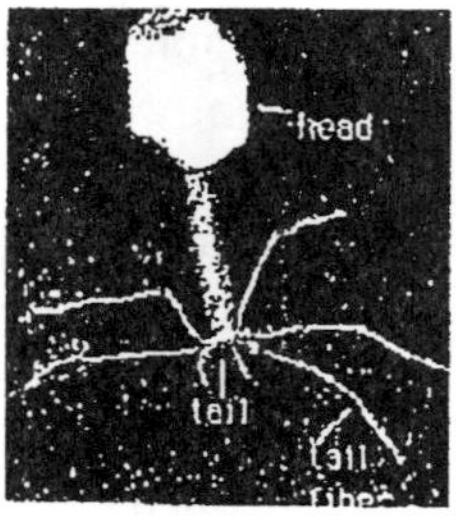

{1000 A}

Fig. 4- Electron micrograph of a T4 bacteriophage[4]

In Fig. 5 we see a cartoon of the T4 bacteriophage illustrating the component structural units. At the top is a representation of the "free" virus particle, showing the sections as seen in the electron micrograph of Fig.4. At the bottom is a representation of the virus particle undergoing interaction with the outside surface of a bacterial cell wall. There is a chemical interaction between the tail and tail fibers of the virus particle with components of the cell wall which creates a hole through the cell wall. This allows the nucleic acid portion of the virus, normally contained within the protective head and sheath, to enter the bacterial cell.

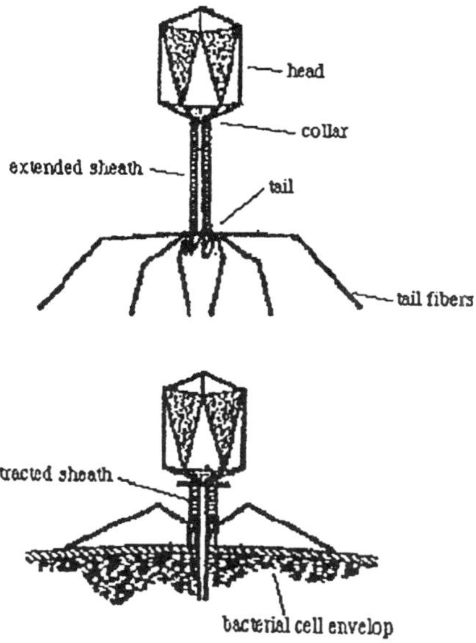

Fig.5 - The T4 bacteriophage as an independent species (top) and as it infects a cell (bottom)[5]

The nucleic acid portion of T4 bacteriophage is double-stranded DNA of approximately 182,000 base pairs. The nucleic acid portion of the virus constitutes approximately 55% of its total weight.

Once the nucleic acid of the bacteriophage has entered the host cell, the ordinary metabolic replication of the host cell ceases and the infecting nucleic acid interacts to express a new metabolism, that reproducing the components of the entire virus. Ultimately, components sufficient for a large number of daughter virus particles, identical to that originally attaching itself to the cell wall, have been synthesized. The nucleic acid of the infecting virus, acting through the metabolic components of the host cell, provides chemical interactions to assemble the individual components. Once the large number of reproductions of the infecting virus have been assembled, the increase in bulk of non-host components filling the available region of the cell causes the cell wall to burst releasing the copies of virus particles to the surrounding medium. These daughter species are now simply macromolecular complexes, having no regeneration or reproduction possibilities until there is association with another bacterial cell.

Let us focus for a moment back on the host bacterial cell. Each original cell has the fundamental characteristics of a "living system". It undergoes both regeneration and reproduction. Suppose we consider again the experiment mentioned earlier with Escherichia coli in which an analogue of sn-glycerol 3-phosphate was added to the growth medium (see Fig. 1). In Fig. 6 we see the phenomenological result shown graphically. The cells enter a growth stasis phase owing to the change in the nature of the cell wall. The immediate capability of the cell for reproduction has been blocked. However, the individual cells remain, and undergo metabolic turnover. The "regeneration" of the cell continues, albeit with a modified rate. If we continue the growth curve to much longer times we find a "recovery" and growth (reproduction) again occurs.

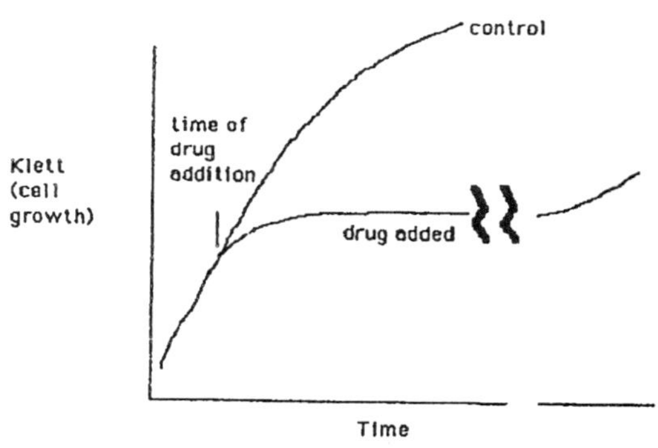

Fig 6 -Effect of (S) -3, 4-dihydroxybutyl-1-phosphonic acid of the growth of *Escherichia coli* cells[6]

If, in the process of treating the bacteria with the bacteriostatic agent, we also add another material, deoxycholate, a bile acid normally present with Escherichia coli existing in the wild, and normally without effect on it, we see a different result, as shown graphically in Fig. 7. Upon the treatment of the cells with both the bacteriostatic agent and deoxycholate there is a sharp decrease in the density of intact cells. That is, there has been "cell death". The modified cell walls undergo a physico-chemical interaction with the detergent deoxycholate, similar to the emulsification of oils by a soap, and the cell wall is destroyed. This disrupts the components within which normally occur the intermediary metabolic processes of the bacterium and regeneration can no longer continue. The reagents of metabolism are dispersed and destroyed, and the individual components no longer exist in a relationship which allows either regeneration or reproduction.

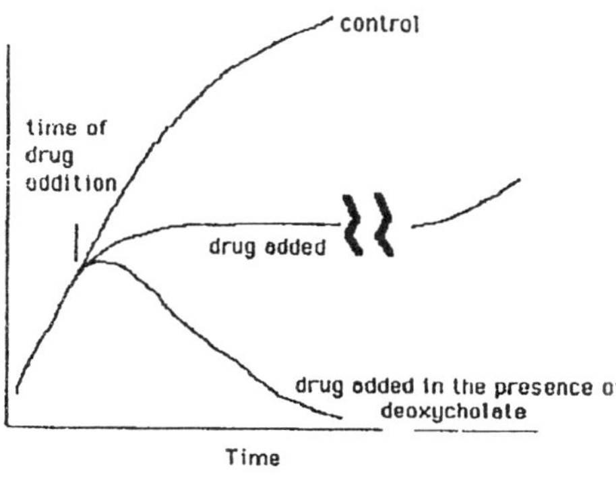

Fig. 7 - Effect of deoxycholate on the inhibited *Escherichia coli* system[7]

To a chemist looking at this or another living system, one sees a delimitation of the concept of life. When the chemical processes of dynamic metabolism are so disrupted or halted to prevent regeneration or reproduction of those specific processes, the system is no longer "living".

While chemists are to a certain extent concerned with the cessation of life on a molecular basis, there is a greater concern with understanding the nature of the normal continuing metabolic processes on a structural-functional basis. Each of the expressions of regeneration and reproduction in a living system is understood to have associated with it a series of highly specific phisico-chemical processes, the specificity of each depending on the molecular structure of the interacting components.

Finally, let us consider very briefly one particular aspect of these interactions which emphasizes the molecular structure aspect of living processes. Thus far we have been concerned only with bacterial systems. Let us now look at the existence of *pheromones* in more complex organisms. A pheromone is fundamentally a chemical, a specific organic compound, which elicits a particular phenomenological response from an organism. Introduction of a particular pheromone into the dynamic metabolism of an organism triggers one of its fundamental life processes. There are a wide variety of pheromones known for the entire range of organisms and various life functions.

We are most familiar with the sex pheromones of insects. For a number of reasons these have been the most extensively studied of the "triggering" agents.

For example, the compound shown in Fig. 8 is the sex pheromone of the silkworm. This is a compound produced by the female silkworm upon reaching a ready state for reproduction. There exists within the male a receptor for this specific molecule which upon interaction with the pheromone results in a mating response of the male. In the absence of this pheromone, the male does not participate in mating. In the absence of a female, but with the male exposed to this compound, it performs mating actions at the site of exposure.

(10E,12Z)-10,12-hexadecadien-1-ol

Fig. 8 - Sex pheromone of the silkworm

Another insect pheromone, one for a different life purpose, is shown in Fig. 9. Oleic acid is a common monounsaturated fatty acid, found in esterified form in virtually all organisms, and used in human food preparation.

Oleic - One of several "death pheromones" of the ant

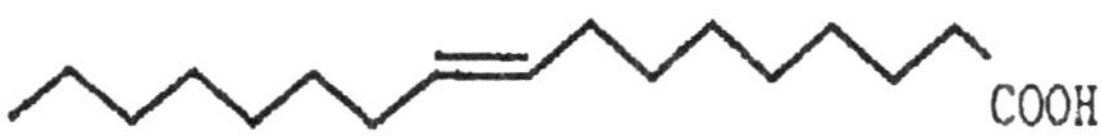

Elaidic acid - An isomer of oleic acid without effect on ant activities

Fig. 9 - Death pheromone of the ant

With ants, however, the free acid is exuded upon the death of the organism, death occurring by whatever cause. Other living ants have a receptor for oleic acid (which also serves for several other related compounds) and activation of that receptor triggers a response for the living ant to move physically the deceased out of the ant hill to a "refuse pile". If we contaminate other materials with oleic acid, even those normally sought as food items for the ant, they too will be moved to the refuse pile and abandoned. The isomer of oleic acid, shown at the bottom of Fig. 9, elicits no such response.

Organisms more developed than insects also act upon stimulation by specific organic compounds.

In Fig. 10 are shown two compounds which, when present together, stimulate aggressive behaviour in the male of the common mouse. This receptor in the mouse associated with aggressive behaviour is apparently more complex than those we have noted in insects. The two species must be present simultaneously to elicit the response. Exposure of the male mouse to this pair of compounds, in the absence of any macroscale aggression demanding situation, results in common aggressive behaviour.

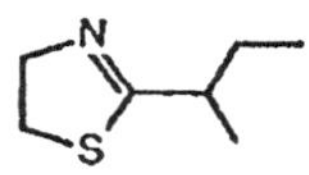

2-sec-butyl dihydrothiazole

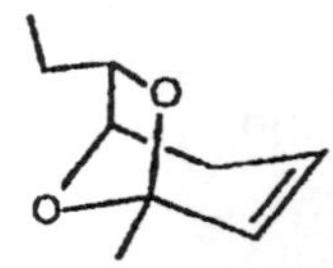

dehydro-exo-brevicomin

Fig. 10 - Male aggresion pheromone of *mus musculus*[8]

Pheromones are also observed to play a role in primate behaviour. In Fig. 11 is shown a group of six carboxylic acids of simple structure which are emitted by the ovulating female rhesus monkey. Upon exposure to these materials, a sexual mating response is elicited in the male rhesus monkey. Artificial exposure of the male to the combination of these compounds elicits a mating response in the absence of females.

CH_3CO_2H $CH_3CH_2CO_2H$ $(CH_3)_2CHCO_2H$

$CH_3CH_2CH_2CO_2H$ $(CH_3)_2CHCH_2CH_2CO_2H$

$(CH_3)_2CHCH_2CO_2H$

Fig. 11 - Pheromones of the rhesus monkey[9]

Finally, in Fig. 12 is shown a compound which appears to serve as a human pheromone. This material, present in human male saliva, appears to elicit sexual stimulation in the human female.

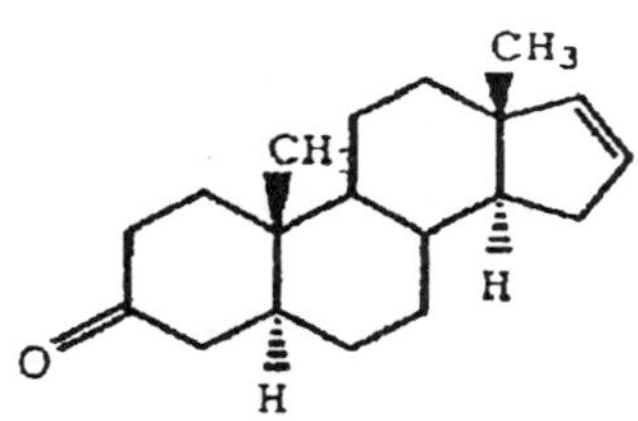

5-α-androst-16-en-3-one

Fig. 12 - Apparent human sex pheomone[10]

For a variety of reasons, not the least of which are the problems involved in performing properly controlled experiments, it is extremely difficult to explore the role of pheromones in humans. However, whether pheromones are involved or not, to a chemist looking at behaviour normally ascribed to "psychological factors", there is the understanding that every "life" action involves a series of particular chemical reactions with highly specific molecular structural requirements. While unable to specify these reactions or their molecular requirements for a given macroscale response, we understand it to exist, and thereby to be controllable.

Department of Chemistry
Queens College of the City University of New York, USA

BIBLIOGRAPHY

1.J.W. Payne, *Advances in Microbial Physiology*, 13, 55 (1976).

2.M. Sheikh, B. Gotlinsky, B.E. Tropp, R. Engel, and T. Parker, *ACS Symposium Series,* 171, 225 (1981).

3.G.Wilke,B.Bogdanovic, P. Heimbach, M. Kroner, and W. Muller, *Advances Chem. Ser.*, 34, 137 (1962).

4.R.C. Williams and H.W. Fisher, *An Electron Micrographic Atlas*, Charles C. Thomas, Springfield, IL, 1974.

5.S.E. Luria and J.E. Darnell, Jr., *General Virology,* 2nd Edition, John Wiley and Sons, Inc. New York, NY, 1968, p. 81.

6.C.S. Shopsis, W.D. Nunn, R. Engel, and B.E. Tropp,*Antimicrobial Agents and Chemotherapy*, 4, 467 (1973).

7.M.M. Connolly, R.Engel, and B.E. Tropp, unpublished results of this laboratory.

8.M. Novotny, S. Harvey, B. Jemiolo, and J. Alberts, *Proc. Nat.Acad. Sci. USA 82*, 2059 (1985).

9.R.P. Michael and D. Zumpe, *J. Endocrinol.*, 95, 189 (1982).

10.S. Bird and D.B. Gower, *Experientia*, 39, 790 (1983).

Christopher Cherniak

Meta-Neuroanatomy: The Myth of the Unbounded Mind/Brain

A picture of human cognitive resources as effectively unbounded pervades mind/brain science to a significant extent, in particular its most concrete level, neuroanatomy. Predominant models of brain structure appear to be profoundly non-quantitative in some respects, not quantitatively coherent. We will focus here on evaluating recent estimates of area of the human cortical sheet, estimates of synaptic densities there, and studies of giant axonic arborizations in the visual cortex. This examination in fact yields some information on actual available cortical connectivity resources that is presently of interest as a basic constraint on models of computation in the brain. Finally, some of the conceptual etiology of the non-quantitative character of brain anatomy will be explored. While the discussion somewhat improves estimates of cortical resources, emphasis will be at the level of philosophy and methodology of neuroanatomy, and on how they can productively shift perspectives that guide scientific practice.

1. «IMPOSSIBILITY ENGINES»

Of several distinct philosophical origins of current tacit ideas of an unbounded mind/brain, the most important may be the Cartesian concept of mind as non-spatial substance. Descartes (1960) gave his key argument in 1641: «I have a clear and distinct

E. Agazzi and A. Cordero (eds.), Philosophy and the Origin and Evolution of the Universe, 219–252.

idea of myself in so far as I am only a thinking and not an extended being.» Since «by 'body' I understand all that...can be located in some place and occupy space,» Descartes drew the dualist conclusion that his mind is not a body, or physical object located in space. Over the next three hundred years, Descartes and many others must then struggle with familiar puzzles of how mind and brain are supposed to be related. But besides this «myth of the ghost in the machine,» there is the prior Cartesian claim that the essential feature of mind is its *non-spatial* character. A second «myth» or tacit picture seems to continue to operate in even the most concrete reaches of mind/brain science, even among anti-dualists: something of the non-spatial character of the Cartesian mind is unthinkingly extended to the brain. (Such a tendency is perhaps only natural, once one opts for mind-brain identity.) The mind, and therefore the brain, cannot be numbered. To the extent that the brain is being tacitly conceived of not as a physical object located in three-dimensional space, its structures will tend not to be thought of as having spatial dimensions. Trends toward an unscaled, non-quantitative «Cartesian anatomy» are then all too understandable.

The pervasiveness of the idea that our cognitive resources have no bounds--in some cases, are actually infinite--is worth emphasizing. At least since the emergence of the formal theory of computation in the 1930's, three levels of abstraction in describing and explaining computational entities have been distinguished: (a) the highest level of the pure function or logical operation, (b) the level of the algorithm or software for executing a given operation, and (c) the most concrete level of the hardware for physically realizing a given algorithm. Interestingly, «impossibility engines» requiring quite unrealistic resources seem to turn up in the models at all levels. (See Table 1.) Much of the basic philosophical conception of a cognitive system consisting of belief and goal structures is drawn from the idea of man as perfectly rational animal with unlimited ability to process information. One of the most striking features of this rationality idealization, which has been tacitly incorporated into many of the models in philosophy, is that it entails the triviality of much of the deductive sciences. It was uncovering such logical impossibility engines in philosophical models of mind (Cherniak, 1986) that prompted the present author to wonder whether similarly extreme unrecognized idealizations might lurk at more concrete levels in cognitive science.

Level of Abstraction	Model
Function (Philosophy)	First-order logic decidability Tautology testability
Algorithm (Cog Psych, AI)	The mind's superprogram "Mentalese" lexicon Maximally interconnected network
Hardware	Neurophysiology "Granny cell" model of pattern recognition Neuroanatomy Area of the cortical sheet Synaptic density of cortex Giant arborizations in visual cortex

Table 1. "Impossibility engines" pervade mind/brain science. Models with unrealistic resource requirements seem to occur even at more concrete levels of explanation.

The next explanatory level of mind/brain science is that of specific algorithms for accomplishing a given operation. Cognitive psychology and artificial intelligence are concerned with identifying such procedures, whether actually employed by human beings or to be executed by machine. Yet the decrease in abstractness here does not seem to yield that much gain in real-world feasibility of the models. One can argue (Cherniak, 1988) that the fundamental framework of computational psychology, the very idea that much of human mentality can actually be represented by some huge superprogram, takes no account of the *size* of this «big text,» nor, consequently, of the issue of the profound unmanageability of such an enormous artifact for the cognitive scientist himself. Similar questions can be raised about tendencies in recent connectionist work on massively parallel and massively interconnected «neural» networks; some of the networks, if they were actually realized in nerve cells, would each require a brain the size of a bathtub (Cherniak, 1990).

The bottom level of concreteness concerns hardware specification; in the case of human cognition, of course, this is the physiology and anatomy of our brain. But, perhaps most surprisingly, when we turn to this most down-to-earth level, a parallel tacit picture of resource unboundedness seems to persist. This blind spot contrasts strongly with the insightful meticulousness that characterizes other aspects of neuroanatomy. The models rather frequently turn out still to be quite unrealistic quantitatively regarding available spatial and temporal resources--for instance, regarding volume budgets of the structures described. Shepherd (1979: 8) notes that Cajal's epochal work (1911), the Big Bang that initiated the modern era of neuroanatomy, included not a single scale marker for any diagram; Carpenter and Sutin (1983), a current standard medical textbook of neuroanatomy, is still sparse, only giving some magnifications for photomicrographs. The one-of-a-kind extended effort toward a quantitative atlas of the brain was Blinkov and Glezer's (1968) compilation, now out of print in English for over a decade. The discussion here will draw heavily upon its hundreds of tables. Concern with size might seem philistine, but a little more attention to *scale* seems worthwhile, in terms of some neuronal bean-counting to check the basic book-keeping. The methodology just consists of navigating a richly structured labrinth of quantitative constraints, a vast spreadsheet, of finding computationally straightforward routes through the maze to some interesting *a priori* anatomical conclusions. Let us calculate.

To sharpen the thesis here, it should be noted that a strong tradition certainly also exists in neuroanatomy of collecting quantitative measurements; the crucial oversight focussed upon below is, rather, at the higher level of simple consistency checking of these lower-level descriptive results for quantitative coherence. A further point in defining the present enterprise: Exhaustive historical review, analysis, and evaluation of the large primary literature of the last century of neuroanatomy on the three specific topics below would be an unwieldy task here. Instead of focussing on practical problems of technique, we pursue a division of labor: our primary objective concerns a methodological meta-issue--the pervasiveness of inattention to basic analytical questions of quantitative coherence in these key neuroanatomical studies. As will be seen in the next section, there is profound disagreement on even the most basic numerical estimates; hence any specific quantitative conclusions based upon them here can be regarded only as provisional starting points.

2. AREA OF THE CORTICAL SHEET

The human cerebral cortex is a complex, convoluted structure, but still, of course, a physical entity entirely located in three dimensional space--not some hyperdimensional Klein's Bottle. Since it is conventionally viewed as the seat of most higher cognitive processes, «cortical imperialism»--a tendency for research attention to focus upon it--is only natural. One of the most basic questions about the cortical sheet concerns its total flat surface area. Table 2 lists several influential estimates of this figure, all except Blinkov and Glezer's published within the last decade or so, many by major anatomists and physiologists.[1] The striking feature of the set of estimates, of course, is that they range over more than *an order of magnitude*. (If the estimate of 4 square meters, given to the author in a recent personal communication by a cerebellar anatomist at a major United States medical school were added to this table, the estimates would range over closer to a factor of 100. Nine square meters is the highest estimate offered by any of the neuroscientists so far queried.) A one square meter or greater estimate is not idiosyncratic; Hubel and Wiesel (1979: 150) had to correct themselves,

«In a 1963 article...one of us gave the area [of the cerebral cortex] as 20 square feet and was quickly corrected by a neuroanatomist friend in Toronto, who said he thought it was 1.5 square feet--'at least that is what Canadians have.'»

Cortex Area (mm^2)	Source
80,000	Rakic (1981: 3)
~120,000	Shepherd (1979: 221)
139,000	Hubel (1979: 150)
160,000	Blinkov & Glezer (1968: 172)
200,000	Colonnier (1981: 126)
220,000	Carpenter & Sutin (1983: 644)
307,500	Hofman (1983: 111)
400,000	Mountcastle (1978: 37)
1,000,000	Anderson & Hinton (1981: 32)

Table 2. The human cerebral cortex sheet: surface area estimates. Recent published figures range over more than one order of magnitude.[1]

It is as if higher-quality published estimates of the height of a normal human adult varied from 1 to 12 feet. A more apposite comparison might be early maps of the New World dating from the voyages of discovery; yet in fact calculations from Mercator's 1587 map of the world yield an area estimate for the North American land mass, despite some *terra incognita*, that appears to differ from current figures by less than a factor of two (Paullin, 1932: 14B). No doubt there are many explanations of such remarkable divergence in ideas about the size of the cortical sheet. Some of the techniques employed differ considerably. But it is also worth noting that both simpler estimations of normal cranial volume and more abstract norms for IQ do not show such extreme variability. Determining the area of a complex structure is a nontrivial task, but still seems feasible in an era, as the platitude goes, when science can land men on the moon. The role of the present discussion cannot be second-guessing in a highly technical field; however, a basic meta-neuroanatomical point still emerges, that there seems to have been little, if any, recognition or discussion in print of these striking disparities. As a bit of informal sociology of the field, a few dozen anatomists have been Socratically queried about these estimates; all expressed surprise about their wide range. Hence, the possibility emerges of considering and explaining this non-recognition in terms of some deep, tacit tendencies toward a way of viewing the brain.

For, the higher estimates of cortex area are quantitatively incoherent in a very concrete sense: As noted, there is no disagreement about normal cranial volume averaging a little below 1.4 liters, nor about cerebral cortex thickness averaging 2 mm or more (with mean thickness running lower at folds). But then

$$1{,}000{,}000 \text{ mm}^2 \text{ cortex area} \times 2 \text{ mm thickness} = 2{,}000{,}000 \text{ mm}^3.$$

That is, such a one square meter «all-cortex brain» alone requires around 2 liters, considerably more than the 1.4 liter total available brain volume. Not much refinement of such calculations is needed to exclude some of the lower cortical sheet estimates: A 400,000 mm^2 cortex 2mm thick requires 800,000 mm^3; on the assumption that all other brain structures combined required a similar volume (in fact, all other cerebral structures alone should require a corresponding volume), total brain volume would again exceed actual cranial volume. If volume of the cerebral hemispheres, minus estimates of the volume of white matter, major vascular system, thalamus and major nuclei is considered, the upper end of the cortical sheet estimates can be further constrained.

Therefore, the explanation for such very large and unacknowledged area discrepancies does not appear to be only, say: «gray-area» indefiniteness regarding what counts as cortical surface in the sulci, or differences in definition or criteria or measurement techniques for cortical surface (e.g., how allowance has been taken for shrinkage of tissue during fixation), or practical difficulty of the measurements, or differences in populations sampled, or in natural variation of cortex area as opposed to brain volume. Something more basic may also be involved, a way of seeing--or tending to fail to see--the cerebral cortex as part of a physical structure, the brain, with dimensional magnitudes that ought to cohere with each other. (The estimates reported in Blinkov and Glezer (1968: 172-3, 346, 374) of around 160,000 mm^2 appear the most likely to be in the correct range (Blinkov and Glezer do not note the diversity of estimates discussed here); they agree well with the recent measurements of Jouandet et al (1989: 95).)

Simple but pervasive quantitative consistency slips elsewhere in neuroanatomy contribute to the mystification of the cortex. Shepherd (1979: 347) had to write,

«It is a commonplace to cite an estimate of 10 billion neurons contained within the human neocortex as evidence for the fantastic overgrowth and therewith unprecedented capabilities for complex functions of this region. Those who cite this figure invariably fail to recall that the number of granulecells in the cerebellum is probably several times this number...»

It is as if there is some tension between cortical imperialism and the idea that the neurons in the seat of higher cognition could be outnumbered by the cell population of an area commonly pictured as merely mediating walking and chewing gum, so to speak. As another instance of flawed arithmetic that fits with the idea of an *idee fixe* of cortical imperialism, Nauta and Feirtag (1979: 96) state that neocortex «is estimated to contain no fewer than 70 percent of all the neurons in the central nervous system» of man and other primates. Shepherd's discussion of the cerebellar granule cells makes clear that, even qualitatively, this is not a fair claim. Individually, cases like these illustrate only the uninteresting point that nobody and nothing (including, inevitably, the present discussion) is perfect; together, they begin to suggest an intriguing pattern of higher-level nonquantitativeness.

3. MICRO-CONNECTIVITY RESOURCES

The examples occupying the rest of this paper focus on small-scale cortical connectivity. Some of the most interesting directions of recent investigation in mind/brain science involve exploration of models of massively parallel and massively interconnected computation in the brain (see, e.g., *Cognitive Science* (1985), and Rumelhart and McClelland (1986)). Attention has only just begun to turn to the fundamental question of the actual neural realizability of these avowedly neural-metaphor models. As mentioned earlier, a number of the connectionist constructs seem in fact to be cognitive impossibility engines. Surely a quantitatively coherent neuroanatomy becomes crucial here; the above inconsistencies in simple estimates of cortical sheet area do not bode well for current approaches to more complex anatomical questions. The most local connectivity constraint is synaptic density, particularly in cerebral cortex. What are the available cortical micro-connectivity resources? Again to shift the perspective, a truism contrary to Cartesian anatomy: Real synapses are not infinitesimal, dimensionless geometrical points or junctions; they each have a small but finite volume. They are not «everywhere-dense» (in the topological sense that, between any two points on the real number line, another can always be interpolated), nor comparable to the dimensionless spheres representing molecules in the classical ideal gas laws. Actual synapses have a maximum packing density. We need to check how many synapses can fit on the head of a pin.

What is the volume of a synapse? It is easier to raise such a question than to settle it. Synapse structure epitomizes the stunning diversity of the nervous system; one can get a sense of the brain as Ultimate Rorschach Blot, with no typical or representative cases, few perceivable central tendencies. Synaptic junctions--often not well- defined--include not only end enlargements in the form of knobs, but also spines, claws, mossy terminals, and many other intricate shapes. To constrain this diversity somewhat, we will tend to concentrate on primate primary visual cortex. In addition, we will proceed by «qualitative arithmetic,» using the minimum computational force necessary, where we seek only rough approximations of the volumes of the complex structures involved. That will suffice for some simple bookkeeping review.

Cragg's (1967) electron-microscope study of cortical synapses yielded average diameters of the order of 0.6 um for mouse and monkey visual and motor cortex, with monkey visual cortex synapse mean diameter of 0.72 um. (Cragg reported there E. G. Gray's similar unpublished results, that rat visual cortex synapses had an average profile diameter of 0.73 um.) In Cragg (1975), one of the most often cited electron microscopy studies of human cortex synapse density, the dimensions of the synapse structures (and synaptic appositions alone) in the plates and tables seem to agree with this quantitative picture. These dimensions are also confirmed by Gray (1959), and Peters, Palay, and Webster's ultrastructure atlas (1970). (For recent studies supporting these results, see also Colonnier's review (1981) of synaptic types in cat and rat neocortex, and Peters' review (1985) of rat visual cortex synapse anatomy.)

Therefore, visual cortex synapse volume--including both pre- and post-synaptic structures--by itself should fall at least in the 0.2 to 0.4 um^3 range. Measurements from Valverde's (1985) remarkably vivid Golgi-method camera lucida drawings of macaque primary visual cortex suggest as much as a mean one cubic micron per synapse estimate. Similarly, in their horseradish peroxidase studies of macaque striate cortex, Blasdel and Lund (1983) report 0.5 to 2 um diameter boutons. (Carpenter and Sutin's (1983: 119) single choice for a micrograph of a presumably typical axodendritic synapse from human cerebral cortex yields a volume in excess of 2 um^3.) It should be emphasized that estimates of overall synaptic volume overhead must include not only the volume of the junction structures themselves in isolation, but in addition the volume of other cell structures that are dedicated exclusively to a particular synapse. For example, one can calculate the average length of fiber associated uniquely with each synapse of a given type; Valverde's (1985) camera lucida drawings suggest a figure of as high as five microns of dendrite and/or axon (see also Blasdel and Lund, 1983: 1398, 1407) per synapse for macaque visual cortex. We can thus suppose that the volume of each synapse plus its uniquely associated structures will average around 1 um^3.

A table of synaptic density estimates can be assembled that begins to resemble the set of cortical sheet estimates of Table 2. Most importantly, there is again the danger signal of estimates ranging over more than an order of magnitude. Toward one extreme, Hubel (1979: 45-6) gives total human neuron and synapse populations that entail roughly 1,000 synapses/neuron; this is for the entire brain, but therefore also includes the extremely high synapse densities observed for cerebellum. Toward the other extreme, Cragg (1975: 85) reported 39,000 synapses/human cortical neuron in frontal, temporal, and parietal areas.[2] In between fall such estimates as Colonnier's (1981: 127) of 15,000 - 30,000 synapses/cortical neuron. Ten thousand synapses/neuron is a rather typical claim for cerebral cortex. Of course, complexity of synaptic structure, and its variability for different human brain regions, makes comparison of different synapse density estimates much trickier than comparison of cortical area estimates.[3]

Nonetheless, some natural questions of quantitative coherence emerge. (See Table 3 for a summary of the following calculations.) To begin, what is the total volume of the average cortical neuron--synapses, dendrites, soma, axons, and all? This crucial constraint on modelling of real-world cortical connectivity is not very readily available. Despite difficulties of generalizing across vast diversity, there seems to be rather good agreement that mean human cortical neuron density is about 50,000 neurons/mm^3 of cortical tissue. The range is from 30,000 neurons/mm^3 for motor cortex (Brodmann area 4), to over 100,000/mm^3 for primary visual cortex, a variation of a factor of 3 (e.g., Blinkov and Glezer (1968: 203, 398), which includes data from Shariff (1953); see also Pope (1978: 16), Peters (1987: 269, 273), and O'Kusky and Colonnier (1982)). Therefore, mean available volume per human cortical neuron would be around 20,000 um^3.

Mean density: ~50,000 neurons/mm3 tissue. Hence, total available volume/neuron = ~20,000 um^3.
Fine-scale vascular system of cortex: ~5% total volume. Extracellular spaces in cortex: ~20%. So, remaining cortical volume: ~75%.
Volume of neurons < volume of glial cells. Hence, neuron volume = ~30% total cortical volume.
Therefore, actual mean volume/neuron = ~6,000 um^3.
Mean soma volume estimate: ~1,600 um^3. So volume of dendrites, axons, and synapses of "typical" cortical neuron = ~4,400 um^3.

Table 3. estimating the volume-budget of an average human cortical neuron: soma, dendrites, axons, and synapses. See text for explanation and references.

Of course, an all-neuron brain would be another impossibility engine. The brain is commonly characterized as over ten times more metabolically active than the mean rate for the human body.

Neurons cannot be dense-packed as a pure matrix, burning immaterial ectoplasm; they require a vascular system, glial cells, and extracellular spaces. As a start on a large issue, let us attempt an estimate of this volume overhead. Blinkov and Glezer (1968: 252, 267) cite measurements that imply the intracerebral vasocapillary system alone (as distinct from larger intracerebral arteries and veins) occupies about 7% of the volume of gray matter. (Cortical capillaries have a internal diameter of about 7 um, with a volume 2 - 3 times larger than in the white matter. These terminal vessels are supplied by intracerebral arteries with diameters of 8 - 20 um, in turn fed by radial arteries with a diameter of 10 - 42 um that sink into the cortex at intervals of around 500 um.) Then there is a corresponding venous system (we exclude extracerebral vessels). Let us suppose that the finer-scale vascular system of the mature cortex by itself could occupy 5% of its volume (Blinkov and Glezer (1968: 287) cite a quite high estimate of 30% for volume of vessels in the cortex).

On extracellular spaces in adult cerebral cortex: Carpenter and Sutin (1983: 22, 134) note that estimates by electron microscopy run lower than estimates from neurochemical techniques, but give a range of recent values consistent with 10% or greater volume. After reviewing conflicts between different techniques, Pappius (1982) concludes that a consensus has emerged, that interstitial space in cerebral cortex tissue constitutes 15 - 25% of total tissue volume. We will assume 20% volume. On glia: Blinkov and Glezer (1968: 239, 252) concluded that neurons occupy a smaller proportion of brain tissue than neuroglial cells and their processes; Carpenter and Sutin (1983: 135) state that mammalian neuroglial cells alone may comprise almost half of total brain volume. With 5% vessel volume and 20% extracellular volume, 75% of brain volume would remain available for neurons and glial cells, which suggests neurons might occupy less than 40% of total brain volume. In fact, ion-uptake studies of Vernadakis (1986: 397) give the neuronal space for adult rat cortex as 35%, which turns out to agree fairly well with Pope's (1978: 18) estimation of the neuronal compartment of human neocortex at 35 - 40% of the fresh volume. Adjusting the latter estimate in light of Vernadakis' study, plus the hypothesis that species with larger brains have a larger glial volume proportion (Tower, 1973, 1978), let us suppose as an approximation that neurons constitute about 30% of adult human cortical volume.

With 20,000 um^3 of cortical volume available per neuron, it would follow that the actual mean volume of a human cortical neuron falls in the range of 6,000 um^3. Of course, within this average value remains room for enormous variance. For instance, Blinkov and Glezer (1968: 387) give calculations of the volume of cell bodies alone of pyramidal cells from seven Brodmann areas that range from 297 um^3 up to 2953 um^3, about an order of magnitude of variation. (Calculations from data compiled by Blinkov and Glezer (1968: 394, 396) suggest mean volume of the soma only of cortical neurons in general might be around 1660 um^3.) A basic coherence question for most customary synaptic density estimates can now be glimpsed. Even just 10,000 synapses, each requiring 1 um^3 volume, per average neuron would result in something worse than an «all-synapse cortex,» for there simply does not seem to be enough room in the cortex for 10,000 um^3 of synapses per neuron, even if all the rest of the neurons' structure were crowded out entirely. The main conclusion here is that, *a priori*, only the lower range of synaptic density estimates for human cerebral cortex appear to be realistic candidates. Once the above soma volume estimate is subtracted from the mean total neuron volume, less than 4,400 um^3 remains for dendrites, axons, and specifically synaptic structures. Therefore, it would be hard to explain a mean synaptic density per cortical neuron that ran much above 4,000. At 4,000 synapses per neuron, hardly any volume would remain for dendrite or axon trunks.

4. LOCAL AND MID-RANGE CONNECTIVITY

The next stage in a connectivity analysis of the cortex is to proceed to extra-synaptic but still relatively local connections--say, in the 5 to 1,000 um distance range from a given neuron. And a natural first step is to ask, what is the maximum connectivity capacity of a single average cortical neuron? In particular, what would be the total length of dendritic and axonic fiber available for an average-volume neuron? If a neuron's processes were infinitely thin wires, to go with geometrical points as synapses, its connectivity could be unlimited; but that is only another impossibility engine. We need first to find out some mean values for diameters of feasible fibers.

Of course, there is a great range. (a) Generally, considering the entire nervous system: Shepherd (1979: 343) describes apical dendrite trunks of sensory cortex pyramidal cells, at 5 - 10 um diameter, as among the thickest dendrites in the nervous system. At the other extreme, thin axons can be found in the cerebellum, such as unmyelinated parallel fibers of granule cells, with a smallest diameter of only 0.2 um, gradually increasing to 1 um (1979: 219); or again, unmyelinated afferent axons of receptor cells in the olfactory mucosa are «extremely thin,» approximately 0.1 - 0.3 um in diameter (1979: 153). Blinkov and Glezer (1968: 170) cite studies of long-range fiber densities in human corpus callosum and anterior commissure that indicate a mean cross-section of 1 um diameter or more, and a recent study of monkey cerebral commissures by Lamantia and Rakic (1990) reveals fine axon diameters of about 0.3 - 0.7 um. (See also Feldman, 1984: 161; Bullock and Horridge, 1965: 150; and Fulton, 1949: 323.) Hillman reports mean minimum diameter attained by dendrites, across a range of cell types, as 0.76 to 1.1 um (1979: 480). (b) Specifically for the primate visual system: Blasdel and Lund found that afferent axon trunk diameters in the white matter of macaque primary visual cortex range between 2 and 5 um (1983: 1400). The very smallest processes in the visual cortex, in the vicinity of synaptic terminals, generally might not go much below 0.3 um diameter; see the high-magnification light micrographs in Shkol'nik-Yarros (1971: ch. 2) for rabbit, cat, dog, monkey, and man (no correction appears to have been made for tissue-fixation shrinkage). For electron micrographs of rat cortex, see Gray (1959) and Peters, Palay, and Webster (1970). Valverde's (1985) camera lucida drawings suggest a rough estimate of near 1 um mean diameter for macaque visual cortex longer-range fibers.

A fairly safe, as well as computationally convenient, guess therefore might be to begin by exploring the supposition that *average* fiber cross-section area in adult visual cortex (areas 17, 18, and 19), with axon generally smaller than dendrite, approaches the 1 um^2 neighborhood. With about 4,000 um^3 available for the dendrites, axons, and synapses of the average cortical neuron, it would follow that such a neuron could have no more than 4,000 um of 1 um^2 fiber, or 6,000 um of 0.75 um^2 fiber. The fact that there might be only half a centimeter or so of intracortical connectivity per neuron in turn has other anatomically interesting implications.

The mid-scale range of cortical connectivity is 1 - 10 mm, still within one Brodmann area. According to the classical picture, the most salient structural feature of the cortex with respect to connectivity is extremely limited «horizontal» connections within the cortex beyond about 1 mm length. Anatomical work of Lorente de No (1949) suggested a predominantly vertical organization of fibers in neocortex, from layer to layer. Physiological studies by Mountcastle (1957) for motor cortex and Hubel and Wiesel (1962) for visual cortex added the idea of a columnar cortical organization, with segregated cylinders or arrays of neurons less than 1 mm diameter having common functional roles, such as edge-orientation detection. What emerges, then, is the prospect of most mid-range cortical connectivity having to proceed relatively indirectly via the deeper white matter, with a potential for some puzzling connectivity bottlenecks there.

However, more recent studies (Gilbert and Wiesel, 1983; Rockland and Lund, 1983; Blasdel and Lund, 1983) with intracellular injection of horseradish peroxidase (HRP) reveal a different picture, of intrinsic horizontal cortical connections over long distances, up to 4 mm or more. (In fact, considerably earlier, Sholl (1955), using conventional Golgi techniques, had reported tangential fibers in cat visual cortex that extend for «several millimeters.») From the perspective of a connectivity analysis of cortex, a basic issue is the density of these longer-range connections. Can they provide a relatively major proportion of cortical connectivity, or is their role apparently less significant? Gilbert and Wiesel (1983: 1116, 1118) provide some information: They report neurons that form widespread clustered axonal projections occur in all layers of area 17 of cat cortex, with at least half of the pyramidal and spiny stellate cells studied of this giant-arbor type. Pyramidal and spiny stellate cells together constitute the preponderance of cortical neurons. In effect, giant-arbor neurons are therefore observed to be distributed on the two orthogonal dimensions of cortical layer and cell type.

The quantitative question here is, what is the maximum feasible density in human primary visual cortex for longer range axonal arborizations? What is the highest proportion of area 17 neurons that could have such long connections? (Table 4 summarizes the following calculations.) The half-dozen cells shown in Gilbert and Wiesel (1983) have axon arbors alone with mean total length exceeding 20 mm. (In addition, it seems a presently open question whether even current HRP techniques typically fill more than, say, about 80% of a neuron's complete giant arborization.) Starting with the earlier idea of an average 1 um^2 fiber cross-section, the giant axon arbor by itself of such a cell would occupy an average volume of over 20,000 um^3. As discussed above, neuron packing density in area 17 is the highest in the human cortex, over 100,000 cells per mm^3; our 6,000 um^3 cell volume estimate was for the all-cortex mean of 50,000 cells per mm^3, and so would be lowered for average area 17 neurons to less than 3,000 um^3. (Glial cell counts/mm^3 (Blinkov and Glezer, 1968: 416, 420) and capillary lengths/mm^3 (Blinkov and Glezer, 1968: 432, 433, 436) for different cortical areas are consistent with a supposition that glial and vascular overhead in area 17 is comparable to other areas of the cortex.) After soma and dendrite arbors are taken into account, the average axonal connectivity available per human neuron would fall below 2,000 um, around a tenth of the giant axon arborization mean. Also, area 17 tissue must include afferent axons from the lateral geniculate nucleus and elsewhere. This suggests that if even one tenth of the primary visual cortex neuron population had such giant arborizations, no volume whatever would remain for axons of the other 90% of the cells. Thus, on the above argument, giant axon arborizations could occur in only a proportion of the cell population below 10%. If the actual proportion of giant-arbor neurons numbered anything like the reported 50%, their dense web would entirely dominate the anatomy of the primary visual cortex--yielding, in effect, an «all-axon cortex.»[4] This seems to open questions about the role of the giant axon arborizations in visual processing.

Density in area 17 > 100,000 neurons/mm^3.
yields mean volume/area 17 neuron <3,000um^3 Less soma and dendrite volume (and afferent arbors)'
For mean 1 um^3 cross-section fiber,[4] yields ~2,000 um of axon length for average area 17 neuron.
On these assumptions, if even 10% of area 17 neurons had giant arbors of 20 mm total length, no volume would remain for axons of the other 90% of neurons.

Table 4. Estimating the total gray matter connectivity-budget of an average neuron in human primary visual cortex. See text for references.

A rather elementary sampling model would explain high observed ratios of neurons with large arborizations. Conventional HRP procedure is first to insert a micropipette electrode into the cortical area being studied until a single cell is penetrated, as indicated by recorded potentials; the cell's receptive field properties are then electrophysiologically mapped (the great majority of striate cortex neurons appear to have classically characterizable receptive fields, so there is not much electrophysiological prescreening), after which the cell is injected with HRP. The probability of such a virtually random microelectrode penetration intersecting a cell of a given size would be a function of the area of the profile of the neuron in the plane perpendicular to the microelectrode track. In this way, a population of giant neurons with, for example, a mean ten times the profile area of the average neuron would, other things being equal, be ten times more likely to be randomly sampled. In fact, this skewness could be affected by the character of the typical distribution of a given cell type's processes--e.g., whether they are densely compacted and overlapping or spread out (as the giant arborizations are). It should also be noted that finer fibers are less likely to be successfully impaled by a micropipette; this will reduce the effective profile area of the giant axonic arborizations. (However, it also appears likely that these cells have larger than average cell bodies.) Thus, with the observed ratio of electrophysiologically characterized giant-arbor neurons of 50%, and a proposed sampling correction of around 10:1, the estimated actual ratio of giant cells would turn out to be only in the range of 5% of the total population, yielding, of course, a much more feasible volume of these cells.

In interpreting this quantitative discussion, it is interesting to consider some lesion experiments by Sperry and coworkers, in which monkey motor cortex (Sperry, 1947) and cat occipital-parietal cortex (Sperry and Miner, 1955; Sperry, Miner, and Myers, 1955) were cut vertically in a cross-hatch pattern. These studies appear to have stood the test of time so far (e.g., Berlucci and Sprague, 1981: 416)). In cat visual cortex, the slices extended 2 mm deep and so severed intracortical connections while leaving underlying white matter tracts largely intact. (Postmortem micrographs indicated no regrowth of intracortical fibers; in some cases mica «insulating» strips had been inserted.) The lesions formed an approximate grid of less than 1 mm spacing--as it turns out, roughly columnar dimensions. It is as if an arbitrary and «impermeable» columnar structure had been imposed on primary and association visual cortex. Yet, after recovery, the cats retained the ability to make fine and complex global visual pattern discriminations at close to the upper limits of their original preoperative capacity. Complete ablation of these cortical areas does destroy these discrimination abilities, so these «micro-ablation» experiments suggest that longer-range direct intracortical connections--such as those revealed in the recent HRP studies--may in fact not be necessary at all for high-level visual performance. There is thus some convergence with the above quantitative analysis of the giant arborizations.

The final stage of a connectivity analysis of the cortex concerns long-range fibers; the mean 170 mm length x 140 mm width of the total human cerebrum (Blinkov and Glezer, 1968: 109) gives some idea of maximum possible connection lengths. Here attention must turn from traditional cortical imperialism to constraints imposed by the cerebral white matter. Long-range connections each cost more in volume than shorter-range ones, and so have to be particularly sparse. An analysis of global connectivity bottlenecks requires, as well as the above quantitative results on local connectivity, some of the formalisms of combinatorial network optimization from graph theory and computational complexity theory, and can be found in another study (Cherniak, 1990).

5. ETIOLOGY OF CARTESIAN ANATOMY

The above discussion suggests that when impossibility engines turn up in neuroanatomy, they exceed actual available resources by about an order of magnitude or two. Relative to higher-level models, this is a moderate overshoot--one of Moliere's «small babies»--but nonetheless an impressive one, given the unavoidable concreteness of anatomy, as opposed to, say, philosophy of mind. Figure 1 impressionistically summarizes resource demands of cognitive impossibility engines at different levels of mind/brain science. The more concrete the model, the less extreme the idealization of resources; but tacit idealization continues, with required resources generally over-estimated, not under-estimated.

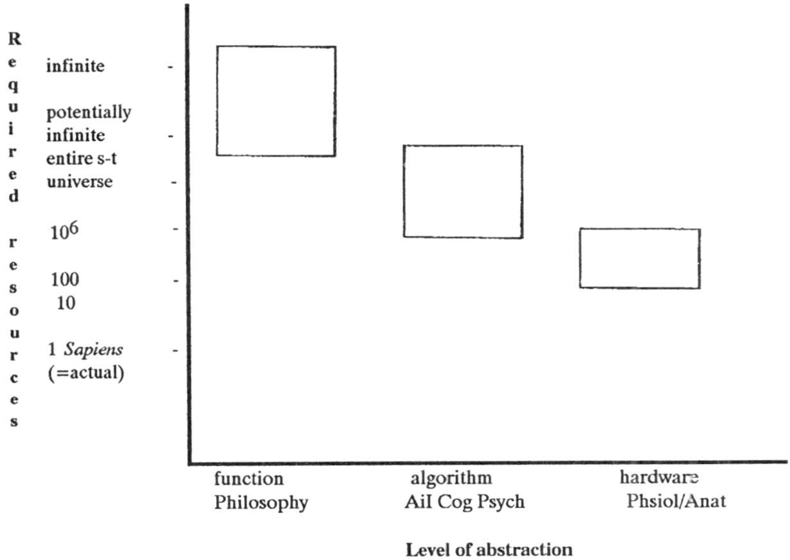

Figure 1. Required resources as a function of level of abstraction. The more concrete the model, the less extreme the resource overestimation; but tacit idealization continues. ("1 *Sapiens*" represents the actual avaible cognitive resources, at any level of description, of one normal human being.)

The picture of prevalent quantitatively incoherent neuroanatomy sketched here seems sufficiently puzzling to invite questions regarding its proper interpretation. In making sense of scientific practice, it is natural, and proper, to appeal to some «principle of (moderate) charity»: A methodological account of the behavior should make it at least minimally reasonable from the practitioner's perspective. Culture-confrontations can occur closer to home than just Whorf contemplating the Hopis--here, in the methodology of a science. Thus, it is no explanatory digression to inquire into the etiology of the idea of brain as impossibility engine. There are, of course, important reasons for the lack of higher-level consistency analysis that arise from actual practice--

for instance, techniques diverging so widely that comparisons become difficult. However, we focus here instead upon some of the theoretical presuppositions that also seem to drive this pervasive inattention to quantitative coherence checking; the rationale for concentrating upon the latter etiology is that sometimes, basic shifts of conceptual perspective can lead to posing usefully different types of internal questions. Table 5 lists several distinct possible philosophical origins of quantitative incoherence here.

1. Cartesian dualism. Mind, therefore brain, is a non-spatial substance.
2. Reifying idealizations. Real neurons are hardware-independent Turing machines, Pitts-McCulloch "neurons".
3. "Qualitative heuristic." To simplify the brain, discard quantitative information.
4. Brain as monad. The brain is a micro-cosmos, comparable to the entire universe.
5. Brain-worship. The last bastion of romanticism/humanism is the quantitative uniqueness of the human species.

Table 5. Etiology of the idea of brain as impossibility engine.

(1) The picture of an unbounded mind/brain was introduced earlier as joining «the myth of the ghost in the machine,» and indeed perhaps arising by a kind of cognitive momentum from Cartesian mind-brain dualism. Hence, a so-called «Cartesian

anatomy» of the brain, on the model of the mind, as an only quasi-spatial substance, with dimensions--and questions of their mutual consistency--lacking or tending to be overlooked.

(2) The notion of an algorithm as something like an abstract mathematical structure, hence without spatial location or magnitudes, can also carry over to the conception of brain as algorithm-follower. In this way, another source of the picture of cognitive resource unboundedness is the idealizations of classical computation theory, that great success story of our century: A basic unifying framework of cognitive science is computational psychology--mind conceived of on the model of hardware-independent Turing machine, with potentially infinite available time and space. A tendency to leap to a corresponding tacit notion of the brain is then perhaps understandable. The main insight of computational psychology, that program, not physical realization, is supposed to be the basic structure, the essence of mind, would thereby reinforce predilections to overlook quantitative anatomy. When, for example, the classical Pitts-McCulloch «neuron» idealization is overextended and reified, taking on a life of its own--with dimensionless, infinitesimal cell bodies and synapses, one-dimensional axons and dendrites (see Figure 2)--its «anatomy» does indeed suggest unlimited spatial resources.

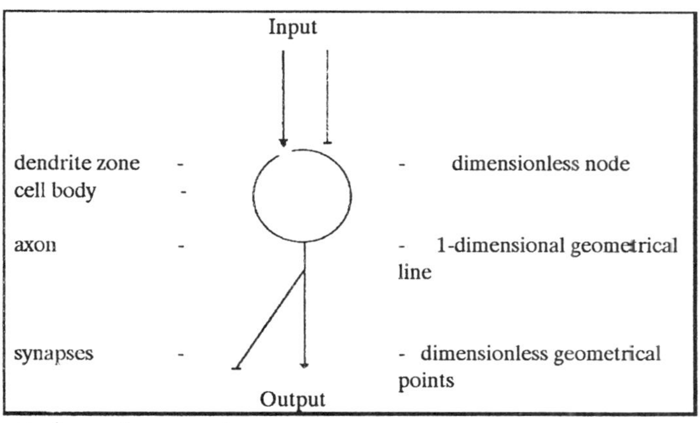

Figure 2. "Anatomy" of the PittsMcCulloch neuron idealization.

(3) Yet another diagnosis of the source of a quantitatively incoherent anatomy might be characterized as the *qualitative heuristic*, a theoretician's quick and dirty strategy of simplification by retaining basic qualitative proportions, but intentionally avoiding and discarding quantitative information. Given the distinctive--indeed, putatively unrivalled--complexity of mind-brain phenomena, the drive toward this simplifying strategy should be uniquely powerful here. The remarkably ubiquitous human tendency to reject event base rate information in probabilistic reasoning, extensively examined by Tversky and Kahneman (1982), for example in their studies of the «representativeness» and «vividness» heuristics, seems a special case of this qualitative heuristic. Another would be the human tendency to simplify quantitatively by «thinking linearly»--that is, treating exponential curves as if they were the more cognitively manageable straight lines of linear functions. We thereby arrive at impossibility engines through our failures to perceive threats of exponential explosion, and limits to their growth. Notwithstanding any cries here to «Bring out the calipers,» sound rationales for some use of the qualitative heuristic must be emphasized--in fact, the methodology of qualitative arithmetic employed throughout the quantitative anatomy of the present discussion is just another instance of this heuristic. After the non-quantitativeness of the standard unbounded-resource models, there is nowhere to go but up. Or down.

(4) Attention to matters of bookkeeping in neuroanatomy can seem mean-spirited or beside the point when the human brain is viewed as micro-cosmos or monad, reflecting--indeed, comparable to--the entire universe. Neuroscientists often assert that our brain (setting aside that of the elephant or whale) is the most complex natural physical structure we now know of in the universe. They also like to point out that, while our brain has 100 billion neurons, our galaxy has about the same number of stars (and the entire known universe contains about the same number of galaxies). Equating brain and universe is the explicit image in an 1862 poem of Emily Dickinson (1965: 632), e.g.: «The Brain--is

wider than the Sky-- / For--put them side by side-- / The one the other will contain / With ease--and you--beside.» If the whole world can be discerned in a grain of sand, seeing it in the brain is a snap.

Toward a purely quantitative demystification of the brain, one can reply that, as noted earlier, the cerebellum--concerned with posture, muscle tone, etc.--contains at least 4 times the neuron population of the cerebral cortex, putative seat of higher mental function. Similarly, the lymphocytes alone of the human circulatory system outnumber the neurons of our nervous system by at least an order of magnitude (Jerne, 1985). Indeed, a few tablespoons of yogurt contain more than 100 billion lactobacillus bacteria. Furthermore, the apparent information-representation capacity of the cortex is in fact by no means unrivalled, much less unbounded (Cherniak, 1988). On the usual assumption that the synapse is the necessary substrate of memory, supposing very roughly that (given anatomical and physiological «noise») each synapse encodes about one binary bit of information, and a thousand synapses per neuron are available for this task: 10^{10} cortical neurons x 10^3 synapses = 10^{13} bits of arbitrary information (1.25 terabytes) could be stored in the cerebral cortex. But the Library of Congress contains 80 million volumes, which (with an average book of 300 typed alphanumeric pages) is:

$$8 \times 10^7 \text{ volumes} \times 300 \text{ pages} \times 16 \times 10^3 \text{ bits/page} = 3.84 \times 10^{14} \text{ bits},$$

i.e., 48 terabytes. On this type of rough estimate, the cerebral cortex could not even contain all the information in the 25 million volume Lenin Library. Such comparisons have an unavoidable ring of silliness, but perhaps also a useful function in at least drawing attention to the conventional tacit «magic meat» perspective on the human brain.

(5) Finally, we must turn to a motivation for the idea of an unconstrained mind/brain that appears to originate in concerns about the relation between the scientific world view and human

values. The issue can be couched as the question of what is the essence of personhood or humanity, traditional answers including: rationality, the capacity for language, a type of functional/computational organization. We can now add another candidate for an answer, that human beings are quantitatively distinct--it is our unbounded complexity of mind and brain that is unique, that puts us on top. What is at stake here is not only the analytic matter of what it is to be a person, but also what it is that gives persons any distinctive moral worth. This is not just species snobbery, drawing a chalk circle around our own kind. Rather, the tacit «what a piece of work is man» agenda seems to be preserving our distinctive moral value in the face of a perceived encroachment of the naturalistic view of the universe and our species' place in it. The real danger is of treating people as no more than just another type of thing, itself an all-too-human predilection. In such a context, brain-worship through quantitatively romanticizing the brain is understandable and in a sense admirable: the problem of the human mind's place in nature is solved by making the brain into microcosmos, comparable to the entire universe. But even as subtext, this still seems incorrect; the puzzle remains of finding sound bases for humanity towards humanity.

***

The above discussion has attempted to show how a type of methodological self-awareness--a bounded-resource philosophical framework--can shift point of view so that different internal neuroscientific issues are considered. A quantitatively self- conscious perspective suggests, surprisingly, that even the most concrete level of mind-brain science, neuroanatomy, like more abstract explanatory levels, needs to focus more attention on questions of quantitative coherence. Analysis of published figures has also in fact yielded a few approximate estimates--with ample but indefinite margins of uncertainty--relating to human cortical connectivity resources which constitute basic constraints on models of massively parallel and interconnected computation in the brain:

(a) Total cortical sheet area falls in the 100,000 - 200,000 mm^2 range. (b) At 50,000 neurons/mm^3, mean cortical synapse density should not run higher than about 4,000/neuron; at 4,000/neuron, an average of approximately 200 million synapses/mm^3 of cortex would be available. (c) The mean total gray-matter connectivity available per cortical neuron will be around half a centimeter of dendrite and/or axon--yielding about 1/4 km of connections/mm^3 of cortex. As emphasized earlier, these estimates can only be a provisional starting point. In this way, some redefinition of research agenda emerges, toward a quantitatively coherent, and therefore computational, neuroanatomy.

Philosophy Department and Institute for Advanced Computer Studies, University of Maryland, USA

Acknowledgements

Persa Batra, William Gasarch, Lydia Goehr, William Hodos, Ayub Ommaya, James Reggia, William Saidel, and Gordon Shepherd helped generously with research for this paper, which was supported in part by fellowships from the University of Maryland Arts and Humanities Research Center and the University of Maryland Institute for Advanced Computer Studies. Presentation of this paper was supported by N.S.F. Grant: INT-8905411.

Notes

1. The figures in Rakic (1981) and Mountcastle (1978) are for neocortex only; Blinkov and Glezer (1968: 374) give neocortex as occupying over 95% of the total cerebral cortex surface.

2. Cragg's influential estimate of cortical synapse density is so high that it begins to approach some of the synapse densities cited for cerebellar Purkinje cells, among the highest in the human nervous system. Some conjecture seems in order regarding just what is the source of the apparent internal disparity between Cragg's synapse size and density estimates. One part of a possible explanation is the fact that Cragg's (1975: 85) mean neuron density of 15,600 neurons per mm^3 of normal cortex runs about 2 to 4 times lower than many conventional estimates (e.g., Blinkov and Glezer, 1968: 398). However, even if neuron volume is adjusted upward for Cragg's neuron density range, his synapse density still seems to leave no volume at all for the rest of the neuron. (Cragg (1975) also unaccountably gives somewhat different synapse densities at pp. 85 and 88.)

3. Uncertainties about the extent of tissue fixation shrinkage, and whether or not it has been taken into account, contribute considerably to the divergence of synapse density estimates. For instance, for tissue fixation techniques for light microscopy, cited estimates of volume reduction range from 45% to 83% (Cragg, 1967: 639), (Blinkov and Glezer, 1968: 17), (Konigsmark, 1970: 318). See also O'Kusky and Colonnier (1982: 281, 286). There are also questions concerning non-uniformity of shrinkage, and the extent of shrinkage of specimens for electron microscopy.

4. As a perturbation analysis, the maximum feasible population of giant arbor neurons should be recalculated for axon diameter ranges that are significantly finer than cortical fiber diameters reviewed above. If *mean* diameter were assumed to be around 0.5 um, yielding a cross-section area of 0.25 um^2 then if 50% of the population of neurons had these slimmer giant arborizations, still no volume («negative volume» actually) would be left for the axons of any other neurons. Even for this different regime of diameter assumptions, a puzzle therefore remains regarding the very great proportion of available cortical volume that so dense a web of giant arborizations would occupy.

References

Anderson, J. and Hinton, G. Models of information processing in the brain. In G. Hinton and J. Anderson, eds., *Parallel Models of Associative Memory*. Hillsdale, N.J.: Erlbaum, 1981.

Berlucci, G. and Sprague, J. The cerebral cortex in visual learning and memory, and in interhemispheric transfer in the cat. In F. Schmitt, F. Worden, G. Adelman and S. Dennis, eds., 1981.

Blasdel, G. and Lund, J. Termination of afferent axons in macaque striate cortex. *Journal of Neuroscience*, 1983, *3*, 1389-1413.

Blinkov, S. and Glezer, I. *The Human Brain in Figures and Tables: A Quantitative Handbook*. New York: Plenum Press, 1968.

Bullock, T. and Horridge, A. *Structure and Function in the Nervous Systems of Invertebrates*. San Francisco: W. H. Freeman, 1965.

Cajal, S. Ramon y. *Histologie du Systeme Nerveux de l'Homme et des Vertebres*. Paris: Maloine, 1911.

Carpenter, M. and Sutin, J. *Human Neuroanatomy*, 8th ed. Baltimore: Williams and Wilkins, 1983.

Cherniak, C. *Minimal Rationality*. Cambridge, Mass.: MIT Press, 1986.

Cherniak, C. Undebuggability and cognitive science. *Communications of the Association for Computing Machinery*, 1988, *31*, 402-412.

Cherniak, C. *Local network optimization in the brain*, University of Maryland Institute for Advanced Computer Studies Technical Report UMIACS-TR-90-90, 1990.

Cognitive Science, 1985, *9*.

Colonnier, M. The electron-microscopic analysis of the neuronal organization of the cerebral cortex. In F. Schmitt, F. Worden, G. Adelman and S. Dennis, eds., 1981.

Cragg, B. The density of synapses and neurones in the motor and visual areas of the cerebral cortex. *Journal of Anatomy*, 1967, *101*, 639-654.

Cragg, B. The density of synapses and neurons in normal, mentally defective and ageing human brains. *Brain*, 1975, *98*, 81-90.

Descartes, R. *Meditations on First Philosophy.* L. Lafleur, trans. New York: Bobbs-Merrill, 1960.

Dickinson, E. *The Poems of Emily Dickinson,* vol. 2., T. Johnson, ed. Cambridge, Mass.: Harvard University Press, 1965.

Feldman, M. Morphology of the neocortical pyramidal neuron. In A. Peters and E. Jones, eds., 1984.

Fulton, J., ed. *Physiology of the Nervous System*, 3rd ed. New York: Oxford University Press, 1949.

Gilbert, C. and Wiesel, T. Clustered intrinsic connections in cat visual cortex. *Journal of Neuroscience*, 1983, *3*, 1116-1133.

Gray, E. Axo-somatic and axo-dendritic synapses of the cerebral cortex: an electron microscope study. *Journal of Anatomy*, 1959, *93*, 420-433.

Hillman, D. Neuronal shape parameters and substructures as a basis of neuronal form. In F. Schmitt and F. Worden, eds., *The Neurosciences: Fourth Study Program.* Cambridge, Mass.: MIT Press, 1979.

Hofman, M. Encephalization in hominids: evidence for the model of punctuationalism. *Brain, Behavior and Evolution*, 1983, *22*, 102-117.

Hubel, D. The brain. *Scientific American*, 1979, *241*, 44-53.

Hubel, D. and Wiesel, T. Receptive fields, binocular interaction, and functional architecture in the cat's visual cortex. *Journal of Physiology*, 1962, *160*, 106-154.

Hubel, D., and Wiesel, T. Brain mechanisms of vision. *Scientific American*, 1979, *241*, 150-162.

Jerne, N. The generative grammar of the immune system. *Science*, 1985, *229*, 1057-1059.

Jones, E. and A. Peters, eds. *Cerebral Cortex*, vol. 6. New York: Plenum Press, 1987.

Jouandet, M., Tramo, M., Herron, D., Hermann, A., Loftus, W., Bazell, J. and Gazzaniga, M. Brainprints: Computer-generated two-dimensional maps of the human cerebral cortex *in vivo*. *Journal of Cognitive Neuroscience*, 1989, *1*, 88-117.

Konigsmark, B. Methods for the counting of neurons. In W. Nauta and S.Ebbesson, eds., *Contemporary Research Methods in Neuroanatomy*. New York: Springer-Verlag, 1970.

Lamantia, A. and Rakic, P. The cytological and quantitative characteristics of four cerebral commissures in the rhesus monkey. *Journal of comparative Neurology*, 1990, *291*, 520.

Lorente de No, R. Cerebral cortex: architecture, intracortical connections, motor projections. In J. Fulton, ed., *Physiology of the Nervous System,* 3rd ed. New York: Oxford University Press, 1949.

Mountcastle, V. Modality and topographic properties of single neurons of cat's somatic sensory cortex. *Journal of Neurophysiology*, 1957, *20*, 408-434.

Mountcastle, V. An organizing principle for cerebral function: the unit module and the distributed system. In G. Edelman and V. Mountcastle, eds., *The Mindful Brain*. Cambridge, Mass.: MIT Press, 1978.

Nauta, W., and Feirtag, M. The Organization of the brain. *Scientific American*, 1979, *241*, 88-111.

O'Kusky, J. and Colonnier, M. A laminar analysis of the number of neurons, glia and synapses in the visual cortex (area 17) of adult macaque monkeys. *Journal of Comparative Neurology*, 1982, *210*, 278- 290.

Pappius, H. Water spaces. In A. Lajtha, ed., *Handbook of Neurochemistry*, vol. 1, 2nd ed. New York: Plenum Press, 1982.

Paullin, C. *Atlas of the Historical Geography of the United States*. Westport, Conn.: Greenwood Press, 1932.
Peters, A. The visual cortex of the rat. In A. Peters and E. Jones, eds., 1985.

Peters, A. Number of neurons and synapses in primary visual cortex. In E. Jones and A. Peters, eds., 1987.

Peters, A., Palay, S. and Webster, H. *The Fine Structure of the Nervous System: The Cells and Their Processes.* New York: Harper and Row, 1970.

Peters, A. and Jones, E., eds. *Cerebral Cortex*, vol. 1. New York: Plenum Press, 1984.

Peters, A. and Jones, E., eds. *Cerebral Cortex*, vol. 3. New York: Plenum Press, 1985.

Pope, A. Neuroglia: Quantitative aspects. In E. Schoffeniels, G. Franck, L. Hertz, and D. Tower, eds., 1978.

Rakic, P. Developmental events leading to laminar and areal organization of the neocortex. In F. Schmitt, F. Worden, G. Adelman and S. Dennis, eds., 1981.

Rockland, K. and Lund, J. Intrinsic laminar lattice connections in primate visual cortex. *Journal of Comparative Neurology*, 1983, *216*, 303-318.

Rumelhart, D. and McClelland, J., eds. *Parallel Distributed Processing*, vols. 1 & 2. Cambridge, Mass.: MIT Press, 1986.

Schmitt, F., Worden, F., Adelman, G., and Dennis, S., eds. *The Organization of the Cerebral Cortex.* Cambridge, Mass.: MIT Press, 1981.

Schoffeniels, E., Franck, G., Hertz, L., and Tower, D., eds. *Dynamic Properties of Glia Cells*. New York: Pergamon Press, 1978.

Shariff, G. Cell counts in the primate cerebral cortex. *Journal of Comparative Neurology*, 1953, *98*, 381-400.

Shepherd, G. *The Synaptic Organization of the Brain*. New York: Oxford University Press, 1979.

Shkol'nik-Yarros, E. *Neurons and Interneuronal Connections of the Central Visual System*. New York: Plenum Press, 1971.

Sholl, D. The organization of the visual cortex in the cat. *Journal of Anatomy*, 1955, *89*, 33-46.

Sperry, R. Cerebral regulation of motor coordination in monkeys following multiple transection of sensorimotor cortex. *Journal of Neurophysiology*, 1947, *10*, 275-294.

Sperry, R. and Miner, N. Pattern perception following insertion of mica plates into visual cortex. *Journal of Comparative and Physiological Psychology*, 1955, *48*: 463-469.

Sperry, R., Miner, N., and Myers, R. Visual pattern perception following sub-pial slicing and tantalum wire implantations in the visual cortex. *Journal of Comparative and Physiological Psychology*, 1955, *48*, 50-58.

Tower, D. General perspectives and conclusions of the Symposium on Dynamic Properties of Glial Cells. In E. Schoffeniels, G. Franck, L. Hertz, and D. Tower, eds., 1978.

Tower, D. The activities of butyrylcholinesterase and carbonic anhydrase, the rate of anaerobic glycolysis, and the question of a constant density of glial cells in cerebral cortices of various mammalian species from mouse to whale. *Journal of Neurochemistry*, 1973, *20*, 269-278.

Tversky, A. and Kahneman, D. Judgment under uncertainty: heuristics and biases. In D. Kahneman, P. Slovic and A. Tversky, eds., *Judgment Under Uncertainty: Heuristics and Biases.* New York: Cambridge University Press, 1982.
Valverde, F. The organizing principles of the primary visual cortex in the monkey. In A. Peters and E. Jones, eds., 1985.

Vernadakis, A. Changes in astrocytes with aging. In S. Fedoroff and A. Vernadakis, eds., *Astrocytes: Biochemistry, Physiology, and Pharmacology of Astrocytes* vol. 2. Orlando, Fla.: Academic Press, 1986.

Mariano Artigas

EMERGENCE AND REDUCTION IN MORPHOGENETIC THEORIES

The origin of the universe and mankind are the two limiting cases within the evolutionary worldview, whose main task consists in providing morphogenetic theories that may explain how new levels emerge out of other more basic ones. In this context, problems about emergence and reduction occupy a central place.

The following reflections refer, first of all, to some difficulties that are found in the already classical analysis of reductionism, and suggest that the problem of reduction may be properly replaced by the analysis of the relations between levels. These reflections are afterwards applied to the examination of some morphogenetic theories, and finally to the problem of ontological emergence, including an evaluation of some ideas about the origin of the universe and mankind.

1. THE POSSIBILITIES OF REDUCTION

Nowadays there is a general agreement about the limited value of classical analysis of the problem of reductionism. On the one hand, this analysis depends too much on the deductive model of scientific explanation, which has been improved by emphasizing the central role of conceptual aspects in science.

E. Agazzi and A. Cordero (eds.), Philosophy and the Origin and Evolution of the Universe, 253–262.

Furthermore, this model does not fit well with actual scientific practice.

Reductionism occupied a preeminent place in the neopositivist program, which aimed at unifying science through the reduction of its different branches to a physicalist language. Although this ideal was soon given up, reduction continued to be treated as logical deduction of laws or theories of the reduced or secondary science out of those of the reducing or primary one. In this line of thought a distinction was made between two types of reduction, namely that called homogeneous or unproblematic, when all terms of the secondary science were already present in the primary one, and other called discontinuous or problematic, when there was at least one new term which was found only in the secondary science. In this second case, the bridge to relate both sciences could be established by means of a logical connection between the terms, or by postulating coordinating definitions, or through factual hypotheses (Nagel [1961], pp. 336-397).

Afterwards some precise arguments were introduced in the discussions, that pointed towards an increasing scepticism about the possibilities of deductive reductions. With respect to homogeneous theories, it was pointed out that, as a matter of fact, reductions by strict logical derivation seldom occurred or that they were even impossible in practice, due to the fact that, when one or several laws are explained by a new theory, the meaning of some basic terms usually change. It was admitted that a weak or instrumentalist reduction was possible, but this kind of reduction was a merely partial and approximate coincidence of results. A strong reduction between heterogeneous theories would also be possible, so that the reduced theory could be retained and even better corroborated; this would happen when some correlational laws are empirically established, or when two classes of entities in both theories are identified. The conclusion was that it would be convenient to substitute the very term reduction by that of quasi-reduction or partial explanation, to stress the difficulties of a full reductionism (Sklar [1967]; Friedman [1982]).

That those points of view are generally accepted could be ascertained at the XIIIth International Conference on the Unity of Science, whose central topic was the problem of reduction and emergence in the main scientific disciplines. There it was stressed that, if actual scientific practice is properly taken into account, what reduction really means is establishing partial connections between different epistemological levels, and it was pointed out at the same time that those connections may adopt very different modalities (Radnitzky [1988]).

What all this means is that it may be safely be said that derivational reductionism corresponds to a philosophical ideal that neither fits well with actual scientific progress nor is practically feasible. Theoretic constructs are formulated with the purpose of solving particular problems in specific scientific areas, and are constructed according to the conceptual and instrumental resources that are available in every moment in time; it seems therefore convenient to replace the traditional problem of reduction between theories by that of establishing relations between fields or areas of investigation , so that epistemological analysis should be focused on concrete relations, that admit a great variety, between problems and solutions that overlap in the development of actual investigation (Darden - Maull [1977]; Artigas [1989], pp. 5-110).

2. MORPHOGENETIC THEORIES

If it is assumed that a morphonegetic theory is one that relates different levels, its possibility will depend on the existence of ordered levels. From a global and quasi-intuitive point of view, these levels are those of physics, chemistry, biology and human sciences, and the problem consists in relating the superior and inferior levels. This problem may be considered from a dynamic point of view, if genetic relations about the origins are considered, or otherwise from a static one; both perspectives illuminate and complement each other.

Difficulties already arise when we intend to establish hierarchical links between the first two levels, those of physics and chemistry. We need to define from the beginning our ideas about the structure of theories and the relations between them. Moreover, every discipline and even every theory is centered on specific classes of problems and use its own intellectual and instrumental resources. Progress towards unification does not eliminate totally such differences, that arise from the limits of our knowledge and from the respective need of adopting partial points of views according to the possibilities of our knowledge. All that may be summed up by saying that there are different epistemological levels (Kanitscheider [1988]), and even a plurality of ontologies that are implied by the different theories and usually are not replaced when some deeper theories are formulated (Rohrlich [1988]). Those reflections are already important when applied to the analysis of the very fist level of physical theories, and even more when we study the relations between physics and chemistry (Primas [1988]).

Much greater are the difficulties when we consider the relations between the physico-chemical and the biological level. Important results have been obtained, such as the knowledge of the structure of DNA and its connections with genetics, the explanation of some functions of hemoglobin through the knowledge of the sequence of aminoacids which determines the structures of the molecule, and the possibilities opened by the techniques of recombinant DNA for decoding the composition of genes and the structure of proteins.

These achievements point out towards the reduction of biology to physics and chemistry. Doubtless they reveal that physical and chemical laws can be applied to a great extent in the realm of living entities. As mechanics showed that celestial bodies obey the same physical laws that are valid for terrestrial bodies, so biochemistry and molecular biology show that living entities obey the same physics and chemistry that other bodies obey. Nevertheless, this does not mean that biology has been reduced to physical science. When we study biological problems we need to use specific concepts and techniques. Explanations expressed through physical sciences neither eliminate nor make useless the epistemological levels of biology (Kitcher [1982] and [1984]; Rosenberg [1985], pp. 69-120).

To exemplify this we may refer to four types of relations between the level of biology and that of physical sciences: the part-whole relations, such as the proof that genetic unities are a part of chromosomes; the explanation of the physical nature of an entity or a process: for instance, the biochemical explanation of the represors postulated in the operon theory; the relations between the structures of entities and processes and the functions adscribed to them: in that way the knowledge provided by chemical physics about the structure of molecules provides an understanding of their biochemical functions; and the relation cause-effect that exists, for instance, in the explanations provided by the theory of allosteric regulation (Darden - Maull [1977]). None of these relations is properly equivalent to a reduction.

The preceeding reflections refer to relations between theories that belong to the same level or to adjacent levels. Greater even are the difficulties when the results of one level are used to explain problems that belong to more distant levels. In this context reference may be made to some theories that usually are labelled as morphogenetic in a special sense, namely non-linear thermodynamics, synergetics, catastrophe theory and theories about chaos.

Non-linear thermodynamics shows that in principle biological processes may be compatible with the second law of thermodynamics: it suggests that in open systems, far from equilibrium, biological structures can arise through amplification of fluctuations that lead to a new state in which dissipative structures would be maintained. Doubtless, this theory is interesting from the morphogenetic point of view, because it points out that a state characterized by a certain degree of organization may be generated out of a less structured state. Nevertheless it can be said that in that case too there is neither logical reduction nor elimination of properly biological levels (Friedman [1982], pp. 28-39).

Synergetics, catastrophe theory and theories about chaos may be considered too as morphogenetic theories, in that they provide explanations about the genesis of new structures; moreover they intend to relate levels that are not only different, but even in many cases are far away from each other. This is why it is very difficult to establish in a rigorous way the general validity of the models that they provide. From the morphogenetic point of view, the main interest of those theories is an heuristic one, as they suggest the existence of similarities in patterns that may be found in different levels.

Finally, morphogenetic theories in a proper sense are those that refer to evolutionary processes. They have a very special epistemological status. By the one hand, there are arguments that support the existence of such evolutionary processes. But, on the other hand, it is difficult to establish with certainty their concrete mechanisms. This is due not only to the limits of our knowledge, but also to the unique character of the processes that must be described. In these circunstances it is evident that differences, even very deep ones, may arise between various explanations. In this context, the realization of those difficulties becomes a guarantee of progress; in fact, whenever partial explanations are considered as if they were complete, this tends to prevent the discovery of better explanations.

3. ONTOLOGICAL EMERGENCE

The partial character of epistemological reductions implies that there are limitations in the possibilities of establishing ontological reductionism. The complete explanation of upper ontological levels resorting only to more basic ones cannot be justified from the scientific point of view. Nevertheless, the quest for the unity of science stimulates searching for relations between the different levels and, in this respect, partial methodological reductionism finds its proper justification. An analogous effect may be attributed to the conviction about the unity of nature , which may be considered as one of the foundations of the scientific enterprise.

Of course, the first condition for the formulation of morphogenetic theories is the existence of the levels that are to be related. In this respect it may be pointed out the basic shortcoming of any physical explanation about the alleged self-creation of the universe (Atkins [1981]; Davies [1983]; Smith [1988]). The problem in this case is not only that quantum gravity is still in its beginnings, nor that there are difficulties in admitting the uncaused character of quantum fluctuations; the main problem is that, if such an explanation were possible, physical concepts should have a significance that transcends the possibilities of experimental method. In fact, physical concepts would not relate two epistemological or ontological levels: they should connect one real level with a non-existing one (Craig [1986]; Artigas [1987]; Carroll [1988]). Moreover, physical method does not allow one to establish that a concrete event has been the absolute beginning of the universe, even in the case that it really would have been [Jaki [1982], p. 260). The problem of the absolute origin of the universe must be formulated within a metaphysical context.

If the problem of the absolute origin of the universe surpasses the possibilities of the experimental method due to the inexistence of the basic level where this method must be grounded, a similar conclussion may be obtained with respect to the complete explanation of man. In this case, the existence of a specific human level is a necessary precondition for the very existence of science. In other words: experimental science cannot deny the peculiarity of man without denying, at the same time, the possibility of its own existence. The ability to formulate questions about the validity of knowledge is a prerequisite if critical argumentation is to have any meaning at all, and this ability can only exist in the context of a subjectivity that, even though is bounded by physical conditions, cannot be reduced to them.

Ontological reductionism, when formulated in the form of a scientistic naturalism or materialist monism, must resort to arguments of the type nothing but. Those arguments, however, lead to blind alleys. If one states that the universe or man is nothing but matter, he is obliged to determine what is to be understood as matter, and this is a task full of difficulties. In fact, sheer matter does not exist, because any material entity is organized and informed. Moreover, naturalism has no scientific or epistemological grounds, unless one denies the reality of the objects that cannot be studied by means of the methods of experimental science; however, this denial leads to a contradiction or vicious circle, and besides it prevents the study of some epistemological and ontological assumptions whose existence is a necessary condition for the possibility and validity of experimental science.

The evolutionary worldview is sometimes used to argue for a substantive monism, that is the thesis that all objects are in the last analysis different forms and manifestations of some original entities (Rohrlich [1988], pp. 297-299). This thesis, however, is a philosophical one, and cannot be justified by means of scientific or epistemological arguments.

Some problems posed by the evolutionary worldview can be classified under the problem of emergence. Of course, emergence can be partly understood by means of the mechanisms that underlie it. However, a deeper understanding transcends the explanations that are characteristic of the experimental method. To recognize this point one has only to consider that any scientific explanation should refer to the existence of laws that are found through all natural levels. Nature considered in its classical sense as an internal principle of activity is a basic prerequisite of science. Experimental science allows us to obtain an increasingly deeper knowledge and control of natural structures and processes; nevertheless, the very existence of natural activity poses problems whose philosophical aspects are not exhausted by scientific explanations.

Faculty of Philosophy, University of Navarra, Spain

REFERENCES

Artigas, M. [1987], «Física y creación: el origen del universo», *Scripta Theologica*, 19, pp. 347-373.

Artigas, M. [1989], *Filosofía de la ciencia experimental* , Eunsa, Pamplona.

Atkins, P.W. [1981], *The Creation* , Freeman, New York.

Carroll, W.E. [1988], «Big Bang Cosmology, Quantum Tunneling from Nothing, and Creation», *Laval théologique et philosophique* , 44, pp. 59-75.

Craig, W.L. [1986]. «God, Creation and Mr Davies» , *The British Journal for the Philosophy of Science* , 37, pp. 163-175.

Darden, L. - Maull, N. [1977], «Interfield Theories», *Philosophy of Science* , 44, pp. 43-64.

Davies, P. [1983], *God and the New Physics* , Dent, London.

Friedman, K. [1982], «Is Intertheoretic Reduction Feasible?», *The British Journal for the Philosophy of Science* , 33, pp. 17-40.

Jaki, S.L. [1982], «From Scientific Cosmology to a Created Universe», *The Irish Astronomical Journal* , 15, p. 260.

Kanitscheider, B. [1988], «Reduction and Emergence in the Unified Theories of Physics», in: Radnitzky, G., ed., *Centripetal Forces in the Sciences* , pp. 89-110.

Kitcher, P. [1982], «Genes», *The British Journal for the Philosophy of Science* , 33, pp. 337-359.

Kitcher, P. [1984], «1953 and All That. A Tale of Two Sciences», *The Philosophical Review* , 93, pp. 335-373.

Nagel, E. [1961], *The Structure of Science* , Routledge & Kegan Paul, London.

Primas, H. [1988], «Can We Reduce Chemistry to Physics?», in: Radnitzky, G., ed., *Centripetal Forces in the Sciences* , pp. 119-133 and 137-142.

Radnitzky, G. , ed. [1988], *Centripetal Forces in the Sciences* , volume 2, Paragon House, New York.

Rohrlich, F. [1988], «Pluralistic Ontology and Theory Reduction in the Physical Sciences», *The British Journal for the Philosophy of Science* , 39, pp. 295-312.

Rosenberg, A. [1985], *The Structure of Biological Science* , Cambridge University Press, Cambridge.

Sklar, L. [1967], «Types of Inter-Theoretic Reduction», *The British Journal for the Philosophy of Science* , 18, pp. 109-124.

Smith, Q. [1988], «The Uncaused Beginning of the Universe», *Philosophy of Science* , 55, pp. 39-57.

Jesús Mosterín

WHAT CAN WE KNOW ABOUT THE UNIVERSE?

0. HOW DID IT COME ABOUT THAT THERE IS SOMETHING TO BE KNOWN?

Immediately after the Big Bang the universe was a very hot and dense gas, nearly homogeneous and in thermal equilibrium. Later on, it fell out of equilibrium. The hot gas expanded and condensed into galaxies, stars and other well structured cosmic systems. The order, structure and thermodynamic information of the universe increased dramatically. Observers and things to be observed became possible. All this would contradict the second law of thermodynamics, if it was not for the presence of the great disequilibrator, namely, the uniform expansion of the universe (of spacetime itself).

In the universe, as in any other system subject to irreversible changes, the entropy has been increasing all the time. Disorder has been increasing all the time. But order has also been increasing. Thermodynamic information has been created all the time. This would be contradictory if we defined thermodynamic order or information as negentropy, i.e., as the negative value of entropy, as Wiener [1961, p. 11] and Brillouin [1962, p. 116, 156] did. Obviously S and -S cannot both increase at the same time. But there is no problem if we define thermodynamic order or information as the gap between the actual entropy and the maximum possible entropy: $Or = S_{max} - S$. As long as the maximum

E. Agazzi and A. Cordero (eds.), Philosophy and the Origin and Evolution of the Universe, 263–289.

possible (or potential - in Leyzar's terminology) entropy increases, both actual entropy and order can continue to grow simultaneously.

The uniform expansion of spacetime is the ultimate source of the disequilibrium and free energy required for building information carrying structures and cosmic systems [Frautschi 1988, p. 16; Layzer 1988, p. 31]. The expansion of the universe and the subsequent creation of disequilibrium has been proceeding at a quicker pace than the degrading and entropic processes pointed towards equilibrium. So the maximum possible entropy of the universe has been increasing more rapidly than its actual entropy, which of course was also increasing at the same time. The net result has been a spectacular growth of structure and order in the universe.

It is amazing how far from equilibrium our universe is. The theory of black holes is extremely simple. Only two parameters (mass and angular momentum) determine uniquely everything about a spherical symmetric black hole. So it is a relatively straightforward task [Penrose 1979; Börner 1988] to calculate the entropy of a black hole which would include all the mass of the universe. The surprising result is that the entropy of that black hole would be 10^{30} times the actual entropy of the universe. This immense distance from maximum entropy emphasizes how much ordered and full of information our universe is. It is this vast reservoir of objective information which makes viable the existence of information-crunching creatures like us.

The thermodynamic order results in the building of conspicuous structures, which give out lots of differentiated and finely modulated signals in all directions. We detect some of these signals, whose form in-forms our brains. It is through the processing of this primordial information that we are able to build our images and representations of the ultimate sender of those signals, the universe.

FOUR SENSES OF "UNIVERSE".

The word "universe" can be used in at least 4 different senses: the perceptible universe, the observable universe, the intelligible (or theoretical or conceptual) universe, and the ultimate universe (or the whole of reality).

1. THE PERCEPTIBLE WORLD.

Each animal has its own world. This world results from applying the animal's inborn processing capabilities to the range of signals to which it is attuned. Each animal only detects a small fraction of the objective information available in its environment, that fraction which the animal is genetically programmed to detect, process and experience. Most of the detectable information is relevant information, information with survival value. The animal receives the signals it is able to detect and submits them to a natural processing which results in certain experiences and perceptions of the world around and in certain effects, movements, secretions and actions. The range of signals the animal is able to detect and the range of information processing programs available to it determine the types of its possible perceptions, experiences, intuitions and imaginations, i.e., determine its perceptible world.

The fox's keen sense of smell detects the fleeing rabbit's molecules left hanging in the underbrush, and from these chemical signals it immediately extracts information about the path followed by the rabbit in its flight. The fox perceives the rabbit's path. The bat has no the slightest difficulty in finding his way in the dark cavern or in the forest at night. It locates its preys or the obstacles to its flight by its ability to emit ultrasounds and to detect their rebounded echoes. Bees see the flower's ultraviolet colours. Some snakes are able to directionally detect the infrared radiation. Some fishes perceive the variations of the magnetic field. All these perceptions produce in the animals which perceive them experiences which we cannot even imagine, as they fall completely outside of our perceptible world. And still less can we imagine the perceptions and experiences of the unknown creatures which possibly inhabit other corners of the universe far away from our solar system.

The perceptible worlds of the animals belonging to the same species overlap to a large extent, so that it makes sense to speak of a species' world. Nevertheless some individuals may have no access to parcels of that world, as it happens with the blind, who do not share in the human visible world, or with the deaf from birth, excluded from the world of sounds.

HORIZONS OF THE PERCEPTIBLE WORLD.

The neural and sensory apparatus characteristic of a species determine the type of perceptions and experiences of which the individuals of that species are capable. It draws a frontier between the perceptions which are possible for those animals and the perceptions which are impossible for them. This frontier marks the limit of those animals' capacity of perception, or, in other words, it constitutes the horizon of their perceptible world. I am not able to have any sensations or experiences, but only those experiences which can be generated by my neural and sensory apparatus, exposed to the kind of stimuli to which it responds. The perceptions and experiences of animals with neural and sensory systems very different from our own are off limits for us, are beyond our experiential horizon. Visual experiences are also beyond the experiential horizon of the blind. My experiential horizon sets a limit to my possible experiences.

HORIZONS OF THE HUMAN PERCEPTIBLE WORLD.

Our neural and sensory apparatus react only to a certain range of stimuli, detect only certain types of signals. These are our only windows to the universe. The sense of hearing is very important in our social life, but its cosmological scope is nil, as the acoustic signals it is sensible to are only produced and transmitted inside the atmosphere. As a matter of fact, all the perceptible information from beyond our own planet which reaches us arrives to us through the sense of sight, which is sensible to the visible light, i.e., to the electromagnetic radiation with wavelength between $4 \cdot 10^{-7}$ m and $7 \cdot 10^{-7}$ m (or, equivalently, with frequency between 4.2

* 10^{14} Hz and 7.5 . 10^{14} Hz). This is the only open window of our perceptible world into the wide extraterrestrial universe. Through it humans have always looked at the starry sky with curiosity and awe. The heavenly bodies, rated as gods in many cultures, have always been honored inhabitants of man's perceptible world and frequent objects of his attention.

NO EXPLORATION.

Other limitations of our perceptible world are due to our specific position in spacetime. Some of these limitations can be overcome through travel or exploration, through locomotion to a different location where new experiences become possible. If, living in a low warm country, I lack the experience of snow, I can get it, making a trip to a high cold mountain. If I have never had the experience of bathing in the salty sea water, because I have always lived inland, I can travel to the coast to get it. But in the case of the cosmos, the exploration is largely excluded.

We can explore our solar system, and we are already doing it. In 1965 the first space walks outside a spacecraft were undertaken. Between 1969 y 1972 several men landed and walked on the Moon. A manned mission to Mars is already in the planning stage, and its implementation is only a question of financing. It is almost sure that we are going to explore the whole of our solar system. It is not impossible that one day we may be able to explore the vicinity of the near stars or even our whole galaxy. But it is practically excluded that we ever explore other galaxies, specially those beyond our local group. Almost the totality of the universe is for ever beyond the scope of our possible exploration.

NO EXPERIMENTATION.

The best way to know the objects, persons or systems of our perceptible world is to test them, to experiment with them, to make them react to factors which are under our control. The degree of an experiment is the number of parameters which are under the control of the experimenter. An observation is (in this sense) an experiment of degree zero, an improper experiment, in which no parameter at all is under control.

Another limitation of our astronomy and cosmology is that experiments are excluded. All we can do is observe passively and receive and process the information naturally transmitted by the observed cosmic systems.

Ruled out the recourse to exploration or experimentation, if we want to go beyond the narrow limits of our perceptible world, we still have the possibility of designing and building extensions of our senses, artificial devices able to detect signals inaccessible to our natural senses.

2. THE OBSERVABLE UNIVERSE.

The observable universe includes not only everything perceptible with the natural senses, but also everything whose signals we detect with the help of the artificial extensions of our senses, specially with the scientific instruments of observation such as telescopes and radio telescopes, photographic plates and neutrino detectors. The observable universe is the universe accessible to observational science and in particular to astronomy.

HORIZONS OF THE OBSERVABLE UNIVERSE.

An observational horizon is a limit to observation, a frontier separating what can be observed from what cannot. Obviously the observational horizon is relative to the observation instruments available at a certain time and to the spatiotemporal position of the observer (or of the scientific community as a whole). Technological progress in observational instrumentation brings about an enlargement of the observational horizon and of the observable universe. The bare passing of time and the eventual changes in space of the observer also bring about changes in the observational horizon.

We cannot draw or specify the horizons of the observable universe on the basis of the bare observation. We do it from beyond, taking into account physical theories and cosmological models. Depending on which physical theories we accept at a certain moment, we will draw the observational horizons one way or another. And, of course, everything we are going to say here about those horizons will presuppose a great deal of contemporary physical theory.

SPECIAL-RELATIVISTIC HORIZON.

No signal can travel at greater speed than the velocity of light. So, here and now it is impossible for us to receive any signals of what happened less than one year ago at a distance from us of more than one light-year, or of what happened less than two years ago at a distance of more than two light-years, etc. Signals can arrive only from events which happened at distances (measured in light-time) smaller than the time elapsed since. Such events are the internal points (of the 4-dimensional spacetime) of the lower hypercone with vertex at our actual position and surface defined by the equation $x^2 + y^2 + z^2 = t^2$. (Remember that we are taking $c = 1$). It is the so-called past light cone. Only from the events located inside our past light cone can we receive signals. From all other events we cannot get any news, as a matter of principle and whatever the progresses of future technology. From the events separated from us by a space-like distance, we will never get news. From the ones located in our future light cone we cannot get signals now, but perhaps we will in the future.

The temporal cuts in the hypercone can be visualized as concentric spheres with center here and now and with radius (in light-time) equal to t, the time elapsed. All the events which happened less than t time ago outside of the sphere with radius t are unobservable in principle (from the center of the sphere, which is where we are). That sphere (or, better, the succession of spheres which makes up the past light hypercone) is the special-relativistic horizon of the observable universe.

We can only receive signals emitted by the cosmic systems at certain moments of their past. So we do not have a world map (in Milne's terminology), a more or less synchronic representation of the universe, but only a certain temporally heterogeneous world picture. We see the near galaxies as they were in their mature age, after several billions years evolution, whereas the signals we get from the quasars were emitted in their infancy. There are no quasars in the spatiotemporal neighbourhood of our galaxy. But, the further and further we observe in spacetime and the more and more we approach the confines of the observable universe, the less and less galaxies and the more and more quasars we observe. It is not possible for us to observe the infancy of the galaxies or the mature age of the quasars. Our image of the observable universe is a mosaic, made up of pieces wildly differing in temporal provenance, an idiosyncratic spatiotemporal cut determined by the peculiar perspective of the observer.

When in 1987 the first signals of the 1987A supernova reached us, what we were observing was an explosion which took place some 170,000 years ago. What is happening there right now is impossible to know, at least to know observationally. The great Magellanic Cloud could have totally disappeared 100,000 years ago and no one could have noticed it yet. As a matter of fact, no one could notice it in the next thousand generations of humans.

The observable universe is always a past universe. The actual universe is unobservable. Our cosmic image is irremediably obsolete. We cannot observe how the universe is today. We can only watch how it was a long time ago (different times ago in different places). The whole universe beyond our local group could have disappeared millions of years ago, much before there were any humans on Earth. We could not have noticed it. It would have been impossible in principle to notice it. We can only observe that there were galaxies beyond our local group several million years ago, but we cannot observe whether they exist now. Those possible actual galaxies are beyond the horizon of the observable universe.

HORIZON DUE TO THE EXPANSION OF THE UNIVERSE.

The universe is expanding. Spacetime itself is in expansion, making galaxies depart from each other at velocities linearly correlated with their distances, at least in the cases of galaxies belonging to different superclusters. Inside the same supercluster it is possible for gravitation to bind the galaxies so strongly together that it more than counterbalances the expansion of spacetime. It is what happens in our own local group, where Andromeda is approaching our galaxy (as shown in the blue-shift of its spectrum), or in our local supercluster, in which our local group is actually approaching the Virgo and Hydra-Centaurus clusters.

The rate of the expansion of the universe changes with time. Hubble's constant H_0 indicates the expansion rate of the universe at the present time, i.e., the recession velocity of far away galaxies in relation to ours at the present time. It is measured in kilometers per second per megaparsec of distance. (1 megaparsec = 10^6 parsecs ; 1 parsec = 3.26 light-years = the distance from the Earth to a star with a parallactic angle of one arcsecond).

We do not yet know the exact value of Hubble's constant, due to the great difficulty in measuring the distances to the far away galaxies with any precision. The estimations range between 50 km/s/megaparsec (Sandage and Tammann) and 100 km/s/megaparsec (de Vaucouleurs). Let us assume the intermediate value of 75 km/s/megaparsec. It would mean that a galaxy at a distance of 10 megaparsecs would recede from us at a velocity of 750 km/s. By the same rule, a galaxy at a distance of 4,000 megaparsecs (= 13 billion light-years) is receding from us at a speed greater than the speed of light c. No material object and specifically no signal can move at greater speed than c. But the cosmic expansion is an expansion of spacetime itself, which is not a signal or a material object, so that it is not subject to the special-relativistic limitation of speed.

The distance c/H_0 is the radius of the so-called Hubble sphere, a sphere around the observer beyond which no signal can reach the observer. Any object located beyond the sphere is receding from the observer at a speed greater than c, so that no signal emitted by it can ever reach the observer. Supposing the previously assumed value of Hubble's constant is correct, it is impossible in principle that signals emitted by galaxies or other cosmic systems at distances from us of more than 13 billion light-years ever reach us. The surface of Hubble's sphere represents an impassable horizon of the observable universe.

THE PHOTON HORIZON.

Most of the signals we get from the cosmos are bursts of electromagnetic radiation, of photons. The electromagnetic radiation arriving on Earth is mainly of two types: (1) galactic radiation, emitted by the galaxies since there are galaxies, i.e., since some 10^9 years after the big bang; and (2) cosmic background radiation, which carries information on the pregalactic epoch, specifically on the time when the universe became transparent to photons, some 10^6 years after the big bang.

In the first half million years after the big bang photons were constantly interacting with free energetic electrons, coupled to them in a gas in equilibrium. When the temperature of the universe (which was descending all the time as a consequence of expansion) dropped below 3000 K, the free electrons began to be bound by the protons, which combined with them to build hydrogen atoms. At the end of the epoch of uncoupling of matter and energy, approximately one million years after the big bang, with a lower temperature and the electrons bound in atoms, the photons stopped interacting with electrons and began to move freely through cosmic space. The universe became transparent to photons. The cosmic background radiation, discovered by Penzias and Wilson in 1965, is the cooled down (till 2.7 K) and fossilized rest of that radiation.

The photon horizon is the temporal frontier separating the regions of spacetime previous and posterior to the time when the universe became transparent to photons. It is only after that temporal horizon that photons became capable of transmitting information about the state of the universe at the moment of their generation. The photon horizon is a fundamental barrier to the observational information we can get about the early universe through electromagnetic signals. Whatever the future progress of signal detection technology, we will never be able to detect electromagnetic messages from the early universe, before it became transparent to photons, just because such messages do not exist.

VISIBLE LIGHT.

The visible light is the electromagnetic radiation with wavelength between 4 * 10-7 and 7 * 10-7 m. It is the only portion of the spectrum which we are naturally equipped to detect and interpret. It is the only window of our perceptible world towards the cosmos at large. Even restricted to this narrow band of the spectrum, our natural capacity to see and discriminate is limited. The observable universe broke the limits of the perceptible world only in the XVII century with the introduction of artificial extensions of our eyes, beginning with Galileo's spyglass and following with a series of ever better telescopes. The substitution of photographic plates for the observer's eye was also a marked improvement in the power of resolution. A further improvement in this direction is the introduction of charge coupled devices (CCD's), capable of recording and imaging light too dim to be recorded by photographic plates. The turbulence of the upper atmosphere acts as a changing lens, distorting the optical images which arrive at the surface of the Earth. This remaining natural limitation will be only overcome with the launching into space of a new generation of telescopes mounted on artificial satellites. In 1990 the Hubble telescope will be put by NASA into a stationary orbit 500 km above the Earth's surface, well beyond the atmosphere's turbulences. The Soviet Union plans to put a 10 m diameter telescope (against the 2.4 m of the Hubble) into space by the year 2000.

All the information about the universe available to humans till the Second World War arrived through the visible light window. Even today it continues to be the main gate of the observable universe and the source of discoveries like the supernova 1987A.

Another potential enlargement of the visible horizon of the observable universe can come from the systematic exploitation of the cosmic gravity lenses. The general theory of relativity predicts that if two distant stars are aligned with an observer, then the effect of the gravitational attraction of the star in the foreground on the light arriving from the star in the background will produce a bright magnified image of the background star in the shape of a ring with the foreground star in its center. When the alignment is not perfect, the image is no longer a ring, but two magnified images of the background star in the shape of two opposing crescents. The same happens wherever two distant galaxies are aligned with the observer. The gravitational lenses can produce images of distant stars and galaxies far surpassing in sharpness and magnification anything obtainable by however advanced telescopes. If the predictions of the theory come out true, the gravitational lenses can provide in the future an important enlargement of the visible (and other electromagnetic!) horizon of the observable universe.

INFRARED SIGNALS.

The infrared radiation is the electromagnetic radiation with wavelength between $7 \cdot 10^{-7}$ and 10^{-3} m. This infrared radiation penetrates into our perceptible world as the characteristic warmth sensation in our skin when we are in the sun (or in front of the fireplace). Besides this solar radiation, infrared radiation coming from other places in our galaxy and from other galaxies also reach the Earth. The observational problem is that most of this radiation is absorbed by the atmosphere, which is opaque to the infrared, except in some small portions of its spectrum.

The development of infrared astronomy in the last 30 years has been made possible by the use of sensible semiconductor detectors, mounted first in airplanes and then in artificial satellites, specially the satellite IRAS, which in 1983 surveyed the entire sky at several infrared wavelengths for a period of ten months. In this way we could get valuable information about zones of space hidden by the galactic dust clouds, which are opaque to visible light, but transparent to infrared radiation.

RADIO SIGNALS.

Radio waves are waves of electromagnetic radiation with wavelength greater than 10^{-3} m (or, equivalently, with frequency less than 10^{11} Hz). In 1931 the Bell engineer Karl Jansky succeeded in detecting for the first time radio waves from outside the solar system. But it was only after the Second World War that the techniques developed during the war were systematically applied to the detection of radio waves from the sky. That was the beginning of radio astronomy, which has experienced a spectacular growth in the last 40 years and has led to the discovery of previously unsuspected cosmic objects, like radio galaxies, pulsars and quasars. Some of the sources of radio have been subsequently identified by optical means; others have resisted till now optical identification. Even the sensational discovery of the microwave background radiation has come through radio wave detection.

We use special instruments, called radio telescopes, to collect and measure the cosmic radio signals. In order to attain much higher angular resolutions than would be possible with single aerials, we use radio interferometers, which are coordinated systems of aerials or radio telescopes located at great distances of each other, but which function as if they were a single radio telescope. In the near future radio interferometers combining earthbound radio telescopes with aerials mounted on artificial satellites will become available and will allow a still better resolution.

Although cosmic objects emit in the whole radio spectrum, the radio waves with wavelength greater than $3 \cdot 10^3$ m (i.e. 3 km) or, equivalently, with frequency of less than 100 kHz never reach the Earth. The photons with such low frequencies get absorbed by the ionized interstellar gases, before they have any chance of reaching our solar system. The information carried by those low frequency photons is lost for ever, at least as far as we are concerned. The observational window opened by radio astronomy has an impassable horizon at the 3 km wavelength.

ULTRAVIOLET RADIATION.

If, starting with visible light, we go on to consider electromagnetic radiation of increasing energy and frequency and decreasing wavelength, we come first to the ultraviolet rays, then to the X-rays and finally to the gamma-rays. Ultraviolet radiation is electromagnetic radiation with wavelength between $4 . 10^{-7}$ and 10^{-8} m. The sun emits radiation in the whole ultraviolet spectrum, but only the ultraviolet with wavelength between $4 . 10^{-7}$ and $3 . 10^{-7}$ m can penetrate the Earth's atmosphere and reach us on the surface of the planet, as the rest of the ultraviolet radiation is absorbed by the ozone layer of the upper atmosphere. This limitation has been overcome by the mounting of ultraviolet telescopes in artificial satellites which orbit the Earth beyond the ozone layer.

Not only the sun, all hot cosmic objects emit radiation in the ultraviolet spectrum. So the development of ultraviolet astronomy has led to a much better knowledge of the interstellar medium and of the gas flows around hot stars. Nevertheless the value of ultraviolet photons as carriers of information about far away cosmic objects is subject to a serious limitation. All the extreme ultraviolet radiation (with wavelength between 10^{-7} and 10^{-8} m) is absorbed on its way to us by the interstellar hydrogen atoms, so that it never reaches us. This is the ultraviolet horizon of the observable universe.

X-RAYS.

X-rays are electromagnetic radiation with wavelength between 10^{-8} and 10^{-12} m. X-ray astronomy began in 1949 with the discovery by Burnight that the sun emits X-rays. This discovery was only possible thanks to the use of rockets as detectors carriers, because the atmosphere is totally opaque to X-rays, which can only be detected beyond 120 Km above the Earth's surface. The satellite Uhuru, launched by NASA in 1970, carried out a complete sky survey and led to the discovery of many discrete X-ray sources. Other satellites followed, specially NASA's Einstein satellite, launched in 1978, carrying a X-ray telescope of very good angular resolution. New ones are in preparation just now, like the German X-ray satellite Rosat, to be launched in 1990.

The X-ray astronomy has opened another very important electromagnetic window and has enlarged the observable universe. The information carried by the X-ray signals has made possible the discovery of many new cosmic objects, which are powerful emitters of X-rays, like X-ray stars, X-ray galaxies, and binary stars one of whose components is a neutron star or a black hole, and has provided fresh news about the mysterious quasars.

GAMMA RAYS.

Gamma rays are electromagnetic radiation with wavelength less than 10^{-12} m. The gamma radiation is made up of photons with energies greater than 10^5 eV. These are the most energetic photons of the whole electromagnetic spectrum. Most of them are produced in the interior of atomic nuclei, when they undergo a transition from a higher to a lower energy level. Their extreme energy makes it possible for them to go through regions of our galaxy and beyond where other less energetic forms of radiation are absorbed. Gamma ray bursts from outside the solar system were first noted in 1967 by satellites designed to monitor the Nuclear Test Ban Treaty. Special gamma ray telescopes mounted on satellites launched in 1972 and 1975 were the real beginning of gamma-ray astronomy, even if the angular resolution of the gamma ray detectors was rather poor. A big increase in accuracy is expected from the Gamma Ray Observatory, to be launched by NASA in 1990, and which will orbit the Earth at a distance of some 450 km.

The detection and interpretation of the gamma rays from outside our solar system opens a new window of the observable universe. Nevertheless this new window meets its own horizon at the wavelength of 10^{-21} m. The most energetic gamma rays of all, those with wavelengths less than 10^{-21} m., are destroyed on their way by collisions with photons of the cosmic 2,7 K microwave background radiation. The two colliding photons give rise to an electron-positron pair, the original incoming photon ceases to exist, and the information it carries disappears [Harwit 1981, p.168].

SUMMARY ON ELECTROMAGNETIC RADIATION.

The electromagnetic radiation which reaches the Earth is the main (and till now almost the only) source of information about the universe available to us. The recent advances in the signal detection technology and the use of artificial satellites as carriers of the instruments have opened wide all the windows to this radiation. Nevertheless there are types of electromagnetic signals which we will never be able to detect, however much our technology may advance in the future, for the simple reason that such signals always get lost in the way from their source to the Earth and never reach us. The only electromagnetic signals from the cosmos which ever reach our corner of the universe are those with wavelength between 10^4 m and 10^{-7} m, and between 10^{-8} and 10^{-21} m. This bandwidth represents the electromagnetic horizon of the observable universe.

COSMIC RAYS.

Not all signals we get from the cosmos are electromagnetic. For example, the so-called cosmic rays are not rays at all. They are mostly charged particles which arrive at the top of the atmosphere from the cosmic space in all directions. Most of them are protons, many are helium nuclei, and there are also heavier nuclei, electrons, positrons, antiprotons and other particles. They travel at nearly the speed of light and have tremendous energies, up to 10^{20} eV. No other particles known on Earth or produced in particle accelerators can ever reach such high energies. Cosmic "rays" are the most energetic particles known. We do not know for sure where the cosmic rays come from: supernova explosions, shocks in their remnants and pulsars are likely candidates. When cosmic-ray particles hit the top of the atmosphere, they collide with atoms of the air, and a shower of new particles (the so-called secondary cosmic rays) is produced. Lots of these secondary particles, specially energetic muons, reach the surface of the Earth. About five of these muons pass through our head every second.

Victor Hess discovered the cosmic rays in 1912 with the help of a balloon. Since then many observations have been made, but the results are rather inconclusive. Most cosmic rays are charged particles, which have been twisted and spread around by the magnetic field of the galaxy, so that it is impossible to know the location of their source, their original direction, the time of their emission or their original energy. Exceptionally they are uncharged particles which have travelled in a straight line; then it is possible to know where they come from. They can also contain interesting chemical information. It is probable that in the future we will be able to squeeze more information out of the cosmic rays which reach our planet, even if the uncertainties about their concrete source will always limit such possibilities.

METEORITES.

Meteorites are chunks of rock which fall to Earth from outer space. Of course they carry information, but their cosmological interest is very limited, as all of them seem to come from inside the solar system, mostly from the asteroids, but also from the Moon, Mars and the comets.

NEUTRINOS.

The neutrinos are electrically neutral particles (hence the name "neutrino", Italian diminutive of "neutral", due to Enrico Fermi), with spin 1/2, insensible to the strong interaction, and with a rest mass either nil or very small (we do not yet know which).

According to accepted physical theory, neutrinos and photons are by far the most common particles in the whole universe. Everything is full of neutrinos and everything is transparent to neutrinos. The whole Earth is transparent to these elusive particles, which are very reluctant to interact with any atom. We ourselves are constantly pierced by billions of neutrinos coming from all sides.

One second after the big bang the temperature would have descended till around 10^{10} K and the density would have decreased enough for neutrinos and antineutrinos to decouple from ordinary matter (with which they were in equilibrium before) and to become free particles. The universe became transparent to neutrinos, so as it would become transparent to photons one million years later. These neutrinos are still there as a sort of cosmic neutrino background radiation, presumably fossilized and cooled down till 2 K. This neutrino background radiation carries information on the very early universe, only one second after the big bang. The problem is that we are quite unable to detect that radiation. There are good theoretical reasons for thinking that this neutrino background radiation exists, but for the moment no one knows how to detect it and reap the information it carries. So, and the postulated neutrino radiation not withstanding, the observational universe becomes opaque behind a temporal horizon situated one million years after the big bang. The neutrino radiation window, which would allow us to look back till one second after the big bang, remains closed.

Besides the cosmogonic neutrinos, which permeate the whole space, we are subjected to a continuous bath of neutrions ofgalactic and solar origin. We think we understand the mechanism of thermonuclear reactions in the core of the stars and specially in the sun. This understanding implies the production of a constant flow of solar neutrinos. Nevertheless the various experiments made by Ray Davis in a deep South Dakota mine for detecting and counting the solar neutrinos have not led to the expected results. This situation is obviously unsatisfactory.

The problem arises from the extremely weak tendency of the neutrinos to interact. The probability of detecting a neutrino is of one to 10^{18}. Because of that, the methods used for the detection of neutrinos are based in the accumulation of enormous amounts of material, in the hope that the very large number of atoms compensates for the extremely minute probability of interaction with any individual atom of the detector. It has even been suggested to adapt big portions of the ocean (suitably provided with arrays of photodetectors for recording the high energy charged particles resulting from the eventual interaction of a neutrino with an atom) as detectors for high-energy neutrinos [Feinberg 1985, p. 148].

We do not perceive the billions of neutrinos which go through our eyes all the time. They do not belong to our perceptible world. Our detection instruments are barely able to detect them. They do not yet really belong to the observable universe. But they fill the intelligible universe. According to our theories, during the big bang one billion neutrinos were made for every proton, neutron or electron. And still today every single nuclear reaction in every star goes on producing new neutrinos, not to speak of the violent cosmic events. The neutrinos were theoretically postulated by Wolfgang Pauli in 1930, but they could only be experimentally detected in 1956 by Cowan and Reines. Davis' experiments have detected solar neutrinos, but less than predicted. It seems that also some of the neutrinos produced in the 1987A supernova explosion were detected, even if there is some polemics about this question. Summarizing, the neutrinos fill our theoretical universe, but they barely show in the observable universe. Probably in the future we will be able to better exploit the cosmic information they carry.

GRAVITATIONAL WAVES.

According to the general theory of relativity the quick accelerations, oscillations or deformations of large masses are bound to disturb and distort the spacetime around them. These disturbances propagate through space at the speed of light. The gravitational collapses which take place at the center of galaxies, the supernova star explosions, the pulsars, black holes and quasars should be rich sources of gravitational radiation, which would carry much information about those phenomena. It is even possible that we are immersed in a fossil cosmic background gravitational radiation, comparable to those of microwaves and neutrinos. This background radiation would carry information on the extremely early universe, when gravitation decoupled from the unified rest force and scattered, perhaps as early as Plank time, 10-43 second after the big bang [Thorne 1987, p. 397]. The detection of such radiation in abundant enough doses could be a potential blow to the inflationary cosmological model (because the later inflation would have diluted the initial gravitational waves), currently favored by many cosmologists and badly lacking in observational checks [Feinberg 1985, p. 142].

Unfortunately gravitational waves are still more difficult to detect than neutrinos. The probability of detecting a graviton (the quantum of gravitational radiation) passing through a detector is only of one to 10^{23}.

Joseph Weber, from the University of Maryland, built a detector of gravitational waves in the shape of cylindric 1.4 ton block of aluminium, which would vibrate with gravitational waves. In order to eliminate local effects, Weber installed two such detectors, separated by a long distance. In 1969 Weber announced the detection of several events of gravitational waves, but other groups trying to replicate his observations in the next years met with no success. His detectors had not enough discrimination capacity to distinguish with any assurance the extremely weak effects of the gravitational waves. At the moment of writing a new generation of gravitational wave detectors is under preparation, based on the laser interferometry. It is possible that they finally open this important new window. For the moment being, gravitational waves have not yet being really detected. They belong still to the intelligible universe, as consequences of the general relativity, and have not yet passed the threshold of our observable universe.

UNEXPECTED SIGNALS.

As far as we know, photons, cosmic rays, neutrinos and gravitons are all the signals from the cosmos reaching the Earth. The sources of these signals make up the observable universe, at least in as far as we are able to detect them. This evaluation of the observable universe is a function of the intelligible universe, i. e., it depends on the validity of certain physical theories and cosmological models. This theoretical situation can change in the future, and so can change the evaluation of the horizons of the observable universe.

It could well be that there are forms of radiation or fields or particles corresponding to theoretical entities postulated by (still) speculative, unmatured and not (yet) widely accepted theories, like the photinos and gravitinos postulated by certain superunified theories. It could also be that our solar system is reached by totally unexpected signals, unknown particles and unsuspected waves or perturbations, not postulated or even imagined by any of today's theories. The point is that we are unable to say anything about such particles or signals, besides recognising their mere possibility. So as not all existing energy is free energy, usable to produce work, in the same way not all existing information is detectable information, usable to produce knowledge.

It is possible to compute the mass of the galactic systems by mechanical calculations. We can calculate the mass of our galaxy from the data derived from the observation of its movements and apparent structure. We can compute the mass of a cluster of galaxies by applying the virial theorem to its movement and size. And we can calculate the deceleration parameter (g_0) of the standard cosmological model from the determination of the distances and recession velocities of the most distant galaxies observed. All these calculations lead us to mass estimations which are several times higher than the masses of the stars and other visible cosmic bodies. It follows that, if our mechanical theories are right, most of the existing mass is missing from our observations. This is the so-called missing mass problem. Most of the mass seems to be dark mass, non visible mass, mass beyond our observational horizon.

We do not know the solution to the missing mass problem. Perhaps there are lots of dark stars. Perhaps neutrinos have some small mass, which, multiplied by their enormous number, could account for a large chunk of the missing mass. Perhaps there are immense quantities of unknown and unsuspected particles of no null mass, making up most of the mass of the universe. If this were so, it would be conceivable that these particles would be carriers of information about unexpected and unheard of cosmic objects and processes.

LIMITS OF OUR DETECTION CAPACITY.

Besides the limits to the signals which reach the Earth, there are limits and observational horizons due to the observer and to the detection instruments. We only have a limited capacity for detecting signals, a limited capacity for discriminating between different signals and a limited capacity to recognize the detected signals. The instruments we use to detect and measure produce their own unavoidable noise, which sets limits to the attainable precision of the observations. These limits often depend on the ratio between the wave length of the signal and the size of the instrument. They are ultimately grounded on the quantum-mechanical uncertainty principle.

3. THE INTELLIGIBLE UNIVERSE.

A theory is an economic way of encoding huge amounts of information into the compressed form of a few concepts and principles. Each theory is like a fisherman's net or a spider web, with which we try to catch an aspect of the universe. When we weave all our relevant theories together, in order to catch the whole universe, what we get is the theoretical or intelligible universe.

The intelligible universe is our maximally comprehensive theoretical construction about the whole of reality, in as far this is scientifically accessible to us. The intelligible universe includes and accounts for the observable universe, but it extrapolates, transcends and goes beyond the observable universe.

It also includes all the relevant scientific laws and theories, and all the entities postulated by them. It encompasses as well the methods of representation as the underlying mathematics and physics and whatever basic presuppositions or conceptual resources are used in the construction.

Some aspects of the observable universe are neglected in the construction of the intelligible universe. And not all theories are equally taken into account in its elaboration. The result of this process of conceptualization is a more or less idealized scientific image of the universe as a whole. This image is a cosmological model. Each different cosmological model is a different image of the intelligible universe, or, if you prefer, is a different intelligible universe.

All cosmological models which are seriously held at the present moment are standard big bang models, based on Einstein's general theory of relativity and on Friedman's solutions to its equations, which incorporate the Robertson-Walker metric (which presuppose a dynamic, spatially homogeneous and isotropic universe). Thermodynamics plays also a major role. And in the newer developments particle physics is continually drawn upon. All of these models postulate a singularity in the past as the origin of the great explosion (big bang) which led to our present universe. The main models of this family are the classical hot big bang model and the inflationary model. They coincide on everything posterior to the first 10^{-30} second after the singularity. They account for the observed expansion of the universe, incorporating Hubble's law; they also account for the microwave background radiation and for the primordial nucleosynthesis, inferred from the actual ratio among the elements, specially between hydrogen and helium.

LIMITS TO CONCEPTUALIZATION. HORIZONS OF THE INTELLIGIBLE UNIVERSE.

A conceptual or theoretical horizon is a limit to what can be known, or conjectured or thought on the basis of a certain conceptual scheme or cosmological model, i.e., making use of the concepts, hypothesis and intuitions of the scheme or model, and of its underlying mathematical and physical theories.

In the Newtonian intelligible universe it is impossible even to raise the question of the curvature of space, as the Euclidean geometry is one of its horizons of intelligibility. In the big bang cosmological model the singularity which makes up the origin of the universe is a horizon of intelligibility of that model. In reaching the point of the singularity the equations break down, their solutions become infinite, the concepts stop being meaningful, and the model does not work any more. It is impossible to go beyond that point, which so constitutes a horizon.

The uniform validity of the laws of physics in all the points of spacetime is a horizon of intelligibility of any intelligible universe. This does not imply that reality itself is like that. Reality itself could be quite different. The laws we know could have only local validity. The speed of light could be constant only in our corner of the universe. The spectral redshift could not be due to the recession velocity. Masses could repel each other. Energy could not be conserved. Of course such an universe would contradict our scientific laws, and it would not be intelligible to us. But it could just be that the real universe is not intelligible after all, in which case we would have no chance of ever understanding it. Nevertheless there is no reason for going on with this line of speculative and futile scepticism. In as far as we want to think about the universe and to conceptualize it, we have to conceive it as an intelligible universe, as a cosmological model, at least provisionally. Our only chance of going beyond the limits of an unsatisfactory cosmological model lies in our building a new cosmological model of broader scope. But with this we will not have left the intelligible universe; we will just have enlarged it.

In our model building activity we seem to have a tendency to extrapolate from the local observations to the whole universe. So the standard big bang cosmological model assumes that the isotropic microwave background radiation we detect on the Earth fills the whole universe, even if there is absolutely no observational warrant for this hypothesis. The same can be said about the relative abundance of hydrogen and helium [Moles 1981].

The perceptible world has a certain stability, as it is anchored in our genetic constitution. The observable universe is an extension of the perceptible universe, and both are obviously linked. The observational results which are not perceptible (for instance, those relating to radio waves) have to be translated into perceptible images (like the audible signals of the radio set or the diagrams or pictures synthesized by the computer) in order to be intuitively assimilated. On the other hand, the perceptible intuitions are often corrected by observational data. Although it is perceptively evident that the sun rises each morning in the east, moves above our heads and sets each evening in the west, we reject that evidence because of other less immediate observational data.

The intelligible universe draws the limits of the observable universe and corrects its data. Today we believe there is more mass around than meets the eye (or the telescope), and this belief is grounded in theoretical arguments. This interface between the observable and the intelligible universe gives rise to frequent discussions and polemics. For example, the astronomer Halton Arp [1987] believes he has good observational reasons to think that the presumedly far-away quasars are excrescencies of much nearer galaxies. If both of them present different redshifts in their spectra, so much the worse for Hubble's law. Arp thinks it is Hubble's law which has to be dropped, not his observations. But the immense majority of the scientific community, committed to a cosmological model in which Hubble's law plays a central role, rejects Arp's conclusion, at least as long as no alternative and more convincing cosmological model is put forward, what Arp does not do.

The observable universe is the touchstone for the intelligible universe, whose first function is to account for the data of observation. New observational data can lead to give up certain cosmological models, as it happened with the steady-state model of Bondi, Gold and Hoyle after the counting of radio sources by Ryle and others (which showed time asymmetries in the number of sources) and, specially, after the cosmic background radiation was discovered in 1965. The same fate could befall the inflationary model, if ever in the future a background radiation of gravitational waves is discovered.

4. THE WHOLE OF REALITY.

The limits of the intelligible universe are the limits of our intellectual and conceptual resources, and they need not be the limits of reality. The whole of reality is everything there is, without any limitation, and with independence of whether it is knowable or thinkable or not. Hence the whole of reality can be properly be said to be the unlimited, the *ápeiron* (the Greek word used by Anaximandros for it).

Anything we are able to think or to say about the universe falls by definition in the scope of the intelligible universe, which is the domain of what can be symbolically articulated and specially the domain of language and mathematics. In as far as reality reaches beyond the intelligible universe, it is an ineffable reality, with which it is possible to entertain a mystical relation, but about which it is impossible to speak.

Karl Jaspers [1935, 2. Vorlesung] has called *das Umgreifende* (the encompassing) "that which encompasses every particular horizon... and which already is not visible as horizon". Benjamin Gal-Or [1987, p. 349 ff.] has called it *Havaya* (the whole, in hebrew). Milton Munitz [1986, chapter 6] has called it the *Boundless* or boundless Existence. This unlimited and encompassing whole, this total and ultimate reality, includes and transcends all perceptible, observable and intelligible horizons.

We cannot perceive beyond the horizon of perception, we cannot make observations outside the observable universe, and neither can we represent, think or understand reality beyond the intelligible universe. Nevertheless, in so far as they go, our perception perceives reality, our observations are real, and the universe we strive to understand is the real universe, reality itself. We only partially achieve what we strive to do. All the same, it is astonishing how much we achieve. The whole of reality is like the able to reach that limit and achieve a total coincidence between his intelligible universe and reality itself. Unfortunately, everything seems to indicate that gods do not exist.

With these last remarks about the ineffable whole of reality I find myself speaking about the unspeakable. This is an unmistakable sign that I have overstepped the mark, and that it is time to close.

Department of Logic, History and Philosophy of Science
University of Barcelona, Spain

REFERENCES

Arp, H., 1987. *Quasars, Redshifts and Controversies*. Interstellar Media. Berkeley.
Börner, G., 1988. *The Early Universe*. Springer-Verlag. Berlin-Heidelberg.
Brillouin, L., 1962. *Science and Information Theory* (2nd edition). Academic Press. New York.
Feinberg, G., 1985. *Solid Clues*. Simon & Schuster Inc. NewYork.
Frautschi, S., 1988. Entropy in an expanding universe. In *Entropy, Information and Evolution*, B. Weber...(ed.), TheMIT Press, Cambridge, Mass.
Gal-Or, B., 1987. *Cosmology, Physics and Philosophy* (2nd edition). Springer Verlag New York Inc.
Harwit, M., 1981. *Cosmic Discovery*. The Harvester Press.Brighton.
Jaspers, K., 1935. *Vernunft und Existenz*. Groningen.
Layzer, D., 1988. Growth of order in the universe. In *Entropy,Information and Evolution*, B. Weber...(ed.), The MIT Press, Cambridge, Mass.
Moles, M., 1981. *Cosmologla y observaciones*. Investigaci6n y Ciencia, Julio.
Mosterín, J., 1989. A round-trip ticket from philosophy tocosmology. In *Foundations of Big Bang Cosmology*, W. Meyerstein (ed.), *World Scientific Publishing*, London and Singapore.
Munitz, M., 1986. *Cosmic Understanding*. Princeton UniversityPress.
Penrose, R., 1979. Singularities and time-asymmetry. In *General Relativity:* An Einstein Centenary, S. Hawking &W. Israel (ed.), Cambridge University Press.
Thorne, K., 1987. *Gravitational radiation. In 300 Years of Gravitation*, S. Hawking & W. Israel (ed.). CambridgeUniversity Press.
Wiener, N., 1961. *Cybernetics* (2nd edition). The MIT Press.Cambridge, Mass.

Massimo Pauri

THE UNIVERSE AS A SCIENTIFIC OBJECT

1. INTRODUCTION

The aim of the present contribution is twofold:
1) to argue in favour of the thesis according to which the Universe *as a whole* cannot be considered as a *scientific object* in any sense that such words have had in the historical development of physics. Although the classical dialectical arguments are not used, the conclusion will be drawn that, essentially, its role is still that of a «regulative framework» having at most the same methodological role as the «idea» in the Kantian sense;
2) to argue in favour of the thesis according to which there is no justification for relativistic cosmology's claim that it has given or is about to give *empirical* answers to the traditional «cosmological problems» (whether the Universe is boundless or not, what its age and destiny may be, how it came about, and so on). And this, not on the basis of philosophical reasons alone but also as a consequence of the weakness of its logical and empirical foundations.

Reading the following quotations will take us to the heart of the matter immediately and give us a feeling for the epistemological condition of contemporary cosmology; they are all taken from works of authoritative members of the scientific community:

i) «The existence of the Universe is clearly its most important characteristic, but I am referring here to the stronger idea that it is meaningful to talk of the Universe *as a whole*, as a single well-

E. Agazzi and A. Cordero (eds.), *Philosophy and the Origin and Evolution of the Universe*, 291–339.

defined concept. This idea is one of the most important, perhaps the most important, scientific discovery of the twentieth century» [1];

ii) «Cosmological theories which pretend to describe the Universe from the *very beginning* (including the quantum-gravitational stage) and up to the present time we shall call *complete*... We assume that in the initial state there was nothing except zero (vacuum) fluctuations of all physical fields including gravitational. Since the notion of space and time are classical, in the initial state there were no particles, no space, no time. Speaking about time, one can say, loosely, that there was a time when there was no time. It is assumed that as a result of a fluctuation there appeared a classical 3-dimensional closed geometry.... It is natural to expect that all the characteristic parameters of the world were Planckian, i.e. classical space-time comes into being at the limit of applicability of classical general relativity.... What are the achievements of the proposed scenario? First, it answers the question about the initial state of the Universe. According to this scenario, the Universe originates, in a certain sense, from nothing.... We assert that the cosmological singularity must be replaced by an essentially quantum-gravitational process which can be called spontaneous birth of the Universe.» [2];

iii) «Present cosmological theories.....have the character of ancient Indian myths, with turtles standing on elephants standing on...Very beautiful fairy tales.» [3].

As one can see, the breadth of such sentences ranges from a «cosmolatry» of a sort, (as Herbert Dingle would have defined it), through «cosmo-mythology and beyond», to the most outspoken, albeit poetical, skepticism.

The epistemological condition *natural* to cosmology, in recent years, has been strongly intermingled with the need of elementary particle physics to find a «laboratory» suitable for its high energy ambitions. On the other hand, one cannot deny that this «cosmic connection» of high-energy physics has given a true «coup de grâce» to the attempt to keep the scientific imagination characteristic of cosmology under control [4], and has made of it a «Science à la Mode». In this situation, I believe I will be allowed to introduce a small contribution in the opposite direction into the overwhelming flow of fashion. The extraordinary lack of critical constraint of certain sectors of the scientific community should induce us to recall the serious difficulties, both philosophical and empirical, which lie at the basis of the demiurgical enterprise of contemporary cosmology, and this even if one runs the risk of repeating, at least in part, well-known arguments. At any rate, my aim here is mainly that of pointing out some peculiar features of contemporary cosmology which seem to have been undervalued or even forgotten also within the cultural community at large.

The conception of the Universe which is implicitly contained in the cosmological perspective presented in i) and ii) will be called *strong*. It is characterized by the consideration of the Universe as a *physical object* which is *given* more or less like an ordinary *thing*, i.e. as an entity of which, for instance, one can say that «it has an age». This point of view presupposes that one is perfectly free to talk, in a non-metaphorical way, of a «material totality of the real» without worrying too much about questions of principle. And, furthermore, it purports to be the true scientific heir of the «cosmological tradition». Substantially at variance with this, of course, is the point of view according to which one merely makes use of the *name* Universe and related mathematical models to organize and unify observational data. In the latter case one has simply to deal with the *largest set* of astronomical objects presently observable, i.e. with the conventional object of astronomy and astrophysics. I will call this second point of view *weak*.

In the following, I will try to show that the historical development of the notion of *scientific object* entails a special relationship with the phenomenological experience of «subjectivity», which becomes emphatically critical precisely within the *strong* cosmological perspective.

The thesis of the *strong* cosmology is in particular associated with the view which holds that the historical formulation of the general theory of relativity (GTR) would have made available, for the first time, a physical and mathematical framework suitable for application, in a rigorous way, to a *whole system* (finite or infinite), and in so doing, it would have changed the cosmological problem from an inconclusive philosophical debate to an empirically meaningful one [5]. This attitude, supported by the impressive improvement of astronomical observations, would lead us to define cosmology as an empirical or even experimental science *to all intents and purposes*, on the same footing as any other branch of physics.

However, it seems to me that attributing the above mentioned role to GTR is both true and paradoxical at the same time. Actually, GTR, *per se'*, not only is not a sufficient basis on which to found cosmology but, a-priori, it even renders problematic and indefinite the semantic content itself of the notion of Universe. As a matter of fact, on one hand, additional assumptions are required for the foundation of cosmology which amount to a strong epistemological and ontological commitment, and, on the other, GTR on account of its technical features, transforms the *definition* of Universe itself into a preliminary problem far from trivial. For instance, the general definition suggested by Bondi [6], according to which «the Universe...is the largest set of all physically significant objects», turns out to be rather ambiguous or, to say the least, misleading. First of all, unless special assumptions are made within GTR, the physical (necessarily local) definition of time cannot be extended to non-local domains, so that there is no concept of a global simultaneity and no concept of a *global (cosmic) time*. Consequently, the notion of an objective sequence of temporal states of objects in space and the very notion of *existing objects* has no universal meaning: the objective distinction between «the already happened» and «what has still to happen» has disappeared. We are confronted here with the fundamental ontological fact that, once the notion of *existence* of physical objects is relativized by relinquishing the Newtonian concept of time (i.e. time *simpliciter*), it looses its essential content.

The possible *physically significant objects* describable by GTR are 4-dimensional objects: they are fields, including the space-time metric itself, or «matter», given essentially in terms of phenomenological aggregates of «genidentical» world-lines (i.e.

corresponding to a space-time sequence of events characterized by a definite and unique *identity*), i.e. they are point particles or classical fluids [7]. There is no consistent theory of quantum gravity (QG), so that all the «matter-energy» described by relativistic quantum theories can be *forced* into the conceptual framework of GTR only in a provisional and heuristic way. Of course, no «genidentical» world-line survives in quantum theories. Furthermore, QG aside, space-time itself should be a deterministic object, a particular solution of the differential equations of GTR, specified in its 4-dimensional *wholeness* by some *initial* or *boundary* conditions. However, as shown by Ellis [8], in actual models of the Universe it is not possible to obtain, even in principle, a complete description of the space-time metric on the basis of observational data, since some elements of it are not directly measurable. In other words, even the knowledge of our «past», including all the information which could have reached us from the hypothetical «beginning», would not be sufficient to forecast the «future» of space-time, unless some additional and not observable conditions be specified. Furthermore, as noted by Glymour [9], independently of the value of the constant k of the relativistic cosmological models, there may also be an intrinsic empirical undecidability as to the global topological structure of space-time. Thus, some basic epistemic *indeterminacy* could exist as an intrinsic feature of the cosmological models. Finally, the theoretical existence of singularities, which seems unavoidable on the basis of Hawking's global theorems [10], clearly limits the validity of the claim that GTR provides us with a self-consistent treatment of the whole Universe. In conclusion, not even the stage of the play is established in its wholeness when the curtain rings up.

At this point, it should be clear that the very notion of «largest set of all physically significant objects» is undefined and that cosmology *in primis* has to tackle the problematic question of defining the semantic content of its object of research. At the very beginning of the inquiry, the Universe can be at most, from the scientific point of view, nothing but a conjecture which expresses, vaguely, the idea of *unity* of the real in the form of *unity* of physics.

2. PHYSICAL OBJECTS

As is well known, it is very difficult to specify what is *physically real*: every attempt of analysis in this direction leads to a «regressus ad infinitum». A satisfactory *formal* definition of *physical object*, together with a characterization of its ontological implications, is also known to give rise to serious difficulties (from Meinong, to Russell, to Quine). Nonetheless, since my purpose here is mainly that of clarifying in which sense the Universe *cannot* be considered a *physical object*, a formal definition will not be necessary. It will suffice to recognize the various stages of formation and transformation of the concept of *physical object* and keep track of the common or *invariant* features in such an evolution.

First of all, obviously, I will not simply refer here to the concept of object as a scientific abstraction of the mere notion of *body*, that is to say a simple stable entity which is distinct *from*, but related *to*, space and time. Indeed, as a consequence of the profound historical evolution that the concept of *physical object* has undergone, that simple abstraction cannot represent, even slightly, the more general content of what *scientific object* means today.

In order to outline the ontological transformations of the concept of *physical object*, I believe it advantageous to identify three main historical periods. The classical period is first; the second originates during the Renaissance with the so-called «mathematization of the secondary qualities» [11] which marks the birth of the modern concept of scientific object. The beginning of the third period, which follows Kant, could be identified with the discovery of Non-Euclidean geometries, the works of Riemann and Clifford and Maxwell's formulation of the electromagnetic theory. I suggest calling this last period a «new mathematization of the primary qualities», since it corresponds to the progressive transformation of the very concepts of space and time into logical-empirical objects themselves and to the first historical appearance of *nomological objects* within the relativistic quantum field theory (RQFT) (I mean *species of objects theoretically defined*): the distinction between primary and secondary qualities (in Locke's terminology) tends to vanish as this new phase unfolds. Within these periods, I believe that at least four or five main steps in the evolution of the concept of physical object should be recognized: the Newtonian mechanism; the formulation of the classical (deterministic) field concept, either as ordinary physical entity or, especially, as metric structure of space-time; the discovery of the quantum *discontinuity*; the formulation of RQFT, and eventually but still in progress, the formulation of a theory of QG.

I will reiterate the essential features of these basic steps. The Newtonian object is characterized by the notion of *simple localization* (Whitehead). The world is decomposable into things; things are aggregates of *material points* (Euler), i.e. entities which always maintain their own identity except for their motion in space with time; mass and force are extra-geometric attributes. The deterministic classical field in vacuum has its own ontological autonomy and identity, not mechanically reducible to material points or fluids; by eliminating and substituting the notion of a *material substratum* of the oscillations, the new concept reduces the Democritean ontological opposition between things and space and sets the stage for a progressive physicalization of geometry. A new *empirical* question of time is asked and answered by the formulation of the restricted theory of relativity (RTR), while a new *empirical* question of space-time is asked and answered by GTR. I will consider later the essential conceptual elements of the relativistic revolution. Some fundamental features of the quantum revolution will be relevant in the following: I intend to discuss them briefly.

As is well known, the *aporias* concerning divisibility or discontinuity have a long history, from Zeno, to Descartes, to Kant: they are essentially connected to the lack of intelligibility of any *actual indivisibility* of an homogeneous extension which is being more or less realistically conceived. Now the *true atom* of contemporary physics is the *quantum of action*, which no longer expresses the indivisibility of a *thing* but rather that of a *process*. A radical conceptual revolution follows from the fact that the *individual quantum jump*, corresponding to the exchange of a single quantum of action, *cannot be described in space and time* since no continuity of physical states exists between the initial and the final one. This has dramatic consequences on the very concept of object: the QT no longer describes objects having a *form* and a distinguishable permanent *identity*, which could be given by an aggregate of events; instead it provides a symbolic description of an essentially *potential* reality. In other words, dissolving the *thingness* for micro-objects - which is a fundamental requisite to empirical macro-physical intuition - is a logical premise for obtaining the maximum of intelligibility of the quantum discontinuity. On the other hand, it is apparent that the consequences of this state of affairs on the concept of space and time, have not as yet been completely exploited. Although the real topological *continuum* can still be utilized in RQFT, the physical role of space-time in it is very peculiar and it is likely to be only provisional. The space-time dealt with in RQFT is a universal deterministic macro space-time (actually the Minkowsky space of the RTR) which is being utilized without any scale limitation from below. As a matter of fact, however, it is introduced into the theory through the group-theoretical requirement of relativistic invariance of the results of measurements with respect to the choice of the macro reference frames in the laboratory. Therefore, this «micro space-time» is anchored to the macroscopic medium-sized objects which asymptotically define the experimental conditions. Its role seems to be essentially that of a *translator* from the symbolic structure of the theory into the language of the macroscopic irreversible traces which set up the experimental findings within macro space-time. However, the limitations posed by QT on the procedures for measuring microscopic space-time intervals (easily reducible to time intervals) are very strong indeed. It can be shown [12] that every possible kind of clock is necessarily a non-microscopic object and, in particular, that even the so-called *atomic clocks* can be considered microscopic devices only on the basis of a rough idealization which conflicts with the fundamental concepts of measurability. Actually, there exists no way of measuring microscopic intervals (e.g. intra-atomic) in a direct operational manner. All measurements are then based on indirect procedures which involve a theoretical interpolation. This, of course, is true for the great majority of physical measurements: the special aspect of the microscopic case, however, is the fact that the theory involved is precisely QT with its probabilistic structure. In addition to the traditional distinction between physical and *phenomenological* time, we have then to consider a further distinction: namely that between physical time, on one hand, and physical *analogical* time (space-time), on the other. The former, as operationally founded on the behaviour of (macroscopic) *clocks*, and the latter as a mathematical structure whose empirical meaning rests on a theory which excludes in principle, except in a mere *analogical* sense, a complete phenomenical representation within the chain of space-time sensible intuitions [13]. I would like to stress my opinion on this delicate epistemological point. Far from worrying about a direct operational justification, what I want to emphasize is that the empirical meaning of the *«analogical»* time (or space-time) is conferred on it solely by the *macroscopic (classical) level of control* of QT and in particular by its being extrapolated from the «macro» time of RTR, which is a well-defined physical concept. A final remark concerns the heuristic attempts to construct a theory of QG. Within the so-called «canonical QG» (in the «metric representation»), we would face an *indeterminism* of a superior order with respect to that met in RQFT. While in the latter theory the «propagators» of the micro-objects have definite causal properties, determined by the pre-assigned micro-causal structure of *analogical* space-time, no pre-assigned space-time structure can exist in the former and consequently neither can any pre-assigned micro-causal structure. Here, the indeterminism is inherent in the space-time structure itself. The deterministic history of space evolving in time has no longer any definite meaning: space-time becomes a purely *virtual* entity. In particular, it is the concept of time which literally does not exist within the «canonical QG» and must be *extracted* from the geometric fabric of the theory only in a classical or semiclassical approximation. It is hardly necessary to remark that all these features have a purely qualitative and heuristic value. I conclude this part with a caveat concerning a very general problem. The *virtuality* of space-time in QG brings to a climax the (macroscopical) paradoxes of QT and, in particular, the puzzles connected to the unrestricted allowability of quantum coherent superpositions of macroscopically distinguishable states. Actually, an indeterminacy of space-time structure at the macroscopic level (i.e.not confined to microscopic or better Planckian scales brackets 10^{-33} cm, by some coherence-breaking mechanism) would result in the most striking of quantum paradoxes and would be clearly unacceptable with respect to a wide range of philosophical positions.

Summarizing, we must now ask which are the main methodological common features shared by the various historical periods and conceptual steps referred to above, as far as the notion of *physical object* or *scientific object* is concerned. I believe we can acknowledge that the description of a conventional physical object in the widest sense, is always based on a scheme which implies a number of *exclusions*. In particular, this description implies the subdivision of the world into three parts: a first part which we can properly define as *the object* (or the physical system), a second part which is *the observer* (or the measuring apparatus) and a third part which is *the remainder of the world*. This last part is considered *irrelevant* to the immediate purpose; its variability, however, determines the *contingent component* of the description through the *initial* or *boundary* conditions (note that having included even space and time within the class of conventional physical objects is not inconsequent, to the extent that we consider them locally). The above separation allows for the formulation of the *possible* time variations of the *object* (laws), in connection with the choice of different relations to *the remainder of the world*. As is well known, this separation is explicitly theorized within the standard interpretation of QT. The distinction between *contingent* components and general laws or *forms* of possibilities implies, furthermore, that a scientific object is never constituted by a single object-like determination.

Actually, it always corresponds to an *equivalence class* of single determinations: this is precisely determined by the *abstraction* from those different relations to *the remainder of the world* which, in the case at hand, are deemed unimportant. Essentially, any object, be it physical or not, presupposes an infinite multiplicity of perspectives. A final pre-requisite for the formulation of the various possible evolutions of *the object* in the form of physical laws, is expressed by the following «Galilean» conditions: i) indefinite temporal repeatability of the whole set of relevant relations between *object* and *the remainder of the world*; ii) irrelevance of the spatial relations between a suitable region, defined by the *object* itself and its local environment, and *the remainder of the world*. These conditions, which extablish the distinction between purely empirical disciplines and experimental science, imply in particular space and - above all - time homogeneity, together with the possibility of repeating the desired world-separation at will.

Clearly, if we take the above features to be necessary elements for a general definition of *scientific object*, the statement that the Universe cannot be a *scientific object*, becomes not only true but indeed a truism. This obvious conclusion does not provide any suggestion as to what could be considered *typical* features of an allowable description of the *totality of material entities*, according to some, possibly reduced, criteria of scientificity. In order to examine this point closer, it seems necessary to probe deeply into the primary «phenomenological» structure of the scientific attitude towards reality in general, as it has been shaped historically, at the beginning of the second period mentioned above. The realm of all possible *scientific objects*, I believe, is precisely determined by this primary structure. This point, however, can be properly dealt with only after having clarified a fundamental aspect of the physical description of nature: namely the general role played in it by the notion of time. This in fact, although deeply connected to the above mentioned *phenomenological* structure, holds a proper autonomous characterization within the theoretical framework of physics.

3. CONSCIOUS SUBJECTS

With this in view, let me return to the relativistic revolution. To set the stage immediately, I will quote Hermann Weyl directly[14] : «The scene of action of reality is not a 3-dimensional space, but rather a 4-dimensional world, in which space and time are linked together *indissolubly*. However deep the chasm may be that separates the intuitive nature of space from that of time in our experience, nothing of this qualitative difference enters into the objective world which physics endeavours to crystallize out of direct experience. It is a 4-dimensional continuum which is neither *time* nor *space*. Only the consciousness that passes on in one portion of this world experiences the detached piece which comes to meet it and passes behind it as history, that is a process that is going forward in time and takes place in space....The objective world simply *is*, it does not *happen*. Only to the gaze of my consciousness, crawling upward along the life line of my body, does a section of this world *come to life* as a fleeting image in space which continuously changes in time».

It is profitable to distinguish and discuss separately the various epistemological levels of the relation between time and space which are only implicit here. Two main orders of questions must be examined: The first has to do with an *intersubjective* separation of time from space in relativistic space-times and with the consequent definition, common to specified classes of observers, of a *physical* (macroscopical) time. The second has to do with deeper ontological questions such as the issue of the *objectivity* of «becoming» and the so-called «arrow» of time. Both orders of problems have a direct cosmological relevance: I will discuss the second one first, deferring the analysis of the first one to a section on relativistic cosmology.

As Grünbaum [15] has lucidly pointed out, the second order of problems shows a three-level structure. Precisely: i) the representation of time, as a scientific entity, by means of a one-dimensional real continuum; ii) the ascription (or non ascription) of *physical objectivity* to the *existential* property of «becoming»; iii) the existence (or non-existence) of irreversible physical processes capable of conferring an intrinsic directionality or «anisotropy» on the real ordered time continuum.

Level i) amounts to a formal problem: by means of a *betweenness* relation, it is possible to define two ordinally opposed time directions on any world-line and therefore to attribute time coordinates t, increasing or decreasing, to the events of the simultaneity classes which constitute the opposed directions. The question as to which direction the increasing (decreasing) coordinates are attributed is completely irrelevant here. At this point, relations such as «earlier than» and «later than» are suitably defined in terms of time coordinatization without any implication regarding the existence of irreversible processes. If such processes exist, then a further differentiation arises between the two time directions: sequences of states of physical systems exist, characterized by, e.g., increasing time coordinates, which do not exist in the direction of decreasing coordinates. In this case, we can say that an «arrow» of time exists or that physical time is intrinsically «anisotropic». It is well known that, until now, this question has received no satisfactory *nomological* answer although the appearance of a time «arrow» in everyday experience is a generalized and unquestionable observational *fact*.

Level ii): within relativistic theories, the time ordering of causally connected events gives rise to an objective representation of evolution in time which appears to be «sub specie aeternitatis»: since it does not make any reference at all to the «now», it is given, as it were, «once and for all» and it holds true, independently of the time at which it is being used. The ascription of «becoming» to events, which originates from the common sense kinematical metaphor of the «flowing of time», does not have any justification within the physical description. Actually, asserting that an event, say E, belongs to the «now», is not determined by some physical attribute of E, nor by objective relations existing between E and other purely physical events. It is instead determined by the observation of something as *existing* and therefore it refers to the immediate awareness of a «conscious subject». The assertion of «nowness», on the other hand, can be transformed a-priori into an intersubjective concordance among a plurality of conscious subjects only if they belong to the same Lorentz reference frame. In this connection, it is clear that a proof, should there be one, that a universal *physical cosmic time* cannot be defined, would render even *logically* inconsistent the ascription of an objective «becoming» to physical events. It is important to point out, however, that the relativity of simultaneity is not a necessary condition to the argument which refutes that ascription. Actually, it is true that a hypothetical subsistence of an objective, *universal and univocal* «becoming» of physical events would validate the assertion that reality, as described by physics, consists of a sequence of states of «nowness» which come subsequently into being, and would imply the existence of a *physical cosmic time*. The contrary, however, is false: a physical definition of *cosmic time*, should there be one, would not amount to ascribing the *existential* property of «becoming» to physical events (it appears that Gödel [16] overlooked the relevance of this logical asymmetry, which strengthens his own thesis, a fact that is rather surprising in view of the position taken a quarter of century before by Weyl, who is the inspirer of the concept of physical cosmic time). As a matter of fact, there is a lot of confusion about this issue in scientific and philosophical literature where, in addition, levels ii) and iii) are often erroneously misidentified [17]. In particular, the alleged assertion of the «becoming» of physical events, is misconstrued as a physical *shifting of the «now»* having the capability of singling out the future direction of physical time as the sense of its *advancing*; conversely, although the physical «arrow» of time does not involve the transient «now», the assertion that there is a factual (though not nomologically explained) anisotropy of time is taken to be equivalent to the claim that there is a «flow» of physical time in the direction of the future. In this connection, it is important to stress that the conclusions drawn up to now, as to the mental nature of «becoming», must be drawn not only for relativistic theories but also for Newtonian mechanics. The circumstance that they have been brought out within relativistic thinking is secondary and clearly due to the peculiar latency of the *observer* within the classical framework. In conclusion, the physical description of reality does not possess a criterion for distinguishing what *exists* from what *does not exist* in the phenomenological sense of the term. It is not possible therefore to confer scientific meaning on the sentence that *past* events «do not exist any longer» while *future* events «do not exist yet». As Russell puts it, «past, present and future arise from time-relations of subject and object, while earlier and later arise from time-relations of object and object». I will say that the description of events furnished by physics is «non-transient». I believe that the «**non-transient**» modality of the temporal evolution presented by actual physical theories is a structural and unavoidable feature of the scientific enterprise, at least in the historical methodological form which was shaped at the beginning of the previously mentioned second period.

Cassirer [18] carried out a very deep analysis of the process through which the Platonic-Aristotelian ontology of substance was surpassed and the atomistic conception of Democritus was reconsidered, giving rise to modern science. I would like to focus the discussion on a single, fundamental point of the Democritean renewal. As is well known, according to Democritus the true reality of Being is not what *appears* and cannot be found within the subjectivity of representations and qualitative perceptions. Instead, it has to be searched for in the *substantial* elements of *matter* (primary objects or *atoms*) in terms of which alone an *explanation* of qualitative perceptions can be obtained. Historically, the resumption of the gnoseological function of this «Democritean circle» has been prepared and accompanied by a fundamental change in the relation between reality and knowledge. The immediate content of perception no longer matched the *internal* conditions of specific substances but it came to be valued according to a mutual *functional connection*, which in turn was susceptible to mathematical representation. In particular, the basic idea of constancy, which classically was grounded on the permanence of substance, was now to be traced to the *functional form*. It is this functional form which confers order on the multiplicity of variable elements of reality and provides the foundation of the «mathematization of secondary qualities» which marks the birth of the notion of *scientific object.*

By abstracting from the *concrete and actual* particularities of immediate experience, a structural *incommensurability*, a true hiatus, is being introduced between the theory and the phenomenical percept. However, far from being a paralysing opposition, this *incommensurability* constitutes - to use an expression invented by René Thom with reference to mathematical thinking - a sort of «aporie fondatrice» of the scientific program. By recognizing the fact that the mathematical analysis of the *qualities* can never be brought to an end, this program is originally grounded on the infinite project of reducing that hiatus. The program of objectivation of the scientific domain implies, together with the removal of the phenomenical percept, a provisional *suspension* of every *experiencing subject.* Historically, this new gnoseological strategy implied an epoch-making transformation of the very conception of *subjectivity*, a transformation which soon became stabilized and shaped many general features of modern thinking. A peculiar and extreme outcome of this development is worth discussing. One fundamental component of this transformation seems to be the predominance of the *principle of representation*, according to which a crucial distinction began to be generally overlooked. I am referring to the distinction between *unreflective* and *intentional* consciousness (which is unobjectifiable and unrepresentable) and the subsequent, objectified, self-consciousness which amounts to a *re-presentation* of the primary experience. Overlooking this distinction makes possible the identification (actually a misidentification) - which takes on different ontological modalities - of the *intro-spective objects* (including *mind*), inasmuch as they are *re-presented*, with the objects of knowledge of other conscious selves. This misidentification stands, in particular, at the basis of monistic (ontological) reductionism, which reaches the conclusion that *even* the world of *physical objects* «is» what is appearing within the immediate subjective experience and ontologically *identifies* them, thus granting the latter a purely *epiphenomenical* reality. In this way it is taken for granted that the scientific program can be brought to an end or even - as it were -

that it is *already* gnoseologically closed and reduced to a mere accumulation of technical improvements. The «aporie fondatrice» would have been thwarted by soldering the «Democritean circle».

In order to clarify the intrinsic limitations of the physical description of the world, I will try to exploit, exemplarily, the internal contradictions of monistic reductionism. In doing so, I will mainly use purely logical arguments and therefore I do not feel obliged to assume here an explicit and articulate philosophical position. I will confine myself to a position of tempered critical realism: ontological value of scientific knowledge but, at the same time, irreducible and foundational role of the *evidence* of the primary phenomenological experience, and, consistently, ontological bearing of the *alter-ego* i.e. of the plurality of conscious subjetcs.

The above mentioned contradictions show up as soon as one considers the consequences of the reductionist thesis on the problems of *free argumentation* and *free will* or, better, *free action*. Concerning the first problem, there exists a well-known argument which possesses, I believe, the logical strength of a «reductio ad absurdum». The following quotation from Eddington may be illuminating in this respect: «Suppose that our brain produces sugar if we say that 7 times 8 equals 56, while it produces chalk if we say, wrongly, that 7 times 8 equals 65. We cannot say, however, that sugar is *better* or *more correct* than chalk...». Precisely within the horizon of the alleged thesis, the supporter of monistic reductionism and his opponent will be ontologically completely *equivalent* as far as the *value of truth* of their argumentation is concerned. Coming now to the question of *free action*, a more detailed discussion is required. Although we all will agree that the Laplace ideal is no longer tenable, I believe that, to the extent to which the scientific description is considered applicable, the conclusion that the physical picture of the world entails a high degree of *macro-determinism* is inescapable. Clearly, the *hole* in the strict causality of nature opened by QT at the microscopic level and the theoretical novelties concerning the phase-space orbit instability of complex classical dynamical systems are completely irrelevant as far as the problem considered here is concerned [19]. As noted by Schrödinger [20]: «without the almost complete precision and reliability of the macroscopic thermodynamical laws ruling the nervous and cerebral processes of the human body and its interactions with the surrounding world, perception and thought would be impossible». In a similar way, Carnap [21] remarked that the existence of a high degree of macro-determinism is an essential pre-requisite to the very possibility of a *free action*. Actually, if a causal regularity did not exist, any meaningful decision would be quite impossible: since this entails a determined preference for a definite course of action with respect to any other, the meaning of any choice would be thwarted by the impossibility of foreseeing the consequences of alternative actions. The previsional capability is guaranteed only if the *choice* gets into a quasi-deterministic causal chain, although the consequences of the choice can obviously be forecasted only within a certain degree of probability, due to the limited information of the acting subject. In spite of the fact that every choice of action belongs necessarily to *one of the causal chains of the world*, Carnap maintains that the choice should not be judged as *compulsory* because, he argues, from a physicalistic point of view, it must be traced to *internal causes*, connected to the character and disposition or psychological history of the acting subject, rather than to a theoretical determination which could make possible an *external* prediction. However, I believe that, in this connection, Carnap overlooks three essential points. First of all, from a *realistic* point of view, if the *free action* is to be traced to *internal causes*(whatever «internality» may mean here), it is necessary that the acting subject be *screened* in some way, *before the choice*, with respect to the former external causal chains in which he is included. In the second place, the choice between alternative possibilities presupposes that they have a *meaning* for the acting subject, a fact that links the present problem to that of the *free argumentation*. As a third point, according to a phenomenological perspective, at least if there are no extreme external constraints, the acting subject experiences immediately a real possibility of alternatives which can even be consciously unsuitable to his own impulse or *character*. Therefore, granting the thesis of an objective causal explanation of *free action* (be it «physical» or «informational», *internal* or not), would entail the admission of a universal mechanism of *illusion* or *deceit*, some astute stratagem of nature which should be continually conspiring against the specific evidence of immediate experience. Of course, the same unreasonable admission would be required as regards the first problem, in connection with the immediate conviction which any conscious subject experiences as to his unconditioned capability of uttering meaningful sentences. It seems to me that, had we to make recourse to psychology (or perhaps psycho-analysis) in this context, it would be to get an explanation for the dogmatic and unintelligible existential tenets of the reductionists themselves.

4. A «MODEL-OBJECT»

On the basis of the foregoing discussion, I believe we can do nothing else than *accept* the fact that, at the present stage of knowledge, we are confronted with an inescapable *fracture* within our comprehension of the total order of reality, which was introduced by the scientific description of the world. The immediate evidence of phenomenological experience and of the *values* deeply rooted in it, together with the support given by independent logical arguments which I deem incontrovertible, impose the reality of *emergent* causal chains determined by *free actions* of conscious subjects [22]. Clearly, within a realistic perspective, one has then to face a problem of consistency between the specific essence of the *emergent* causal chains and the constraints specified by physical laws. My point, however, is that this problem *must* have a solution. The nature of this solution should in turn reflect the primary structure of the *suspension* of subjectivity which is immanent in the foundation of the scientific attitude towards reality. First of all, it is necessary to admit that every conscious subject is given an intrinsic, unconscious, capability of carrying out a *synthesis* of the interaction between his own body and the physical environment. Furthermore, this synthesis - as seen by us conscious beings

- should have the effect of transforming *symbolically* the *material effectiveness* of the previous causal chains, consistently with all the local physical laws. I am not interested here in finding a thorough and convincing *explanation* of this meta-empirical connection: consistency does not mean *explanation*. As already mentioned, I am not assuming here a detailed philosophical position, especially not a Cartesian *dualistic* one. I am simply claiming that an intelligible unity of the real requires this solution to exist, and I regard this to be sufficient for my present purpose.

Now, one may ask what is the connection arising a-posteriori, between the *emergent* causal chains and the space-time description provided by physical theories. An original aspect of the problem of consistency shows up here: the subsistence of *emergent* causal chains is simply **incompatible** with the «non-transient» description of time evolution, i.e., with the scientific description «tout court». Far from representing a logical contradiction, this asymmetric situation reflects precisely the aporetic nature of the «Democritean circle». In other words, while it is imperative to push the domain of the physical description as far as possible, a *total* scientific description of the real is not allowed. In fact, were one to try, by a purely hypothetical argument, to go beyond this structural limit, and to get a (incomplete anyway) physical description of a world including conscious subjects, one would have to *update* the space-time description in correspondence to *every emergent* causal chain. I would say consequently **that time is inherent in the emergent structure of the real as «becoming», but it is necessarily represented in «non-transient» modality as soon as it is objectified within the space-time description of physics. It could be said moreover that the inherent temporal structure of the real shows a particular kind of «anisotropy» which cannot be described within the nomological structure of physics. «Becoming» and «anisotropy» turn out to be strictly related at the level of the «real» temporal dimension of the world, as opposed to what happens at the physical level of description**[23]. It is also manifest that the **plurality of conscious subjects** plays a fundamental role here: it constitutes an essential bridge between the apprehension of subjective «phenomenological» time and the ascertainment of the **inherent temporal reality** of the world. On the other hand, this reality is in itself intrinsically open and cannot be given an objective and univocal formal structure. Therefore no implication at all can follow concerning the necessity of a universal (cosmic) **«real» time**.

The fracture thus introduced into the scientific description of the world comes unavoidably to the fore within the cosmological perspective. As a matter of fact, it seems that without having used any dialectical argument, it is nevertheless natural to derive a conclusion which, at least for my present purpose, is equivalent to that drawn by Kant in his first antinomy, precisely that it is impossible to think of the space-time totality of material entities as *closed* and therefore it is impossible to understand the Universe as a *scientific object* anyway, let alone a *thing*. Consequently, it is also natural to consider the notion of Universe as a «regulative framework», which has the same methodological effectiveness of the Kantian «regulative idea»: namely that of unifying theories and producing increasing empirical knowledge. This is precisely what I meant by *weak* cosmological perspective. In this sense, every possible cosmology originates from an *abstraction* which is essentially different from that which lies at the basis of any other *scientific object*, including the «analogical» objects of QT. Actually, within conventional physics, the *emergent* causal chains stay constitutionally *outside* the scientific description, or they take part, at most, as *measurements*. Therefore, conscious subjects, although *provisionally suspended*, are in some way tacitly accepted as *potentially* present in the background of reality. When it comes to the «totality of the real», however, separating out the «material entities», or smoothing out the *emergent* causal chains within a physical cosmological approach, leads to a peculiar kind of *inverted* statistical description which artificially assumes, as a real ontological possibility, not so much a *provisional suspension* as a *radical ablation* of any subjectivity. In essence, this fictitious separation is ontologically equivalent to an implicit acceptance of ontological reductionism (or of a rigid «dualism» as well). The obvious circumstance that the *emergent* causal chains are cosmologically irrelevant from a physical point of view, cannot modify the particular epistemological condition of the intellectual operation involved here [24]. I suggest calling the resulting *cosmological referent* a «model-object».

I will try now to enucleate, in a sketchy way, the main features which a scientific theory of this «model-object» (MOT) may have, as well as its epistemological limitations in comparison with conventional physical theories. These features follow mainly from the fact that the «model-object» happens to be *defined*, or better, *constituted* by the theory, in a quasi-Kantian sense. Here, however, I will not probe deeply into all the implications of applying the scientific method to a *construct* like the «model-object». In particular, I will not claim that all the weaknesses which I recognize in the logical and empirical structure of relativistic cosmology should be traced to the specific epistemological nature of the «model-object», although this will occur occasionally. (After all, all species of cosmologies must confront the difficulties resulting from that specific nature). For these reasons, the two argumentations set out in the Introduction will remain, in the present work, relatively separated.

First of all, either a MOT presupposes some «a-priori» principle of cosmic uniformity which guarantees the cosmological extrapolation of the local physical laws, or it should produce some difficult justification for this extrapolation in case this uniformity cannot be assumed. A MOT must presuppose, in addition, that the local laws be known to a high degree of approximation, since any variation which is deemed small, according to local criteria, can imply a catastrophic change at cosmic level. In particular, every *evolutionary* MOT must presuppose that a *global (cosmic) time* be empirically definable, in order to formulate the very concept of cosmic evolution; however, at the same time, it must be such that, within it, one could recover locally, both in space and time, the Galilean conditions of repeatability, i.e. spatial and temporal homogeneity. Even more, in any evolutionary MOT, it is necessary to *assume* that the local laws (derived on the basis of temporal homogeneity) can be *extrapolated in time*. In other words, it is necessary either to assume that the local laws remain *identically* valid, irrespective of the change of cosmic time (which is itself the source of *inhomogeneity* by definition), or to *fix* their variation in cosmic time *arbitrarily*, since this operation is *totally out of the empirical domain*. And this even under the extreme conditions of «matter» considered by evolutionary cosmologies. In particular, the cosmological applicability of GTR involves the hypothesis that its equations retain their validity in spite of the evolution of the «model-object» they rule, even independently of the circumstance that such equations predict (under certain conditions) precisely this evolution. Actually, this hypothesis is not so much an internal coherence condition of the theory as an epistemological pre-requisite. As a matter of fact, if a *strong* cosmological perspective is adopted, the *hyper-extrapolation* which is implicit in the assumption of *evolution-invariance* of the physical laws is such as to confer on them a *meta-empirical* status. The laws themselves must be thought of as *external* rules, that is to say, as a-priori structures which are ontologically pre-existent to the concrete and actualized Universe. In this same perspective, one should then be obliged to explain *why* the laws happen to be what they are in the actual Universe, a question which becomes even more critical in all speculations concerning the «birth», the «death and resurrection» of Universes, and so on [25]. Pushing the argument a little harder, one could even say that a *meta-empirical* status of the laws would require them to be *exact* laws and such as to realize «eo ipso» the unity of physics. Thus, in principle, cosmology should present itself as a *total knowledge* of the Universe, a condition which, in a certain sense, is immanent in every monistic gnoseology. On the other hand, if one correctly adheres to the more critical position I am defending here, there is no risk of ontological commitment of this sort. The proposed *weak* interpretation of the Universe, associated to a MOT, has another point of advantage in connection with the circumstance that applying the physical laws to the Universe entails the choice of *initial* or *boundary* conditions. Actually, the fact that only *certain* conditions are realized in the actual concrete, unique Universe, implies that the non-realized ones are totally deprived of any empirical meaning; in this way, however, the empirical meaning of the whole theory could be seriously questioned. On the other hand, in a MOT, due to its *constitutive* character, the *essential* features (as opposed to the *accidental* or *contingent* ones) are obviously selected by their relation to the MOT itself, rather than by an independent criterion. Therefore, from this point of view, the notion of «model-object» appears to be useful in saving the empirical meaning of the laws which are being utilized in its *constitution*. (Note, in this connection, that an apparent way out of this difficulty is that offered by the inflationary theories according to which the observable Universe is only one (a *bubble*) among an infinite number of others which are *actually realized*. In this way *all* the possible *initial conditions* could be considered as realized too. The epistemological situation, however, is even worse, since the meta-empirical nature of laws is still there and the whole Universe has become unobservable in principle except for our *bubble*; interestingly enough, some proponents of these theories explicitly describe them as «metaphysical (rational) cosmology»).

Another important difference between MOTs and conventional theories appears in connection with historical development. The infinite perspective of the scientific program and the necessarily finite amount of experimental evidence in any given historical situation, results in the obvious fact that any scientific theory is a *limited* theoretical structure, which has a recognized *domain of validity*. In spite of the fact that the historical development of scientific theories does not correspond to a progression of theoretical *inclusions* and widenings of the domains of validity, it is nevertheless legitimate to speak of scientific progress in a cumulative sense [26]. The important point to stress here is that, far from being a weakness of the scientific enterprise, the limited applicability of any specific theory appears to be a *necessary condition* for the existence of an *objective truth content* (although partial) of the theory itself. And this in the sense that only under this *boundedness*, can it be *historically stabilized* within the limits of a specific *domain of validity* and with reference to a specific *domain of objects. Limited applicability is thus also a necessary condition for the overall growth of scientific knowledge.* Furthermore, with reference to the so-called *discovery context*, in every conventional scientific domain, whenever one experimentally checks a new theory or a new physical hypothesis which go beyond the old knowledge, one has to assume in some way that the approximation guaranteed by the known laws is such as to be relied upon for interpreting the empirical chain going from the object-phenomenon to the observer or to the measuring apparatus. Only in a subsequent stage, is it possible in general to get a new consistent synthesis [27]. Now, from this point of view, it seems essential that the object-phenomenon be a limited part of the material world and that the experimental checks be indefinitely repeatable. In spite of the well known complexity [26] of the relations among *observational data* and *theoretical sentences*, there can be no question that no *observational datum* could ever be referred to the «model-object» as such. There will always be strong theoretical interpolations which contain fundamental a-priori assumptions. As Bondi [6] explicitly recognizes, while observational astronomy is already far less reliable than experimental physics, cosmological data, with one notable exception (namely the measurement of the isotropy of the background radiation - see, however, section 5), are among the least reliable in astronomy. In addition, a MOT is based on extrapolations of many local theories which may even be in some way «incommensurable» [26] among themselves (as happens e.g. in the case of GTR and RQFT whose «marriage» is only provisional and heuristic). Thus, even if the *interpolated* experimental data support all the extrapolation hypotheses, one cannot expect in general that they will be able to prove that the hypotheses are correct in a definitive (stable) way. Furthermore, «past» experiments - in the real sense of the temporal dimension of the world - are impossible; more generally, from the uniqueness of the «model-object», i.e., the lack of repeatability, it follows that cosmic experiments are necessarily made in terms of *correlations* among different points and instants; this implies that the MOT should have been given *before* one can even speak of an experimentally meaningful object (remember the quasi-Kantian *constitution* of the «model-object»). Therefore a MOT cannot be methodologically assimilated to some of the great known conventional theories. For example, it does not seem possible to assert that, given a MOT, another MOT could be *theoretically* more inclusive than the former one. Being theoretically more inclusive means that at least the class of the *observational predicates* i.e. the *domain of objects* of the new theory contains that of the previous one [26]: when this happens, the objects of the two theories can even be put into correspondence between each other by some semantical *projection* (or *amplification*), and in this case one can speak of a new theory with an enlarged *domain of validity*. In a MOT, however, the *domain of objects* is almost completely *theory-dependent*, due in particular to the heavy burden of a-priori assumptions on which it is based. Therefore one cannot say that this domain can ever be widened unless in the mere quantitative sense of an enlargement of the observational boundary. Substituting one MOT for another MOT is tantamount to giving a different definition or *constitution* of the «model-object». In this sense, a-priori, all possible MOTs should be considered *incommensurable* among themselves. This conclusion means that it is not possible to characterize a specific *domain of validity* of a MOT and, consequently, that a MOT cannot be *historically stabilized*. One could even conceive that new observational data will prove in the future that no species of MOT whatsoever could be acceptable with respect to reasonable criteria of scientificity. As a matter of fact, after approximately seventy-five years of passionate debate and research work, and in spite of impressive observational achievements, one cannot yet assert that the theoretical rank of cosmology within scientific thinking has been settled: cosmology still suffers from precisely the same epistemological diseases as at its birth.

In conclusion, it seems to me that the proposed terminology is sufficiently justified: instead of having a model (i.e. a provisional and pragmatic subsidiary scheme) for a theory, we have here a *theory for a model.*

5. COSMIC TIME

Let me begin by making some general remarks on the problem of the *physical definition* of time, within GTR in particular. First of all, it should be stressed that a scientific, empirical, definition of time has to be originally *local* in any case. The *stipulation* of temporal congruence is always based on procedures which never lead to measuring a *pure* time; they determine instead relations within a natural process which is essentially periodic. Therefore, all the procedures presuppose the individuation of *identical* states and, consequently, an intrinsic degree of approximation. On the other hand, the actual possibility of this individuation rests on prerequisites of (*weak*) cosmological flavour: there must be relatively autonomous and physically isolated *subtotalities* in which, within a controlled approximation procedure, a *stability of recurrence* can be recognized. In conclusion, the basic definition of *physical time* (actually *proper time*) is founded on a local procedure which is valid under *conventional* physical conditions. It is obvious that all these pre-requisites loose their meaning completely if applied to the *totality of material entities*. The notion, sometimes referred to, of a would-be universal time defined by the *evolution* of the Universe, considered as a clock in itself, is a true «petitio principii», unless a physical correspondence with a *locally defined* time can be established.

Now, let me consider the question of the *extension* of this local definition within GTR. As is well known, a synchronization of clocks for arbitrary observers is generally not allowed, although this should not be construed as preventing this possibility for particular solutions of the GTR. The negation of this possibility in general is strictly connected with Weyl's previously quoted assertion according to which there is, a-priori, no global *separation* of time from space. A number of different cases can occur: i) relativistic space-times in which an enduring physical 3-space cannot be defined, so that they cannot be given a global consistent time direction (in the sense of coordinatization) [28]; ii) relativistic space-times which do not possess a family of global space-like hypersurfaces, so that a one-dimensional time cannot be defined:

they violate the so-called *chronological condition* and may contain closed time-like curves which, formally at least, imply acausal features; finally: iii) space-times in which an enduring 3-space and a one-dimensional time t can be defined, but in which the world-lines of the space points are not the orthogonal trajectories of a one-parameter family of time slices of the manifold. Then there is no notion of time that allows for the spatial distance within the 3-space to be interpreted as the *space-time distance simultaneous with respect to t*, and consequently no global simultaneity either. The connection of these space-time features with causality has been exhaustively investigated by Hawking and Ellis in their well-known book. In particular, Hawking [29] was able to prove that a necessary and sufficient condition for the existence (let me say: theoretical definability) of a global cosmic time (i.e. the global orthogonalization of time and space) is the so-called *stable causality* condition: this requires that the *chronological condition* be satisfied in a neighborhood of *deformed* space-times adjacent to a given one. Of course, the question here is not the mathematical result but its actual relevance for the physical Universe. Hawking's argument is based on variational considerations and implies viewing GTR as the classical limit of some QG theory. Properly speaking, a satisfactory theory of QG does not exist. The only explicit mathematical results so far obtained [30] imply that the spatial metric (not to mention the space-time metric) cannot even be a quantum observable; time, for its part, can be defined only within a classical approximation, as already said. Thus the argument is rather speculative. Finally, as pointed out for example by Earman and Kanitscheider [31], rejecting closed time-like curves (*chronological condition*) on causal grounds, is not so convincing from a general point of view. As is well known, Gödel [32] has worked out an explicit solution of GTR in which there are closed time-like curves (and, of course, no cosmic time); running across them, however, requires unphysical velocities, and it would be interesting to see if they violate the Sakharov acceleration limit as well [33]. It seems therefore that we should praise Hawking's theorem as an important mathematical result whose empirical relevance, however, is still to be questioned. On the other hand, given the implicit mathematical background of Hawking's theorem, if cosmic time were not empirically definable, no necessary unpleasant *empirical* consequence would follow as to the chronological condition and the cosmological applicability of GTR.

Let me turn now to the logical and empirical foundation of *cosmic time*. Any relativistic Cosmology has to solve a number of fundamental theoretical problems. The most basic of these is the construction of a cosmic «substratum». Furthermore, this construction has to carry out the peculiar task of adapting a theory like GTR, which is founded on the generalized equivalence of all observers with respect to the formulation of the *laws*, to a situation which, instead, is characterized by the *real motions* of a *unique* system like the Universe. In particular, any evolutionary relativistic cosmology requires the «substratum» to be compatible with a proper definition of a *global (cosmic) time*. Actually, the most important of the conceptual revolutions brought about by GTR is the dissolution of the universal chrono-geometrical «substratum» provided by the Euclidean-Galilean framework. Space-time can no longer be, a-priori, a differentiable and metric manifold of *free-mobility* in which a universal notion of crono-geometrical *congruence* objectively resides. Chrono-geometrical *congruence* and *causal* structure depend now (at least partially) on the (space-time) distribution of matter; on the other hand, specifying the distribution of matter-energy requires in turn a space-time framework: so all should be given simultaneously and consistently. Finally, according to the original epistemological inspiration of GTR, the whole program should be realized in a *relational* way. Since now space-time coordinates have no objective meaning, *bodies* and light-rays first of all have to define, i.e. to *individuate*, points and instants, by conferring their identity upon them by means of coincidences, thus enabling them to *serve* as the *loci* of other bodies and events (as Grünbaum puts it).

The conventional solution to all these problems is based on a set of geometrical assumptions which, in their essentials, can be traced to a pioneering work of Weyl [34] («Weyl postulate»). The universal «substratum» is defined by a specific structure of *virtual* (not in the quantum-mechanical sense!) trajectories of *fundamental* particles which *relationally constitute* space-time. Precisely, Weyl assumed that there is an expanding time-like geodesic congruence of *virtual* trajectories having a common origin in the past [35]. As one can see, matters are turned around with respect to the standard approach: a geodesic is a geodesic of some metric; here a particular geodesic structure is assumed in order to *construct* a metric having certain desired properties. Actually, in the present case there is one and only one (up to spatial rotations) geodesic passing through each point of space-time. This means that at any point of space-time, the *virtual* matter passing through it possesses a unique velocity: this velocity in turn represents precisely the *relational expansion* of the «substratum» in the form of a hydrodynamical «motion». At this point, *fundamental observers* (FO) are ideally attached to every fundamental particle. The idealized nature of the «substratum» reveals itself in the further hypothesis that the FO form a continuous set. The specific chrono-geometrical properties of the «substratum» allow for a global orthogonalization of time and space and consequently for a definition of *cosmic time* having the property of being measured by the *proper time* for every FO comoving with the «substratum». In other words the peculiar «motion» of the «substratum» acts as a universal *synchronizing* mechanism for the FO's clocks: in this way the physical meaning of *local time* is transferred to *cosmic time*. At this point the so-called Cosmological Principle (CP) (spatial isotropy with respect to any point and consequently global spatial homogeneity) can be enforced by requiring that any pair of FO which are simultaneous with respect to the cosmic time see the Universe identically, i.e. homogeneous, isotropic and expanding at the same rate. In this way, relativistic equivalence is partially recovered, for the particular class of observers represented by the FO, as cosmological equivalence in the Machian sense of identity of relation with the global structure of the «substratum». Of course, the *proper time*, as measured by the clock of any other observer, *does not fit with the cosmic time* and it is subjected to relativistic effects. As is well-known, on the basis of the above framework, from GTR equations and a description of «matter» which presupposes the knowledge of an equation of state for it, the usual Friedmann-Robertson-Walker space-time metric follows: it is (locally) characterized in terms of a constant k and a cosmic *scale* function $R(t)$, where t is the cosmic time. Its global topological structure, however, is not entirely determined, and, what is more, as already mentioned [9], can even be found empirically undecidable. Thus, the conclusions about a closed or open Universe may turn out to be entirely a matter of philosophical faith even within the currently accepted relativistic cosmology. The assumed models of «matter» give rise to the *dust-filled* and the *radiation-filled* Universes, for the so-called *matter-dominated era* and the *radiation-dominated era* (terminated when the scale factor $R(t)$ was more than 1000 times smaller than its present value), respectively. Finally, it can be shown that, within a reasonable approximation, it is possible to recover the conditions of temporal and spatial homogeneity, locally, with respect to the inertial frames which move with constant velocity and without rotation with respect to the FO.

These are the chrono-geometrical premises. The empirical connection is set as soon as one identifies fundamental particles and FO with the *dust grains* of the *dust-filled* Universe and these, in turn, with the actual galaxies or centers of mass of clusters of galaxies. In order for this to be consistent, one has to assume first that, in *every region* of the actual Universe, the distribution and motion of matter are such that there exists a *minimal scale* suitable for defining an *average* which reproduces isotropy and homogeneity on the large scales and, second, that the *average motion* of these idealized averaged regions adjusts itself, within a good approximation, to the motion of the FO. That means, in particular, that the so-called *peculiar* motions of galaxies and clusters are unsystematic and very small relative to the geodesics, so that they can be neglected with respect to the overall cosmic flow of the «substratum». This is necessary not only for the chrono-geometrical identification, but also in order to approximate the galaxies contribution to the energy-momentum tensor by a smooth fluid with the same velocity and negligible pressure.

Let me assume for a moment that all of these empirical conditions are met. There are nevertheless a lot of conceptual problems. First of all, I observe that the assumption of a global spatial isotropy and homogeneity is in itself a strong extrapolation. In fact, every methodology for measuring cosmic distances presupposes a *model* of the «substratum»: model-dependent effects in the observations can be detected over the background of experimental errors more or less beyond the order of magnitude of the Gigaparsec. The assumption of isotropy and homogeneity beyond this scale is equivalent to assuming precisely what grounds the observations that should be used to check the model; then there is, at the very least, a big consistency problem. As already pointed out, this should be considered as typical of a MOT. This extrapolation is currently considered well-founded on the basis of the observed very high isotropy of the microwave background radiation (MBR) and of the quasi-perfect blackbody structure of its spectrum. The argument is that, since the MBR is gravitationally coupled to the expansion dynamics through its own gravitational field, if it is assumed that it has travelled freely towards us from a distance of about 10^{27} cm., remaining isotropic, one must conclude that on a large scale the space-time metric must be of the Friedmann-Robertson-Walker type (i.e. spatially isotropic and homogeneous) at least since the cosmic time when the MBR was emitted or last scattered. However, *this argument is inconclusive*. Actually, Ellis [36] has devised an interesting counter-model of the Universe in which there is no expansion, no homogeneity, no *cosmic time* and still there is isotropic MBR with respect to any observer which is situated outside a hot fireball encircling «the singularity» of the model. What is accounted for in terms of cosmic time variations in standard relativistic Cosmology gets interpreted in terms of spatial variations in this static model. The theoretical importance of this work is not so much that of providing a «true» model of the actual Universe as that of disproving the belief that the isotropy of the MBR might be a compelling argument in favour of the standard view.

Let me turn now to cosmological expansion. Since the «substratum» is *relationally constituted*, the physical meaning of its *expansion* is entirely based on the ascertainment of the *expansion-invariance* of the local reference *standards*. The question is currently answered on the basis of the observation that both atomic or nuclear *standards* and the local gravitationally-bound systems (planetary systems, double stars, even galaxies) do not expand. From this point of view, as noted by Wheeler, the expansion would be an «effect» of the large-scale homogeneity and isotropy. Thus, far from being a dynamical gravitational effect, the overall expansion should be viewed as a *de facto* behaviour of the «substratum» (realizing a global «inertial framework») which is merely *compatible* with the equations of GTR. In this perspective, the alleged *expansion-invariance* of the local gravitationally-bound systems would be the result of a dynamical action *against* the expansion of the «substratum». I don't believe, however, that the question is so simple. Given the fact that not even the two body problem can be solved explicitly within GTR, the question has been dealt with on the basis of various approximations. In a first work of Einstein and Strauss,[37] the Schwarzschild metric is replaced with continuity within empty 4-dimensional spheres which are cut out from the «substratum»; it is then shown that the overall expansion has the effect of expanding the matching boundaries of the spheres without modifying the internal metric. A more refined analysis by Noerdlinger and Petrosian,[38] however, shows that if one correctly looks at the «substratum» as an omni-permeating medium, the time evolution of gravitationally-bound systems depends in a negligible measure on their initial dimensions (i.e. on their being local systems !), while it depends strongly on the energy density of the «substratum» itself: in particular, for low densities, the clusters of galaxies do not expand while for high densities practically all the gravitationally-bound systems expand. Expand with respect to what? Of course, with repect to atomic and nuclear *standards*. The invariance of the atomic and nuclear *standards*, however, is an empirical *fact* which is valid under the *present* physical conditions and not a phenomenon which can be theoretically justified by an integrated theory of gravitational, electromagnetic and nuclear systems: it is in turn merely *compatible* with GTR in view of the *incompleteness* of GTR itself with respect to the description of matter; from the point of view of the *conceptual foundation* of the «substratum» this is not totally irrelevant, in particular for its so called «early phases». Given the fact that the *analogical* Minkowsky structure of RQFT has been experimentally tested up to about 10^{-16} cm. and that any further extrapolation of it to the physical conditions prevailing in the «early phases» of the Universe is merely speculative, it seems that there are serious conceptual problems for the very notion of expansion , at least close to the «beginning». I believe that this situation reflects in some way the confusion connected with the rather obscure *relational epistemology* which accompanied the birth of GTR. The main point is that the «Weyl postulate» (or its modern equivalents) has an ambiguous conceptual status: it is neither a *law*, since it is not contained in the equations of GTR for which it even presupposes the explicit solutions, nor a general *principle*, since empirically it can at most be realized *on the average*. In essence, this issue is just another logical component of the riddle of the «initial» conditions.

A third important point, actually the most important one, concerns in fact the empirical (and consequently the ontological) status of cosmic time. This question should be scrutinized by dividing it into various «temporal sections», each involving a different theoretical and empirical commitment; precisely: a section corresponding to present-epoch *«macro»-cosmic time*; a section characterized by the «analogical» or *«micro»-cosmic time* implied by the cosmological utilization of RQFT (specifically in the context of the so-called GUTs) in connection with the «early phases» of evolution; finally, a would-be section defined by QG regime.

First of all, even if we assume that all the empirical conditions mentioned before are met, it remains true that the definition of the *«macro»-cosmic time* is intrinsically *statistical* and approximate and depends on the particularities of the contingent distribution and motion of matter. Thus, cosmic time has originally the same theoretical status as the *temperature* of a gas. Furthermore, *cosmic time* is measured only by the privileged class of the FO, while for a generic pair of observers the ordinary relativity of simultaneity survives; thus its epistemological value seems reduced nearly to a question of *descriptive simplicity*. Now, what is the bearing of this state of affairs on the other «sections» of the question? I would like to observe that, quite in general, the conceptual framework of relativistic cosmology is rather peculiar: its fundamental notions are first grounded *statistically*, then they are *geometrized* in an exact way, as a first approximation with respect to the local irregularities, whose description is being delegated, as a second approximation, to the conventional physical laws (these irregularities, however, may include black holes and infalling *standard clocks* whose *proper time* stops at the singularity and *gets out* of cosmic time (!), although after an infinite span as seen by FO). Once clothed in this new, more fashionable suit, these notions are extrapolated backwards in time *deterministically*, up to a situation in which the empirical connection is totally lost. Actually: i) there are no «ideal» FO measuring *proper time* and no «genidentical» *dust grains* falling along the Weyl geodesics in the *radiation-filled era*: the notion of «substratum» gets «analogical» and becomes weaker and weaker. ii) No classical and «external» standpoint can be introduced in order to confer empirical meaning on the «analogical» *micro-cosmic time* which is supposed to replace the *macro-cosmic time* when relativistic quantum effects predominate; on the other hand it appears that what is being extrapolated to microscopic scales is itself a concept lacking a satisfactory physical foundation. Furthermore, it is implicitly taken for granted that QT still applies in a cosmological framework. This assumption, however, contradicts the standard interpretation of the theory [39]; still worse, were it to apply as such, its «macroscopical» paradoxes would turn into «cosmological» paradoxes. Finally: iii) no possible concept of *time* whatsoever can be formulated *when* one supposes that quantum-gravitational effects should come on the scene. Perhaps contemporary «cosmo-mythology» reaches a true climax here: not only should the Universe be an *object*, it should also *have been* a *micro-object when* there was no conceivable time. It seems to me that the physical meaning of *cosmic time on the whole* is, to be generous, rather elusive. This may be the reason why people start bravely jumping «beyond», to see what happened «before» the Big Bang.

Lastly, let me reiterate that all these conceptual problems exist *whether* the empirical conditions required for the definition of a *macro-cosmic time* at the present epoch are met or not ! By now, however, these empirical conditions are seriously questioned. So, what about the «age» of the Universe?

6. EMPIRICAL EVIDENCE

Let me begin by referring to a recent brilliant article by Burbidge [40] who looks critically at the modern observational evidence concerning cosmology. As Burbidge reminds us, the most important observational pillars of relativistic cosmology are:

a) the redshift-distance relation;

b) the existence and properties of the MBR;

c) the relative abundance of the light isotopes.

Furthermore, if one accepts the relativistic cosmological models, then the following additional elements become important:

d) the determination of the deceleration parameter $q(o)$;

e) the determination of the «age» of the Universe, and the ages of the stars and the elements.

Finally, a concluding touch could be reserved for:

f) the inflationary theories.

I believe the best thing to do is to report synthetically the relevant excerpts from the conclusions of Burbidge himself:

a) «The *direct* test has never been successfully carried out...There is a considerable amount of evidence suggesting that large parts, if not all, of the redshifts of many of these objects (quasars) are not due to the expansion of the Universe... Finally, patterns in the values of the observed redshifts, involving peaks and periodicities in the redshift distribution, are present, and these are not expected in normal cosmological models...Thus it appears that some galaxies may also have intrinsic redshift components... This result may have tremendous repercussions. For example, if appropriate parts of the redshift of galaxies in the extensive surveys were interpreted in this way, the large-scale structure would disappear....What it (all this) does mean, of course, is that discussions of the Universe at high redshifts which are based on observations of quasars and related objects cannot be taken seriously... This may lead us to open Pandora's box. This in turn might lead to new cosmological models. Only the future will tell»;

b) «there is the alternative possibility that the MBR did not arise in a hot Big Bang. In this connection....the result that the energy density in the MBR is the same as that found in the starlight in our own galaxy....remains one of the strangest result of cosmology» (remember in this connection the theoretical counter-model of Ellis !);

c) «the relative abundances of the light isotopes are *compatible* with Big Bang nucleosynthesis.....however if it could be shown that the fraction of the observed helium which is primordial is less than 22%, it could not have been made in a Big Bang»;

d) «none of the observational methods of determining $q(o)$ leads to results which we can take very seriously»;

e) «the current prejudice (to accommodate the ages of elements, stars and globular clusters) is to believe in an Einstein-de Sitter model with critical density. However....there is no real evidence that the critical density of matter is present»;

Finally, what about inflation?

f) «Since there is no way of testing the inflation hypothesis by direct observation, it has always seemed to me that it also is an idea with only a metaphysical basis. It is unfortunate at least that the inflationary idea has been so widely accepted primarily because it is an elegant mathematical solution to a problem (I disagree with Burbidge about this being the main reason of acceptance!) , without sufficient attention being paid to the fact that it solves a problem which may not be a real problem at all, and which certainly cannot be proven to be one. Yet another aspect of the bandwagon approach to cosmology is the fact that the only good account of the anti-inflationary thesis, by Rothman and Ellis (1987), [41] has been relegated to a popular journal.... Any conclusions which stem from the inflationary ideas, for example that the density of matter equals the critical density, and therefore that there is a large amount of non-baryonic matter present, should not be taken too seriously unless or until some direct astronomical evidence for its presence is found.... What bearing does all of this have on the observed Universe? In my view, it has none.»

I would like to add a single comment to the excerpts from Burbidge. If the redshifts turn out to be no longer a reliable confirmation of an expanding Universe, the Big Bang model will be find itself bereft of the phenomenon it was invented to explain.

Let me now examine more closely the empirical conditions which I described before as essential in connection with the definition of *cosmic time*. Actually, in recent years, astronomers have found astonishing observational data concerning the large-scale structure of the Universe [42].

First of all enormous structures have been charted which cannot be accounted for by the standard theory of galaxy formation nor by the *ad hoc* introduction of exotic «dark matter». These structures are nearly two and half billion light-years long and, inside them, galaxies are packed more than twenty times as densely as they are outside. To grow to such a scale under the force of gravity alone would have taken a time span of the order of 100 billion years, i.e. five times longer that the Big Bang model allows, so it may be that recourse should be made e.g. to plasma physics in order to explain them [43]. Similar clustering has been found in the distribution of quasars within eight billion light-years of the earth. The most recent observations are still more impressive: it seems that the further the astronomers look, the bigger the structures they find. It appears in addition that the variety of structures is so great that the simple classical Lambert-Charlier hierarchical model of the Universe cannot describe their complexity: in order to render the complexity of the hierarchy, a dozen or so hierarchical levels would have to be introduced. It has even been suggested that recourse should be made, at least for the intermediate cosmic scales, to a *fractal distribution* [44]. In any case, it has become quite clear that the present day distribution of matter is highly non-homogeneous out to a distance of several hundred million light-years, and it seems probable that the inhomogeneity extends out to billions of light years and characterizes the entire observable Universe. Thus, if the Universe is crowded and irregular on this scale, there is no empirical justification for the assumption of homogeneity made by the CP; on the other hand, the same high isotropy of the MBR becomes now a problem in its own right.

The second important point concerns the large-scale motion of matter[45] Over the last decade, astronomers became aware that the outward motion, or «Hubble flow», which is allegedly generated by the Big Bang is not the only large-scale dynamic in the Universe. Actually, observations have revealed that big structures, corresponding to a large number of galaxies extending over hundreds of Megaparsecs, have independent and ordered motions which are at odds with the direction and speed of the Hubble flow. These *peculiar* velocities are of the order of 500/1000 Km/sec. and represent a deviation from the *ideal* expansion of the «substratum» that had been neither seen nor even imagined before. Therefore, in any case, the Hubble flow is now manifestly non-uniform.

What is the result of all this? The result is that there is no sufficient empirical evidence - to say the least - supporting the Cosmological Principle and, above all, no evidence that the integrability conditions which express the mathematical definability of *cosmic time* can be considered to be empirically met at all. This means in turn that the theoretical construction of a «substratum» associated to the existence of a *cosmic time* does not even apply *approximately* (as it should be anyway) to the actual observable Universe. It is no longer possible to say with a sufficient degree of empirical support that the galaxies' (or clusters') *trajectories* are really orthogonal to a family of global hypersurfaces which should define time. The real observers are then no longer guaranteed to be in condition to synchronize their clocks as the *ideal* FO of the «substratum» are supposed to do. Therefore only a concept of *medium time* for local regions of the present epoch could survive. Still less justifiable, of course, is any backwards extrapolation in *cosmic time* to the «early phases» of the Universe, let alone to quantum gravity regimes. The claim of relativistic cosmology to have been able to give *empirical* answers to the traditional cosmological problems seems very weakly based indeed. What is more, it should be realized that the vanishing of the concept of *cosmic time* could make the whole theoretical structure of relativistic cosmology collapse completely.

The epistemological conclusion, in this case, would be that the seeming universality and uniqueness of the fundamental temporal relations of «earlier than» and «later than», which are so deeply rooted in the immediate awareness of our consciousness, properly reflect some basic aspect of the *real temporal structure of the world* but can be *objectified* in the nomological space-time structure of physics only *locally* and *relativistically*. After all, this should not be viewed as the only enigmatical issue concerning locality and non-locality in the physical description of the world.

ACKNOWLEDGEMENTS

Heartfelt thanks to my friend, prof. Antonio Scotti, for reading the manuscript and for many stimulating discussions.

Physics Department - Theoretical Division
University of Parma, Parma, Italy

NOTES

1. See Sciama 1973.

2. See Grishchuk and Zeldovich 1982.

3. This sentence was uttered by Nobel Laureate H.Alfven (paraphrasing William James), as quoted in Rothman 1989.

4. It is amusing to recall that, according to what Whitehead believed in 1929, «Cosmology...restrains the aberrations of the mere undisciplined imagination», (see Whitehead 1929).

5. In what follows I will confine the discussion to evolutionary relativistic cosmology. At present, the other cosmologies are not «à la Mode»: if a hot initial state does not exist, the cosmic laboratory for testing speculative high-energy theories does not exists either. Of course, the range of relativistic cosmologies is not restricted to the evolutionary theories; actually, as noted by (Kerszberg see Kerszberg 1989), «the word *non-static* has always referred to something basic alien to GTR as a physical theory». Of course, the purely philosophical arguments apply to any species of scientific cosmology.

6. See Bondi 1960.

7. As is well-known, the very measuring devices which must provide its empirical foundation are not described by the theory: *standard clocks* and *practically rigid rods*, or equivalents, have a purely plausible existence from the theoretical point of view. Therefore, the theory is intrinsically *incomplete* as far as the description of matter is concerned.

8. See Ellis 1979.

9. See Glymour 1973.

10. See Hawking and Ellis 1973.

11.On this point see for example Husserl 1959.

12. See Salecker and Wigner 1958.

13. A new, more profound, detachment from the Kantian conception could be noted in this connection: we have here «concepts» of space and time which, although empirical, cannot be traced to a synthesis of any sensible intuition, if not in an *analogical* sense alone.

14. See Weyl 1949.

15. See Grünbaum 1971.

16. See Gödel 1949, I.
17. See, for example, Merleau-Ponty 1974, Whitrow 1980 and Kerszberg 1989.

18. See Cassirer 1906.

19. I disregard here the *macroscopical* paradoxes of QT, already mentioned. As to *microscopic* indeterminism, I believe that the dualistic, *quantum-based*, way out of the «subject-object» dilemma held by Eccles (see Eccles 1986) is exceedingly rigid from a philosophical point of view and not properly justified scientifically.

20. See Schrödinger 1945.

21. See Carnap 1966.

22. Here I will not extend the discussion to include the problem posed by the existence of living beings in general. I think this problem too is to be treated within the perspective of the consistency between the physical description and the existence of emergent causal chains.

23. Of course, I am not claiming here that a purely physical «anisotropy» of time does not exist or that it could not be represented within the nomological structure of physics. On the contrary, it must be stressed that the «anisotropy» of time is a fundamental epistemic *pre-requisite* for the foundation and the empirical control of every physical theory (including the time-reversible dynamical descriptions). Without the development of macroscopic irreversible traces, any interpretation of an ordered physical state as a true manifestation of «past» events, and, in particular, any cosmological inference, would become impossible. Whether this *pre-condition* will remain *external* to scientific formalization or will eventually be grounded into the theoretical structure of physics, as well as the possible forms this achievement might take, are open questions. In this connection, it could be reminded that, according to a line of thought going from Kant to intuitionism, there should be a temporal psico-genesis of the numerical concepts. In this way, the «unreasonable effectivness of mathematics in natural sciences» (as Wigner puts it) would appear to be connected to the «phenomenological» notion of time. A bold attempt to ground the «anisotropy» of time, as a particu-

lar *pre-condition* of knowledge, by introducing a «tense» temporal structure at the basis of physical theories has been made by von Weizsäcker (see von Weizsäcker, 1971 and 1988). My point of view, in this regard, is that were an attempt of this nature to attain a true success, it would amount to the most radical revolution of physical thinking since its formation during the Renaissance. I will not pursue any further here the analysis of the possible relations between the «anisotropy» of the real time, resulting from the existence of *emergent* causal chains, and the would-be nomological explanation of the anisotropy of *physical time*. Incidentally, it may be interesting to note that a seeming (formal and indeed solipsistic) «imitation» of the epistemological situation originated by the subsistence of *emergent* causal chains is provided by the so-called «many-worlds interpretation» of QT (see De-Witt and Graham 1973). Although QT is the only physical theory which, through the *state vector reduction* process, takes notice of the fact that the «conscious» observer is himself a component of the «objective» world (at least according to its standard interpretation), it does not describe the observer himself within its conceptual framework. I think that if the *macroscopic* paradoxes of QT are eventually settled within a coherent quantum theory of measurement, the resulting description would refer entirely to physical micro-*objects* and measuring apparatuses: i.e. it should be compatible with *microscopic* non-separability without any ambiguous intervention of the observer as a «conscious» subject. It is also worth noticing that, precisely in connection with the undefined intervention of the observer within the current theory, *time* plays a *privileged* role in it with respect to *space*. This circumstance gives rise to one of the main difficulties met by any attempt to construct a theory of QG (see Hartle 1989). Finally, quantum *indeterminism*, being connected to a peculiar transition from the «potential» to the «actual», could appear to be capable of conferring the property of «becoming» to the events. That this is not the case, has been exhaustively argued by Grünbaum (see Grünbaum 1971). Of course, I do not wonder at this conclusion, in view of the fact that QT satisfies all the general methodological constraints of scientific theories.

24. Note that my conclusions are in some way exactly the contrary of those which are sometime *metaphorically* (and naively) derived from the point of view of an extreme reductionistic monism. For example when it is asserted that the probabilistic character of the elementary laws should be traced to an incompleteness of the information due to the fact that our own brains are *not yet* included into the physical theory, taking for granted that *were* they included, a completely deterministic description of reality would follow.

25. The «boundary of the boundary is zero» *principle* of Wheeler is clearly an attempt to answer this problem. The same is true of the purely «Riemannian» path integral approach taken by Hartle and Hawking (see Hartle and Hawking 1983) which tries to solve the problem of fixing *boundary conditions* by specifying that there are *none*, so that the Universe would be created *ex nihilo* on geometrical grounds.

26. See on this point the recent re-examination of the questions of «commensurability» and «cumulability» of scientific theories made by Agazzi (see Agazzi 1985).

27. Incidentally, this is one of the reasons by which, «strictu sensu», a scientific theory cannot even be *falsified* if one is not allowed to rely on another theory which can be considered *verified* to all the relevant implications.

28. As Schrödinger argues, however, this could be compatible with assigning a consistent physical interpretation to these manifolds see (Schrödinger 1956).

29. See Hawking 1968, and Hawking and Ellis 1973, (Chap.6.4).

30. See, for example, Ashtekar 1988.

31. See Earman (1970),(1972) and Kanitscheider (1982).

32. See Gödel (1949), II.

33. See Sakharov 1966, and Unruh 1976.

34. See Weyl 1923.

35. The modern expositions of relativistic Cosmologies (see for example Wald 1984) no longer rely on the Weyl postulate explicitly; their geometric assumptions, however, are more or less equivalent to it, at least insofar as the logical and empirical difficulties of the subject are concerned. Actually, the difficulties of interpretations have even grown up. In modern relativistic Cosmologies people speak confidently of a Big Bang also for an open Universe. The conceptual situation in this case is even worse than in closed Big Bang (see for example Rothman 1989).

36. See Ellis 1978.

37. See Einstein and Strauss 1945.

38. See Noerdlinger and Petrosian 1971.

39. As is well known, QT depends on the existence of a classical background to which all questions of observation must ultimately be referred: thus it divides the world necessarily in two parts, even though it does not determine the separation line theoretically. Since in cosmology there cannot be any «external» and classical vantage point, QT cannot apply, at least in its currently accepted form. It is therefore not surprising that the so-called «quantum cosmologists», when they take an explicit stand, adhere to the unorthodox «many-worlds interpretation» of Everett and De-Witt (see De-Witt and Graham 1973). However,...,let me quote a shrewd excerpt from John Bell (see Bell 1981): «Everett's replacement of the past memories is a radical solipsism - extending to the temporal dimension the replacement of everything outside my head by my impressions, of ordinary solipsism or positivism. Solipsism cannot be refuted. But if such a theory were taken seriously it would hardly be possible to take anything else seriously. So much for the social implications. It is always interesting to find that solipsists and positivists, when they have children, have life insurance».

40. See Burbidge 1989.

41. See Rothman and Ellis 1987.

42. See, for example, G.Chincarini and G.Vettolani; and M.Geller, J.P.Huchra and V. de Lapparent, in Hewitt, Burbidge and Fang 1987.

43. See, for example, Peratt 1990.

44. See Giavalisco, Calzetti and Ruffini 1989.

45. See, on this point, Rubin and Coyne 1989.

REFERENCES

AGAZZI, E.:

«Commensurability, Incommensurability, and Cumulativity» in Scientific Knowledge, Erkenntnis, 22, (1985), 51-77;

ASHTEKAR, A.:

New Perspectives in Canonical Gravity, Bibliopolis, Napoli, 1988;

BELL, J.S.:

«Quantum mechanics for cosmologists», in *Quantum Gravity 2*, C.Isham, R.Penrose and D.Sciama, eds., Clarendon Press, Oxford, 1981; reprinted in J.S.Bell: *Speakable and unspeakable in quantum mechanics*, Cambridge Univ.Press, Cambridge, England, 1987;

BONDI, H.:

Cosmology, Second edition, Cambridge Univ. Press, 1960;

BURBIDGE, G.:

«Modern Cosmology: The Harmonious and the Discordant Facts», Int.J.Theor.Phys., 28, (1989), 983;

CARNAP, R.:

Philosophical Foundations of Physics, Basic Books, 1966;

CASSIRER, E.:

Das Erkenntnisproblem in der Philosophie und Wissenschaft der neueren Zeit, vol I, Bruno Cassirer, Berlin, 1906;

DE-WITT, B. and GRAHAM, N. eds.:

The Many-Worlds Interpretation of Quantum Mechanics, Princeton Univ.Press, Princeton, N.J., 1973;

EARMAN, J.:

«Space-Time, Or How To Solve Philosophical Problems and Dissolve Philosophical Muddles Without Really Trying», J.Phil., LXVII, 9, (1970),259;

«Implications Of Causal Propagation Outside The Null Cone», Australasian J.Phil., 50,3, (1972), 222;

ECCLES, J.C.:

«Do mental events cause neural events analogously to the probability fields of quantum mechanics?», Proc.Roy.Soc.Lond., B227, (1986), 411;

EINSTEIN, A. and STRAUSS, E.G.:

«The Influence of the Expansion of Space on the Gravitation Fields Surrounding the Individual Star»s, Rev.Mod.Phys., 17, (1945), 120;

ELLIS, G.F.R.:

«Is the Universe Expanding?», Gen.Rel.Grav., 9, (1978), 87;

«Cosmology and Verifiability»,Qu. Jour Roy Ast. Soc., 16 (1975), 245;

«Limits to verification in Cosmology», Ann. N.Y. Acad. Sci., 336, (1980), 130;

GIAVALISCO, M., CALZETTI, D. and RUFFINI, R.:

«The Fractal Distribution of Matter in the Universe», in *Proceedings of the V Marcel Grossmann Meeting on General Relativity*, D.G.Blair and M.J.Buckingham eds., Part B World Scientific, Singapore, 1989;

GLYMOUR, C.:

«Topology, Cosmology and Convention», in *Space, Time and Geometry*, P.Suppes ed., D.Reidel Publ. Co. , Dordrecht, 1973;

GÖDEL, K.:

«Theory of Relativity and Idealistic Philosophy», in *Albert Einstein: Philosopher-Scientist*, P.A. Schilpp ed., The Library of Living Philosophers, Evanston, Ill., 1949;

«An example of a new type of cosmological solutions of Einstein's field equations», Rev.Mod.Phys., 21, (1949), 447;

GRISHCHUK, L.P. and ZELDOVICH, Ja.B.:

«Complete Cosmological Theories», in *Quantum Structure of Space and Time*, M.J.Duff and C.J.Isham, eds., Cambridge Univ. Press, Cambridge, 1982;

GRÜNBAUM, A.:

«The Meaning of Time» in *Essays in Honor of Carl G.Hempel*, N.Rescher ed., D.Reidel Publ.Co., Dordrecht, 1971;

HARTLE, J.B.:

«Time and Prediction in Quantum Cosmology», in *Proceedings of the V Marcel Grossmann Meeting on General Relativity*, D.G.Blair and M.J.Buckingham eds., part A, World Scientific, 1989;

HARTLE, J.B. and HAWKING, S.W.:

«Wave Function of the Universe», Phys. Rev., D 28, (1983), 2960;

HAWKING, S.W.:

«The Existence of Cosmic Time Functions», Proc.Roy.Soc.A , 308, (1968), 433;

HAWKING, S.W. and ELLIS, G.F.R.:

The Large Scale Structure of Space-Time, Chap.6.4, Cambridge Univ.Press., Cambridge, 1973;

HEWITT, A., BURBIDGE, G. and FANG, L.Z., eds.:

Observational Cosmology, D. Reidel Publ. Co., Dordrecht, 1987;

HUSSERL, E.:

Die Krisis der Europäischen Wissenschaften und die Traszendentale Phänomenologie. Eine Einleitung in die phänomenologische Philosophie, Nijhoff, The Hague, 1959;

KANITSCHEIDER, B.:

«L'influenza della teoria comtemporanea della relatività sulla filosofia», in *Il Problema del Cosmo*, G.Toraldo di Francia ed., vol II, Istituto della Enciclopedia Italiana, Roma, 1982;

KERSZBERG, P.:

«Of Learned Ignorance in Relativistic Cosmology», in *Proceedings of the V Marcel Grossmann Meeting on General Relativity*, D.G.Blair and M.J.Buckingham eds., part B, World Scientific, 1989;

MERLEAU-PONTY, J.:

Cosmologie du XXe siècle, Gallimard, Paris, 1974;

NOERDLINGER, P.D. and PETROSIAN, V.:

«The Effect of Cosmological Expansion on Self-Gravitating Ensembles of Particles, Astrophys».J., 168, (1971), 1;

PERATT, A.P.:

«Not with a Bang», The Sciences, January-February 1990;

ROTHMAN, T.:

Science à la Mode - Physical Fashions and Fictions - The Garden of Cosmological Delight, with ELLIS, G.F.R., Princeton Univ.Press., Princeton, N.J., 1989;

RUBIN, V.C. and COYNE, G.V., S.J, eds.:

Large-scale Motions in the Universe, Princeton Univ.Press, Princeton, N.J., 1989;

ROTHMAN, A. and ELLIS, G.F.R.:

Astronomy Magazine, 15, (1987), 6;

SAKHAROV, A.D.:

«*Maximum temperature of thermal radiation*», JETP Lett.,3, (1966), 288;

SALECKER, H. and WIGNER, E.P.:

«Quantum Limitations of the Measurement of Space-time distances», Phys.Rev., 109, (1958), 571;

SCHRÖDINGER, E.:

What is Life?, Cambridge Univ.Press. Cambridge, 1945;

Expanding Universes, Cambridge Univ. Press, Cambridge, 1956;

SCIAMA, D.W.:

«The Universe as a Whole», in *The Physicist's Conception of Nature*, J.Mehra, ed., D.Reidel Publ.Co., Dordrecht, 1973;

UNRUH, W. G.:

«Notes on black-hole evaporation», Phys.Rev.,D14, (1976), 870;

WALD, R.M.:

General Relativity, The University of Chicago Press, Chicago, 1984;

von WEIZSÄCKER, C.F:

«The Unity of Physics», in *Quantum Theory and Beyond*, T.Bastin ed., Cambridge Univ.Press, Cambridge, England, 1971;

«Classical and Quantum Descriptions», in *The Physicist's Conception of Nature*, J.Mehra ed., D. Reidel Publ. Co., Dordrecht, 1973;

«Exerience and the Unity of Physics», in *The role of Experience in Science,* Proceedings of the 1986 Conference of the Académie International de Philosophie des Sciences, E. Schebe ed., Walter de Gruyer, 1988;

WEYL, H.:

«Zur Allgemeinen Relativitätstheorie», Phys.Z., 24, (1923), 230;

Philosophy of Mathematics and Natural Science, Princeton Univ. Press, Princeton, N.J., 1949;

WHITEHEAD, A.N:

The function of Reason, Princeton Univ. Press, Princeton, 1929;

WHITROW, G.J.:

The Natural Philosophy of Time Clarendon Press, Oxford, 1980.

Erhard Scheibe

GENERAL LAWS OF NATURE AND THE UNIQUENESS OF THE UNIVERSE

Dedicated to Peter Mittelstaedt
on the occasion of his 60th birthday

It seems a generally acknowledged view that physics is confined to the investigation of events that can be reproduced. «The natural scientist - says Pauli[1] - is concerned with a particular kind of phenomena . . . he has to confine himself to that which is reproducible... I do not claim that the reproducible by itself is more important than the unique. But I do claim that the unique exceeds the treatment by scientific method. Indeed it is the aim of this method to find and to test natural laws...» Here for Pauli as for everybody else a natural law is a statement expressing a regularity more or less directly related to repeatable events. And one may add that it is not only the possibility of testing that is responsible for our demand of reproducibility. Rather it is the very fact of regularity expressed in it that gives a natural law its dignity and makes it a subject worth studying on its own account.

The characterization of physics by the natural law and the reputation that physics thus understood has gained during the last centuries has often been felt to be a difficulty for cosmology, evolutionary biology and other kinds of natural history. For in these disciplines the typically historical element and with it the unique event and the unique development becomes the primary

E. Agazzi and A. Cordero (eds.), Philosophy and the Origin and Evolution of the Universe, 341–360.

subject of investigation. The opposition is expressed - to give but one quotation - by Friedrich Hund[2] by saying: «One may characterize physics as the doctrine of the repeatable, be it a succession in time or the co-existence in space. The validity of physical theorems is founded on this repeatability... By contrast 'cosmology' is the doctrine of the unique universe, of its special, perhaps historically, developed features.» Now history in a general sense must not be a stumblingblock to repetition and reproduction. Strictly speaking all events are unique, and in the sense in which they can be repeated they are repeated in the course of time. It is only when we come to more and more extended processes such as biological evolution or the recession of the nebulae that we cannot hope to become witnesses of a repetition. And it would be such processes that we had to face if we wanted to save the scientific status of the disciplines mentioned in any direct way.

In this paper I shall not take this direct way. Rather I shall investigate the premise of the foregoing argument, i.e. the claim that in physics proper we really have the situation of repeatable events and natural laws expressing regularities between such events. And the particular aspect under which I want to analyze this claim is the fact that all events, processes, objects etc. that have ever been made the subject of an empirical investigation are events, processes, objects etc. *in one and the same universe*. Thus I want to challenge the common view of the lawful character of physical theory by taking seriously an aspect of uniqueness that, although it may be very weak, has some obvious relevance to our theorizing in the natural sciences.

There are two main reasons to be suspicious about the regularity view (in a wider sense) of physical theory if we introduce the aspect in question. One is that we cannot a priori exclude a thoroughly holistic structure of the universe, and we cannot do this even after having accumulated hundreds and thousands of empirical evidences to the contrary. This is certainly an extreme position but it has been taken even by physicists. Schrödinger, for instance, asking how we can come to make precise predictions about the future behaviour of a physical system argues[3] that «it may be, and if we are entirely strict about it, it certainly is the case that we are forced to extend the system considered to the entire universe.» The second reason that may raise doubts about our subject is methodological in nature. It is that the usual formulation of a physical theory does anything but invite us to believe in the regularity view in any innerworldly sense. Rather the theories are formulated primarily as statements about a single physical system, and their generalization to universal statements about a whole class of systems, although it lends itself to a possible-worlds interpretation, does in general not give the slightest hint to find an interpretation within one, namely our, universe.

I

In the first section I will explain in greater detail what I mean by generalities of the 1st kind. The main thesis, coming in two parts, is here: First, a physical theory is essentially a theory about one single physical system. What a theory says - what makes us recognize that in a given case we are faced with quantum mechanics and not with electrostatics, with thermodynamics and not with acoustic etc. - these contents of the theory, I say, concern one single physical system. Insofar a theory, if viewed as a statement, essentially is a singular statement. However - and here comes the second part of the thesis - already this statement, singular with respect to the physical system, contains two obvious generalities: the concepts in which our system is described and the quantifiers - the universal and existential quantifiers - applied to the concepts in order to bring about the statement in question. These generalities are the ones I want to call *generalities of the 1st kind*. Obviously, they are system-internal generalities and are not used to express the eventual universal validity of the theory. This applies even in the case of probabilistic theories, e.g. quantum mechanics. This is again obvious if we advocate the view that probability statements are about single systems. But even if one is not willing to view probability statements as statements about single systems one has to admit that the theory then is *about* one single ensemble of physical systems in the usual sense. What the theory says is then said about this ensemble and is certainly not a universal statement with respect to the individual systems. For according to the very advocates of the ensemble view a probabilistic theory does not make any statements about single systems. Precisely for this reason it cannot make a universal statement about a single ensemble.

Among the evidence for the fact that most people, contrary to what I have just been saying, like to view theories as being universal in the first place there is the fact that they emphasize the exceptional situation occurring in cosmology: In cosmology - it is usually said - we meet with the serious obstacle that our theory is about one system only simply because the universe as a whole is given to us only once. To me this seems to be the wrong kind of emphasis because it favours one component of theorizing - universality - to the exclusion of another one that, as we shall see, is equally important. I would, therefore, rather begin my analysis with the remark that in physical theory, although we are never concerned with the universe *in toto*, we always conceive of the actual system of our interest *as if* it were the whole universe. Thus, on the one hand I take it for granted that even in physical cosmology we never make the whole universe the object of our theoretical investigation. The conception of the universe as the unrestricted totality of everything existing may be an interesting conception from a philosophical point of view. In physics it would be of no use whatever. There a drastic selection takes place in every case, and the amount of what is selected usually is negligibly small when compared with what we omit. The selection is made under various viewpoints: we idealize, we neglect, we isolate, we simplify, we abstract. In every case this means that we pass from a larger whole that really is a piece of nature to some fraction of it, and it is only this fraction which we are going to deal with.

On the other hand, it is important to realize that what is omitted in this way - what is not taken into account in our theory

- is so radically wiped out that we cannot but view the product of our selection as being a world of its own a complete substitute for the actual universe. May be that the latter still plays a role in the background and that it is re-introduced in part when we apply the theory. The theory taken by itself does not know about this. The object of electrodynamics as defined by Maxwell's equation is a field and charged matter and nothing else. Quantum mechanics of the hydrogen atom has as its object one hydrogen atom (or an ensemble of such) and nothing else, and so on. In each of these cases we act *as if* the object of our theory be the total universe although we know that this is not the case and sometimes mitigate the situation by introducing more complex systems. The method of the «as if» might be called after its inventor the *Galilean method* [4].

The fact that it works is a highly non-trivial fact about the universe that we shall keep an eye on: We can successfully investigate parts of the universe without considering everything. And actually we do so already in our daily life. The Galilean method, however, deserves to be studied with special care.For it involves the farreaching and intricate concept of a physical theory, and therefore we should now have a brief look at the logical structure of our theory concept[5].

In physics we attempt to describe physical systems by means of mathematical structures. In this way physical laws, obeyed by those systems, can be expressed by statements about the describing structures. Let me exemplify this procedure by the theory of a particle moving in a central field according to the laws of newtonian mechanics. In this case the structure being used to describe the behaviour of the particle consists of four parts: absolute space, absolute time, a field of force as well as the orbit and mass of the particle. Correspondingly, our theory is made up of euclidean geometry of space, a corresponding degenerate geometry of time, general newtonian mechanics and a special force law. And all this is usually formulated in mathematical terms wellknown in this case even to the beginner.

In general the statement of our theory is of the form

$$(1)\ \Sigma\,[\,X_1,\ldots,X_\mu;\ s_1,\ \ ,s_\nu]$$

where Σ usually is a conjunction of statements concerning more and more of the elements X_μ and s_ν , of the describing structure. These elements are sets later on to be identified with the extensions of the basic concepts describing the system in question. We distinguish the socalled principal base sets X_μ. from the typified sets s_ν. The nature of the elements of the former can only be known from without our system. By contrast, the elements of the latter are known from within in the sense that they are the product of one or the other of a class of universal constructions from the principal base sets. In our example, X_1 and X_2 would describe space and time respectively, s_1 and s_2 the distance in space and time, and so on. The elements of s_1 would then stand for triples consisting of two points in space and one number such that the number is the distance of the two points. On the other hand, the question what a point in space (or time) is could not be answered in this way. And this is the general situation whether we are dealing with point mechanics, continuum mechanics, electrodynamics, quantum mechanics, gravitational theory according to Newton or Einstein or what not: In each and every case the theory is given by a statement of the form (1) where the structure $< X; s>$ stands for a physical system and Σ for what the theory says about the system.

In this situation we first meet with a system-internal *conceptual generality* (of the 1st kind) : The given physical system is described by means of concepts the extensions of which are the sets X and s. Accordingly, for these concepts as well as the ones defined by them it is mandatory that they refer to one well determined system although in a general discussion as the present one this determinateness is only assumed. Thus, for instance, the concept that a point P_1 has distance r from another point P_2 refers to a well determined space, and unless this space is given we do not know what is meant by that concept. Likewise, in our illustrating theory the concept that the particle has velocity v at time t means nothing unless a well defined orbit is given, and so on.

Trivial as it is the matter has to be emphasized in view of the concepts of the 2d kind to be introduced later on. The unity of these concepts will be given not by one single system but by one theory. They will be concepts, for instance, of a space or of a particle orbit. Accordingly, they will refer always to a whole class of systems and thus will have,so to speak, a generality of a higher order. But plain generality we already find in the concepts describing any single system. For its description we cannot but introduce possibilities that certainly are not realized: There is an infinity of points P_1 and P_2 not having a given distance r, and similarly an infinity of velocities which our particle does not have at a given moment. This conceptual overproduction not only is a fact but a necessity: We simply do not know a method to describe an individual object without introducing more theoretical elements than correspond to what is actually there.

We find a corresponding situation if we now turn to the propositions that are made in physical theory about a single system, i.e. the propositions of which Σ in (1) is built up. Being the axioms of a theory these propositions are not singular statements by which we are informed which is the distance of two given points or which velocity our particle might have at a given time. Thus although they *are* statements about a single system they are *not* singular in the usual sense. Rather we here meet with a propositional generality (of the 1st kind) already on the level of one physical system and precisely corresponding to the conceptual generality mentioned previously. If, for instance, our theory includes a theory of space then, being a theory about a particular system, it necessarily must refer to a particular space. Any theorem about this space, e.g.the triangle inequality, then is a case in point: It says what it says by essentially using quantifiers binding the variables of the terms in which the theorem is formulated. The same holds for our particle theory, for instance, with respect to its equations of motion: They are differential equations submitting the position functions to certain conditions to be satisfied at *every* time. Likewise a field equation would have to be valid at *every* point in time *and* space, and so on. The typical situation as to the axioms of a physical theory is that once we have introduced concepts *and* want to make a general use of them quantifiers are unavoidable and then represent the propositional generality of the 1st kind that was to be introduced.

II

In the previous section it was argued 1) that a physical theory, if viewed as a statement, is a statement about one individual physical system and 2) that already this statement, although being singular in this sense, involves two generalities one conceptual, the other one propositional - generalities of the 1st kind as I called them. If we now turn to the problem of *universal laws of nature* - the main theme of this paper - an entirely different kind of generality comes into play. At any rate this is my second thesis, and I am somewhat puzzled that this thesis does not appear in the relevant literature with sufficient clarity[6]. As distinct from generality of the 1st kind it is essential for the generality *of the 2nd kind* that it concerns a certain *totality of physical systems* or-as philosophers are used to say - of objects. It is the kind of generality that philosophers have in mind when they talk about the universal validity of a law of nature. And, of course, also physicists do not restrict the meaning of their theories to singular statements of the form (1) . Somewhat more modestly they speak of the domain of validity of a physical law. This indicates that the generality of the 2d kind also comes in two parts: one *conceptual*, the other one *propositional*. Insofar as we make general use of it a theory is a concept: some physical systems fall under this concept and others do not. Secondly, if we want to express universal validity of the theory we would have to say something like

$$(2)\ \Lambda y \,.\, y \in Y \rightarrow \Sigma\,[\,X(y); s(y)].$$

Here Y is the domain of validity and $<X(y);s(y)>$ is the structure describing the system $y \in Y$. (2) then says that all systems belonging to a certain domain Y satisfy the axioms of our theory. And this statement will now be the major subject of discussion on the new level of generality as had been statement (1) on the lower level.

A first point to show us that proposition (2) is different in kind from (1) is that the demarcation of the domain Y cannot be produced by an ostensive act as it could be done in the case of (1) where only one individual system had to be pointed out.The only alternative then seems to be a conceptual description. But this in turn cannot be given in the language in which Σ is formulated and defines a certain range of structures as its models. To give but one example, gases are described by their pressure, volume and temperature. If we now want to use van der Waals' equation (as Σ) in a universal statement (2) then, even if we take the risk to claim the equation for *all* gases, we still would have to say what we mean by a *gas*. It would not suffice, as is usual, to restrict generality by restricting our parameters to certain intervals, e.g. to low pressure. In the last analysis the characterization of a gas in the *premise* of (2) has to be given in a language *different* from the one used in the conclusion: it has to be characterized by the way a system is given to us or is produced or something of the sort. By contrast we were not forced to do this in the case of stating the singular version (1) of our theory.[7]

For a second consideration that may clarify the situation I want to compare the statement (2) with the philosophical folk formulation

$$(3)\ \Lambda y.Py \rightarrow Qy.$$

of a law. This frequently discussed version is very likely to mislead us because in it only one of the two kinds of propositional generality occurs, and it is usually left undecided which one. From the point of view taken in the present approach it is immediately clear that quantification in (3) is of the 2d kind if (3) is meant to be a law and the range of the variable y is a class of objects obeying the law. Moreover, propositional generality of the 1st kind does not occur in (3) because the extremely simple description of the objects provided for by (3) is not in need of it. If, on the other hand, (3) is not viewed to be a law the quantification may very well be of the 1st kind. Y being the universe of discourse with respect to which (3) is interpreted anyway, in the case of a law the elements $y \in Y$ would be the physical systems. In the other case it may be that a »closed universe« showing lawful behaviour is only reached in form of the entire set Y as, for instance, in geometry.

(3) being the analogue of (2) it may be asked what the analogue of the quasi-singular statement (1) is in the philosophical folk case. Evidently, it must be

$$(4a)\ Py \rightarrow Qy,$$

and this is the moment to re-emphasize that it is this statement and not (3) which conveys the important information. (4a) can easily be rewritten as a species of structures in the sense of (1). It is then given by

$$(4b)\ \Sigma\ [\{P,P>\},\{Q,Q>\};\ s_1,\ s_2] \equiv$$

$$s_1 \in \{P,P>\} \ \wedge\ s_2 \in \{Q,Q>\} \ \wedge\ (s_1{=}P> \ \vee\ s_2{=}Q).$$

where P> and Q> are the negations of P and Q respectively. Our system then is a system that, since only two predicates are available for the description, can assume only four states *a priori,* only one of which (namely $s_1 = P$ and $s_2 = Q>$) is excluded by the law. It is as if we would restrict the investigation of a circuit having resistance R to *one* value U_o and I_o of potential and current respectively, these values obeying Ohm's law

$$U_0 = R.I_0.$$

Proposition (4) expresses this connection for *one* system only, and the generality of (3) consists by no means in admitting also other values of potential and current. Rather it refers to other *systems* - other circuits (with resistance R) - stating for these the *same* connection between the two given values U_0 and I_0. The philosophical folk case simply is degenerated into two properties P and Q, and all possibility of variation concerns the systems, *not* the physical description.

Thus we see that my present point cannot be illustrated in full by the simple universal implication (3), and I mention this case only because, as the philosophers of us know, whole books are filled with considerations on laws of nature without transcending their folk version (3). However, as soon as we turn to real life examples from physics we can easily recognize the symbiosis of the two kinds of generality, propositional as well as conceptual. Take our standard example of a particle moving in a field according to classical mechanics. The usual physical concepts of this theory as, for instance, the concept of the position of the particle at any time are concepts of the 1st kind used to describe one concrete system to which our theory is applied. But there is also the concept of a system satisfying the theory in question, i. e. the concept of a particle moving in a central field, and so on. And this concept is of the 2d kind. With it not the system but, in a sense, the theory is described. Correspondingly, in the axioms of the theory we have quantifications with respect to the concepts describing the system, e.g. we require the equations of motion to hold for every time point. This requirement concerns one system only, and its generality is of the 1st kind. But again there is also the claim that what our theory says about any one system holds for every system belonging to the domain of validity of the theory - an obvious claim of the 2d kind.

The distinction in question is particularly perspicuous in the case of frame theories of physics like classical hamiltonian or quantum mechanics. For these theories it is important that such parts of the describing structures as the phase space or the (quantum mechanical) state space are variable. The concepts of these spaces are therefore genuine concepts of the 2d kind.By contrast a symplectic metric on a given phase space or an expectation function on a given quantum mechanical state space are concepts of the 1st kind. It is true that these concepts can be *defined*, so to speak, with generality of the 2d kind. But the very definitions then show that *before* we can speak of a particular metric, a particular expectation function etc. a phase space, a state space etc. must already be at hand. The question: which is the expectation value of observable A in state s? cannot be answered unless a particular quantum mechanical structure has been given. By contrast, the question whether a given structure satisfies the axioms of quantum mechanics not only can but *must* be answered without referring to a second structure. It may be added that from a purely set theoretical view point the elements X_μ and s_ν of a given structure are always *sets* in the proper sense whereas the model class of a theory is always a genuine *class* (and not a set). But in physics as we shall see in the last section it is seldom a whole model class that matters, and the distinction of concepts of the 1st and 2d kind has also different roots. One last consideration may show this.

In connection with propositions (1) and (2) the concept of a physical system will allow that numerically different real systems are described by the same mathematical structure. This has to be admitted if only because we might *never* attain absolutely complete descriptions. However, we are free to require that systems with different descriptions really are different.Thus if we are given different data for the states of two gases we are entitled to infer that the given data refer to two (numerically) different systems. We then count two gases as different even if merely two different states of the same material object are prepared. This, of course, is but another way of expressing what has already been put into the notation $X^{(y)}$ and $s^{(y)}$ in (2): In the context of one theory to every system a unique description is assigned. This convention is perhaps less innocent than it looks like. For it is meant to imply that different descriptions are even *incompatible*. If system y is described by structure $<X^{(y)};s^{(y)}>$ then we cannot without leaving the theory, *add* to this structure in order to get a more complete description. The presence of several systems in the sense of the quantification in (2) is, therefore, altogether different in kind from the presence of the elements of any of the sets X_μ and s_ν making up our structure and (possibly) quantified over in (1). There is *no competition* between, say, the points of space or time as there *is* a competition between the various descriptions offered by a theory as possible descriptions of a real physical system.

We have, then, no difficulty in conceiving of any two parts of a physical system - parts in a very general sense - as »belonging to the same world«. The way in which systems are described by structures clearly shows the cooperative role that the various *elements* of any set belonging to the relevant structure play in building up this structure and, with it, the system. On the other hand, it has still to be clarified what it means that the various systems of which the universal statement (2) speaks all »belong to the same world«. The statement (2) by itself offers no hint whatever to answer this question. On the contrary, we have just seen that there is a certain competition between these systems which thus may even be unfavorable to their coexistence in a common world. Although everybody believing in laws of the form (2) tacitly implies that the systems in question do belong to our universe the prevailing interpretation almost seems to contradict this implication. According to the usual understanding the occurrence of any two systems submitted to our theory according to (2) amounts to what is most frequently called by such terms as »repetition« or »reproduction«. Indeed it is the age-old methodological requirement of the repeatability or reproducibility of every experiment that stands behind a universal law like (2). We are here not dealing with a repetition in the system-internal sense, it is not the question of a periodic motion - no two swings of the pendulum. In the context of lawlike behaviour repetition of a first instance of some law means a second independent instance of the law - instance or counterinstance - but at any rate a new system with different initial conditions. The system-external generality of (2) sometimes is even raised to the metaphysical level that the systems comprised within Y are so many different worlds. This, in an obvious sense, is not realistic. But there is a realistic approximation that I will call the laboratory view of lawlike generality. It is the view that we are able to produce (or: re-produce) in our various laboratories approximately independent systems with different descriptions but all obeying the same law according to (2). In a sense we can, therefore, practically realize different independent worlds within our universe.

III

However, we should take our problem also as a matter of principle, and in this case it is easily seen that a certain dilemma is coming up. The general proposition (2) - as a 2d kind generality - has no more the innerworldly character of the quasi-singular proposition (1). Precisely if the latter refers to such a world substitute as a physical system is taken to be, it is unclear within what new world a proposition (2) is to be understood. As we have seen, to a certain degree we want to conceive of the systems $y \in Y$ referred to in a law (2) as being mutually independent possible worlds. Only this, after all, explains our amazement about the regularity expressed in the law. On the other side we know that in any case of a physical theory the systems to which it refers are to be met with, if at all, as parts of *one and the same*,namely our *universe*. This, however, is *not* expressed in the theory (2) as it *is* expressed in (1) for the corresponding elements.

The question, therefore, poses itself whether a reformulation of (2) as an innerworldly proposition (1) is possible. The physicist probably would find this question to be of little importance. If experiments of a given kind have been repeated at various places on earth at different times, why demand that, apart from the experiment itself, i. e. its *kind*, also this general fact or even its extrapolation to future experiments of the same kind should be given a separate innerworldly formulation? It is obvious that for the working physicist the interesting part of the task is completed with the description of the experiment itself, i. e. its kind,even if he would admit that it were of no interest to him if it could not be repeated. However, a philosopher might wish to confront this laboratory view - as I called it - with the *cosmological view* according to which we do not satisfy ourselves with pragmatic excuses but *insist* on a strict innerworldly formulation of the universal implication (2).

To see whether the cosmological view of a general law of nature can be strictly maintained let me recall how systems composed of several independent subsystems are treated in physics. In the simplest case the connection between the three descriptions may be indicated by

$$(5a)\ \Sigma_1[sy_1] \wedge \Sigma_2[sy_2] \equiv \Sigma_{12}[sy_1,sy_2].$$

The independence wanted is here expressed by the fact that the statement Σ_{12} about the total system $<sy_1, sy_2>$ is the bare conjunction of two statements, each referring to one subsystem only. Whereas in (5a) all three statements may still be different, in the attempted reformulation of (2) the two statements Σ_1 and Σ_2 would have to be identical. There are even theories Σ for which we have

$$(5b)\ \Sigma\,[sy_1] \wedge \Sigma\,[sy_2] \equiv \Sigma\,[P(sy_1,sy_2)]$$

if (sy_1,sy_2) is a suitable description of the total system. Hamiltonian mechanics is a case in point and with some qualifications also quantum mechanics. The more general conjunctive decomposition (5a) can be found also in the universal parts of a physical theory. What Newton says in his »Principia« on space and time can easily be reconstructed in this way (with space as one and time as the other member of the conjunction).We know, of course, that this was not the last word on the matter and also for systems in the proper sense we feel that we cannot stick to such decompositions. But before we move on it is important to state that the innerworldly reconstruction of a law (2) that we are after would adequately be prepared by the procedure indicated.

However, let us now turn immediately to certain restrictions to which the formation of the product of independent systems is submitted. The most important restriction - the one on which perhaps all the others are dependent - is the uniqueness of space and time. The most obvious assumption that we make in the innerworldly description of several physical systems is that all systems are to be met with in one and the same spacetime and that therefore spacetime must be a common element of the conjunctive members in (5a).. Already Kant said:[8] ». . . if we speak of diverse spaces, we mean thereby only parts of one and the same unique space . . . Space is essentially one; the manifold in it, and therefore the general concept of spaces, depends solely on limitations.« Today we differ from Kant in several respects. Space has to be replaced by spacetime, and there is a general concept (of the 2d kind) of spacetimes. Together with other qualifications this concept is basic for general relativity. Still we do not consider theories of the universe or any physical theories in which the universe of discourse would be described by a structure containing *several* spacetimes.

Now the uniqueness of spacetime has consequences for the presentation of its material content. At the dawn of modern physics Kepler's three laws did not yet allow to recognize this. But they are a good starting point for showing the difficulty we have to cope with. Kepler's first law, for instance, can be spelled out for any planet without taking into account the existence of the other planets. In these statements space and time as well as the sun are common elements. Apart from them the worlds separate, and we have as many physical systems as there are planets. The law that *all* planets move in ellipses can essentially be expressed by a (finite) conjunction with identical predications. However, as we know since Newton this reduction is only an approximation that eventually becomes grossly false, for instance, in the case of a system consisting of the sun, the earth and the moon. The essential insight was that, since all celestial bodies exist in the same world, they may interact with each other such that only their totality makes up a closed system whose behavior as a whole follows a law. In fact, the matter stands even worse: The mutual gravitation in a system of bodies, according to Newton's theory leads strictly speaking to a totally irreducible system of equations of motion: If a system of bodies moves according to these equations no subsystem does. As a consequence, it is hopeless to look for an innerworldly reformulation of (2) if Σ essentially is given by Newton's equations. Within one and the same space-time strictly speaking at most one gravitational system could be realized.

The Kepler/Newton case illustrates a general reciprocity of *lawfulness and interconnectedness* in nature. Lawfulness in the standard form (2) demands strictly independent instances of the law. In searching for laws the point just is to find such independences. Of course, these independences go together with *internal* dependences as they constitute the contents of the respective law. At the same time they mark the limits of the latter. As long as we have reason to assume that laws in this sense are realized in nature - strictly realized - there is no *total* interconnectedness in the universe. On the other hand, the realization in one and the same universe, as it will be required even by a modest empiricism, constantly draws our attention to the possibility to have missed some dependence. And the discovery of any one in the context of an accepted law inevitably will destroy the law. The increase of discovered dependences cannot but lead to a decrease of laws in the usual sense as something fundamental. The physically interesting form of a theory is *not* the conjunction in (5a) but

$$(5c)\ \Sigma_1\ [sy_1]\ \wedge\ \Sigma_2\ [sy_2]\ \wedge\ \text{int}\ [sy_1, sy_2]$$

where *int* is an interaction term. But if (5a) is dismissed then so is (2) in any innerworldly interpretation. The interest in natural law and in causal connection, although going together in some sense, conflict with each other as soon as we widen the horizon, and in extreme cases it may follow from a theory (1) that its generalization (2) has at most one realization.

Since we know quantum mechanics we have difficulties not only with the interactions but already with the product formation in (5b). Let us consider, for instance, the quantum mechanics of free electrons. The state of an electron is given by its Ψ-function that determines for every observable its expectation value in the given state. According to the theory there is a whole Hilbert space of states. Now let Ψ_1 and Ψ_2 be any two of them. According to the laboratory view the pair (Ψ_1,Ψ_2) again determines a possible description of the situation. In a concrete case we would say that we have prepared both states independently of each other. According to the cosmological view, however, this conjunction is by no means the most general description of the situation: If we take seriously that the particles belong to the same world we have to treat the situation as a 2-particle-system. We must pass from two 1-particle ensembles to one 2-particle ensemble. For the latter, however, the pair (Ψ_1,Ψ_2) is a correct description only in exceptional cases, the socalled separable cases. In general the two subsystems are *inseparable*, and our knowledge about them is not maximal. Rather the information about the total system concerns many *correlations* between observables of the two subsystems. Again the possibility of an innerworldly reformulation of (2) is paralysed from the outset.

Against this argumentation it may be objected that the difficulties for a cosmological interpretation of (2) in connection with interaction and inseparability do not have any *practical* importance. All fundamental interactions have a finite range allowing for practically independent and yet internally interacting systems. Similarly, we can prepare practically separable quantum mechanical systems showing all the features of inseparability internally. And both possibilities are in accordance with the respective theories. All this has to be admitted, let alone the overwhelming number of cases where we find the independence in question not by looking at the theories but simply by experience.

On the other hand, we have to remember that we are investigating a matter of principle. Theories about gravitation and the mechanism of compound systems are of a *fundamental* character.

Such theories are not proposed only to say afterwards that they are to be taken *cum grano salis* anyway. And if such theories show us that the galilean method, successful as it is, in the last analysis not only misses the factual constitution of the universe but also violates its laws then this deserves to be recognized and understood. The question whether there are any universal laws of the form (2) that are not only approximations but strictly valid is a matter of principle. If the question should be answered in the negative this would mean that the theories from which we can derive the laws in question as approximations cannot themselves be of this kind. And we would then be faced with the question of *what* kind they are after all.

This paper does not allow to touch on this question - let alone to answer it. I may only remind us of one important feature of physical theories in their singular version (1): Together with assumptions contingent upon the theory we can infer other contingent statements from the theory. The best known examples are given by socalled initial conditions. Now whenever a real system is a candidate for satisfying the theory at least some of those additional premises will be true of the system. The conclusions drawn from these premises together with the theory are thus open to test without going beyond our system. It is true that in this way only a fraction of the theoretically possible initial conditions (in the general sense) can be put to use. However, as is shown by celestial mechanics we sometimes had to be and actually were content with this situation. and the amount of available evidence is restricted anyway. On the other hand, if one day we should come to the conclusion that the regularity view can only be maintained as an approximation this would be a most interesting turning point in methodology. We should neither be afraid of it nor loose sight of its possibility even now.

Heidelberg Universität, Germany

NOTES

1. W. Pauli *Aufsätze und Vorträge über Physik und Erkenntnistheorie.* Braunschweig 1961, p. 94

2. F. Hund: *Geschichte der physikalischen Begriffe.* Mannheim 1972, p. 274: see also E.P. Wigner: *Symmetries and Reflections.* Woodbridge, Conn., 1979, p. 3. no. 1; G. Vollmer: *Kann es von einmaligen Ereignissen eine Wissenschaft geben?* In: Ders. : Was können wir wissen? Bd. 2. Stuttgart 1986. 53-65

3. E. Schrödinger: *Ueber Indeterminismus in der Physik.* Leipzig 1932, p. 2

4. See the quotations in *Galileo. Man of Science.* Ed. by E.McMullin. New York 1967. Pp. 329 f., 356 f.

5. For the following view on theories see E. Scheibe: On the Structure of Physical Theories. In: *The Logic and Epistemology of Scientific Change.* Ed. by I. Niiniluoto and R. Tuomela. Amsterdam 1979. 205-24

6. See, for instance, C.G. Hempel: *Aspects of Scientific Explanation.* New York 1965. Pp. 264 ff., 335 ff. , 354 ff.; E. Nagel: *The Structure of Science.* New York 1961 , Ch. 4

7. The problem how the domain of application of a physical theory can be described is treated in W. Stegmüller: *The Structure and Dynamics of Theories.* New York 1976. Ch. IX. 4 and S

8. I. Kant: *Kritik der reinen Vernunft*, B39.

Bernulf Kanitscheider

THE ANTHROPIC PRINCIPLE AND ITS EPISTEMOLOGICAL STATUS IN MODERN PHYSICAL COSMOLOGY

I. FROM THE COPERNICAN TO THE ANTHROPIC PRINCIPLE

1. A Thousand Years on the Way to the Copernican Principle.

It was a long way from Greek geocentric cosmology to the modern centerless universe which is linked to the name of Copernicus.

Aristotle was the first to find a physical cosmology based on mathematical principles. His model of the universe was built out of the homocentric spheres of Eudoxos and Kallippos and was strongly devoted to a cosmic center. The geocentric picture of the Aristotelian world had its origin in dynamical foundations. He based his astronomical system on a dualistic physical ontology. The so-called sublunar sphere consisted of the four elements earth, water, air, and fire. The natural motion of these terrestrial elements was «up» and «down» and they had an intrinsic tendency to find their proper places according to their weight. On the contrary, the supralunar sphere was built entirely out of the fifth element, and its natural motion was the endless circular revolution around the center of the world occupied by the center of the earth.

E. Agazzi and A. Cordero (eds.), Philosophy and the Origin and Evolution of the Universe, 361–397.

More or less, this situation did not change even by the refinement of Ptolemaic astronomy, and during the medieval times the geocentric world picture with its natural place fitted very well into the cosmological ideas of Christianity. Thomas Aquinas showed how the Aristotelian universe could be adapted to the Christian thought by relatively slight modifications. Christian revelation was tightly connected with the conviction that mankind does indeed occupy the center of the universe. The displacement of the cosmic center could get on only step by step.

In 1543, Copernicus shifted it to the sun. Here it remained up to the times of William Herschel. At the end of the 18th Century, Herschel found that on astronomical reasons it should be removed to the core of our galaxy. But even that position was overthrown by Harlow Shapley in 1922, and at last Walter Baade discovered in 1952 that our milky way is a normal spiral galaxy not remarkably larger than the customary type. Hence, we can recognize in the history of cosmology the steadily growing conviction that mankind does not occupy the center of the universe.

Already at this point we should make an important distinction: cosmology may convince us that, from a global view-point, our kind of intelligent life has no special *location* in the universe, but in a local perspective we are surrounded by characteristic types of objects that might be unique. Our planetary system, the special location where the only known biological evolution took place, is perhaps distinguished in so far as peculiar constraints led to the building site of the biosphere. Current theories on the origin of life give no hint on the intrinsic properties of possible organisms that might be engendered by other planetary sites. The reason is, besides the low state of theoretical knowledge on the origin of life, that cosmology gives us only the main outline of the cosmic picture and ignores the irregularities of the local detail. Even a globally homogeneous universe without cosmic center and without outer edge can have a physically privileged place, where evolution took place. Cosmology makes spatiotemporal assertions, it is not concerned with estimations of complexity and a fortiori not with statements of value. Accordingly, it is possible that the human neural network is the system with the highest complexity in the entire universe and the only system that engenders values. This is fully compatible with our living in a typical location of the universe.

2. The Copernican Principle in Modern Cosmology

Modern cosmology in the time of Einstein and Hubble brought about the idea of the isotropic and homogeneous universe. This is relevant to our purpose in the following sense: the Copernican principle only makes a claim on the likelihood that *we* have a special location in the universe. Or, in the words of Hermann Bondi: «The Earth is not in a central, specially favoured position.»[1]

Such a principle leaves undecided the question whether or not a cosmic center does exist somewhere. The standard class of models (FRW) is quite explicit on this point. It takes into account only the subclass of spacetimes, i.e. exact solutions of Einstein's field equations, that comply with the boundary conditions of homogeneity and isotropy. But there is a snag in these restricting constraints. Observation leads unequivocally to an isotropic distribution of matter and radiation. Local irregularities set aside, beyond 100 Mpc galaxies are scattered evenly throughout 3-space up to the horizon. Radioastronomers have found that the very distant radio sources are distributed isotropically around us. The same is true for radiation, e.g. the X-ray background and foremost the cosmic microwave relic radiation, the remnant of the fireball state, which today, after having expanded adiabatically for $15 \cdot 10^9$ years, is measured with a temperature of 2.9 K, show the same feature. Recent measurements reveal that the isotropy of the 3 K-radiation amounts to about 1 part in 10 000, $\frac{\Delta T}{T} \leq 10^{-4}$

Now comes the snag. It lies in the fact that humans are immobile observers. For obvious technical reasons we cannot explore the vastness of space in order to test the homogeneity of 3-space. But by looking around carefully we can only ascertain local isotropy. Local isotropy means rotational symmetry around our special point of observation. However, to establish homogeneity, we need global isotropy, that means the same symmetry around any point whatever of the cosmological substratum.

It is a well-known theorem of differential geometry[2] that exact spherical symmetry around any point entails that the universe is spatially homogeneous. Such a spacetime admits a six-parameter group of isometries. Its surfaces of transitivity are spacelike three-surfaces of constant curvature.[3] In more colloquial terms: any point of 3-space in a homogeneous universe is physically equivalent to any other point on the same surface.

Hence, in order to fill the gap between local and global isotropy we need a bridging law. The lawful connection fulfilling this requirement is the Copernican principle. The name of this principle is clearly a misnomer, since Copernicus believed that the sun occupies the central place in the universe. Apparently, it perpetuates the belief that, indeed, a center does exist somewhere. Names, however, are of no importance, and we should pay more attention to the *trend* starting with Copernicus (with the ancient paragon Aristarchos). According to it we have to realize that we are living on a medium sized planet revolving round a normal main sequence star that is located on the rim of an average spiral galaxy which is in turn a member of a local group of galaxies. If we do so, we can make use of the term «Copernican principle» in the sense that our position in space is not distinguished in any way.

Since we have to deal with counterarguments against the Copernican principle in due time, we should have in mind the many epistemological advantages of a homogeneous 3-space. This high symmetry makes cosmology a much easier task. A universe containing one or many special locations with physically distinct properties could not be dealt with in a comprehensive way. The customary inference from a cosmologically relevant sample to the whole spacetime (even beyond the event horizon and even if its global topology were non-compact!) would be invalid.[4] In a hierarchical universe which is maximally inhomogeneous, no part, howsoever large, is significant for the universe as a whole.

Needless to stress that Einstein's field equations do not demand anything like homogeneity in the distribution of matter or radiation nor does this basic law require constant curvature for 3-space. Besides the simple Robertson-Walker spaces there is a large class of solutions in which the requirement of isotropy is dropped, but spatial homogeneity (in accordance with the Copernican principle) is retained. Even absolute rotation and shear could be included in the large-scale description, if astrophysical data indicated such complications. The fact that no large-scale anisotropy of this kind has ever been detected indicates that we are living in a universe with very special initial conditions that are friendly to cosmologists.

So homogeneity is highly desirable from an epistemological point of view, and the Copernican principle - sometimes called the principle of cosmological democracy - has a clear-cut function if it can be rationally defended.

But the Copernican principle included an emotional component; it brought about an abdication of a cosmic privilege rooting in theology. It is on a par with the abdication physics and biology have remorselessly forced on us, when, by Darwin's evolutionary theory, the species of man was filed in the long range of living organisms. Maybe the Copernican movement evokes a feeling of injured vanity and is therefore feared and resented in some corners of the intellectual community up to the present time.

3. The Roots of the Anthropic Principle and its Ramifications

The origins of the countermovement against the Copernican principle are deeply rooted in 19th Century's physics. Ludwig Boltzmann, the founder of statistical mechanics, was led to explain the direction of time by a natural line of arguments coming close to a concept which nowadays is called the Weak Anthropic Principle (WAP). His attempt to infer the thermodynamical arrow of time from mechanics forced him to assume either that the entire universe is at present in a highly improbable state or that our observable region is a tiny part of the whole universe which, globally, is in thermodynamic equilibrium.[5] Relatively small regions will deviate on account of the unavoidable fluctuations from thermodynamic equilibrium. The reason, why a living being finds himself in such a corner of the universe, for which a significant deviation from equilibrium defines an arrow of time, can be understtod, if we remember that only on a slope of the entropy curve life, consciousness, and intelligence can evolve. Therefore, it does not come as a surprise that in our local cosmical environment time has the direction we actually observe, because open systems need thermodynamic unequilibrium as a necessary condition for the origin, evolution, and maintenance of their life.

For the moment it is worthwhile to notice that Boltzmann had to make his choice, either to accept very improbable initial conditions, for which in principle no further experience could be given, or to include man and his thermodynamical presuppositions in a physical argument. Even at that time Boltzmann's colleagues were reluctant to accept the anthropic fluctuation interpretation, and in modern times it was severely criticized by Karl Popper, who accused Boltzmann of having violated the objectivity of the direction of time.[6]

Even if reproaching Boltzmann with subjectivity is historically not entirely correct[7] - because Boltzmann defended a realistic and objectivistic theory of knowledge - we can recognize that already since their rise anthropic arguments ran great risk to be interpreted in the direction of an epistemological idealism.

In modern times, we can observe various roads to the Anthropic Principle (AP). Science (not so much the humanities) can be regarded as the search for invariant aspects of the phenomena expressed in fundamental laws. Every law, however, contains constants the exact values of which remain more or less unexplained. Some of these constants of nature turn out to be tightly fixed by the necessary condition that they enable the existence of intelligent beings. This condition acts the part of a selection effect and characterizes the type of universe that can be inhabited and of which scientific knowledge is possible.

If we regard these constants that are inevitably contained in every law of nature, surprising coincidences show up which can be classified as follows:

i) absolutely senseless or chance coincidences. For an example we may consider the ratio of the mass of the nucleon to the mass of the electron, $\frac{m_N}{m_e} = 1836.1515$. It resembles largely to the result of $6\pi^5$, which is 1836.118. Speculations on a hidden physical meaning of this coincidence probably lead nowhere.

ii) rather contrived numerical relations[8], the physical meaning of which can be assessed hardly today, like the mass of the nucleon, of the Δ , Σ and Ξ hadron which can be ordered progressively, $m_N : m_\Lambda : m_\Sigma : m_\Xi = 1:2^{1/4}:2^{1/3}:2^{1/2}$, or the so-called mass splitting coincidences $\frac{m_n - m_N}{m - m_N} \simeq \alpha$ which connect the masses of some hadrons with the fine structure constant α.

They are not just a result of mysterious combinatorial juggling, but at the moment no deeper meaning can be grasped.

iii) very surprising coincidences of the type that, in 1937, caused Dirac to introduce his large number hypothesis (LNH). Dirac was bewildered by a prevalence of dimensionless large numbers resulting from the ratio of atomic and cosmological quantities. More exactly, he felt that there was a need for an explanation of the chasm between the numbers built out of physical constants like $\frac{m_N}{m_e}$ and $\frac{e^2}{\hbar c}$ which are within a few magnitudes in the order of unity, and the really large numbers like

$$N_1 \frac{\text{age of the universe}}{\text{atomic light crossing time}} = \frac{t_0}{e^2/m_e c^3} = 6 \times 10^{39}$$

$$N_2 \frac{\text{electric force between proton and electron}}{\text{gravitational force between proton and electron}} = \frac{e^2}{G m_N m_e} = 2{,}3 \times 10^{39}$$

To explain the surprising coincidence of the two large numbers N_1 and N_2, he put forward his famous LNH: «Any two of the very large dimensionless numbers occurring in nature are connected by a mathematical relation in which the coefficients are of the order of magnitude unity.»[9]

It might be, of course, a matter of debate, whether the occurrence of two large numbers of that kind is *really* an item to be explained causally. Should we really regard it as an explanandum that points to a hidden nomological structure? Superficially, the two categories of numbers, the microphysical and the cosmological one, are quite unrelated according to current physics. If the hypothesis «$N_1 = N_2$ up to some trivial numerical factors of the order of unity» is taken seriously, we get dramatic consequences for gravitational theory in general and cosmology in particular. Since N_1 includes the Hubble age t_0, any number of the order of 10^{40} should be time-dependent and, accordingly, all numbers of the order of $(10^{40})^n \alpha t^n$. The question as to why physics contains these large numbers at all is answered by referring to our cosmic age. When the universe was young, these numbers were small, but now they are large, and they are getting still larger in future.

Since $N_1 \alpha N_2$, one of the constants of N_2 had to be time-dependent, too. With respect to well tested results of local quantum mechanics, Dirac chose the very consequence of the LNH that gravity must weaken with the passage of cosmic time, namely $G \alpha t^{-1}$. Besides this novel non-Newtonian and non-Einsteinian gravitational theory, where $G = G(t)$, Dirac's approach led to the unusual prediction that the number of particles in the universe N (which is of the order of 10^{80} within the Hubble radius cH^{-1}) must increase with the square of t, $N \alpha t^2$. This consequence, of course, leads to a head-on collision with energy conservation, if the universe is finite; a conflict that can only be avoided if the universe is infinite and N therefore not defined.[10]

Numerology of the Diracian type led many even renowned astrophysicists to delve into strange speculations. Pascual Jordan[11], e.g., extended violation of energy conservation to the realm of stellar masses[12].From here, there is only a small step left to stars popping out of nothingness. Philosophers of science have always criticized the idea of continuous creation of matter[13] as an untestable process. The spontaneous genesis of entire stars ex nihilo is without doubt beyond the fringe.

On the other hand, varying G-theories act a part in a kind of subculture in physics up to the present. Many spatiotemporal variations of the fundamental «constants» were tried out, but physics is a strongly interrelated network, and every time one's proposal seemed to work another physicist could show up an unintended consequence which is contrary to the observed facts.

A devasting critique of the incorporation of Dirac's $G \alpha t^{-1}$ into stellar structure and planetary dynamics has been put forward by Edward Teller. He showed that on account of the sun's luminosity being higher in the distant past and the earth's orbit being smaller, the surface temperature of the earth would have been so high that in pre-Cambrian era the oceans must have been boiling. This argument already includes a kind of feed-back to the existence of life. Our knowledge of the very early microbiological form of life constrains the terrestrial surface temperature at that time. We urgently need research programs on a self-consistent set of variations of all fundamental parameters[14], including constants defining the strength of the known forces.

A new cosmological perspective came to the fore, when several scholars, included the famous J.B.S. Haldane, realized[15] that cosmological models in general and the variable G-model in particular could have great importance for the origin and maintenance of life.

Of course, a cosmological model that is selfreferentially inconsistent in so far as it excludes man, the formulator of the problem, cannot be a viable scientific approach. Dirac's original model was of that kind. If N_1 equals permanently N_2, then G does vary in a way that eliminates the creatures, who observe the coincidence $N_1 = N_2$.

Therefore, a theory was strongly needed in which the gravitational constant G varies at a much lower rate. In this situation, the so-called scalar tensor theory of Carl Brans and Robert H.

Dicke came to aid. Its construction was motivated by the endeavour of bringing Einstein's theory of gravitation into accordance with Mach's principle. The field equations[16], with an additional scalar field ϕ , hadn't any longer exact solutions for empty space, as Einstein's original equations did. A low variation of G, of course, can be brought into accordance with the original evolution of terrestrial life, but as the permanent validity of $N_1 = N_2$ is no longer upheld, the original intuition of Dirac is destroyed.

R.H. Dicke turned Dirac's approach upside down. For a rather long time, he was busy with a research program connecting biological factors and large number coincidences. Finally, he found the missing link between the apparent coincidences and the necessary pre-conditions for the existence of observers. When comparing N_1 with N_2, he argued, we should concentrate on the question why, today, we observe $N_1 = N_2$. In the early universe, when the world was young and hot, number N_1 was small, but there was nobody in the quad to notice this value. In the far future, when the universe will be old and the stars burnt out and partly collapsed to black holes, N_1 will be large and again nobody will be there to observe this value. Therefore, the discordance of N_1 and N_2 at very early and very late times is unobservable. Only within a limited epoch of cosmic time, when the astrophysical conditions are favourable to intelligent life, observers will exist and then be very suprised at the coincidence of N_1 and N_2.

Dicke realized later on that his argument in the above version was incomplete resp. contained many suppressed premises.[17] In order to make the causal claim more explicit we remember that according to current biochemical knowledge life is built upon elements heavier than H and He. Heavy elements, however, are not primordial, as the standard hot big bang model tells us, they are «cooked» in massive stars and enrich interstellar material via supernova explosions. Carbon as the basic material for *our* kind of intelligent life cannot be produced in a universe that recollapses long before the first generation of stars ended their life. So, surely, our carbon based life could not come into existence before the age of the universe t_0 is beyond the main sequence stellar age t_{ms}, $t_{ms} \approx t_0$ No wonder that t_0, the cosmic age of today, is $\alpha\ t_{ms}$ and as a consequence we observe $N_1 = N_2$ at present.

Important to stress that Dicke's argument does not presuppose, as Barrow and Tipler pointed out, that only carbon sustained life is physically possible. Even if many chemical bases for the origin of life reveal to be possible, «the fact remains that *we* are a carbon-based intelligent life-form which spontaneously evolved on an earthlike planet around a star of G 2 spectral type, and any observation we make is necessarily self-selected by this absolutely fundamental fact».[18]

It remains a matter of speculation, whether other material sites like interstellar clouds have enough inner complexity to bare living structures. Selection is a key word in this context, and Dicke's biological «explanation» of the coincidence $N_1 = N_2$ can be regarded as a reduction to an artificial selection effect. Whether it should be called a genuine explanation, this will be focussed in the following philosophical icial selection effect. Whether it should be called a genuine explanation, this will be focussed in the following philosophical section. In any case, Dicke's turning Dirac's LNH around is the First modern use of the Weak AP that establishers a link between a global property of the universe, hamely its age, and certain necessary conditions for a special subsystem, i.e. a living organism. The stronger versions will be analysed in due time. Already now it should be stressed that the argument only yields a *logical* connection between the self-awareness of organic matter and a certain constellation of crucial parameters, namely if (but not iff) matter has begun to contemplate itself, then $N_1 = N_2$. Dicke's argument, however, cannot be constructed as a sufficient condition for the existence of life. Carbon, oxygen, silicon are presuppositions for the genesis not only of life, but also of anorganic complex structures like mountains. Physically possible are universes with many rocky planets in it, yet no intelligent life. Nobody looks around in this possible place for habitation to be aware that $N_1 = N_2$. Already at this stage I should point out that there is a conceptual uncertainty in the demarcation between the WAP and its corresponding strong version (SAP). Many authors included myself[19] have taken the borderline between both principles coinciding with the distinction between necessary and sufficient conditions favourable for the existence of observers. Taken in this way, the SAP holds that the above mentioned situation of a mindless universe is impossible; the evolution of matter to higher states of complexity cannot stop at the anorganic chemical level. Every universe that is physically possible at all must be cognizable and in the formulation of Brandon Carter «admit the creation of observers within it at some stage».[20] In other words: a certain cosmic coincidence like $N_1 = N_2$ has to lead *inevitably* to life, mind, and consciousness. However, John Leslie reminded me in a personal communication, that there might be a more restricted interpretation of the difference between both principles, the distinction being that the *weak* principle states that we must be observing a life-permitting location, while the *strong* takes this «location» to be a life-permitting «universe». In this case, where the difference only concerns the domain of influence, both principles act as a necessary presupposition.

There is, however, the still stronger Participatory Anthropic principle (PAP) asserting that «observers are necessary to bring the Universe into being», which takes over the rôle of an assertion stating a mind-dependent reality. It is, of course, a matter of debate, whether we really need the PAP to understand coincidences which otherwise, if we were restricted to the use of the SAP, would escape our comprehension. It is philosophically important to realize that some authors make us believe that the stronger versions of the Anthropic Principle can be interpreted semantically only as design arguments that means, in a teleological framework. Although we will be occupied with the epistemological evaluation of the various AP's later, it suffices at the moment to note that most scientists do not intend such a usage. To avoid idealistic and premature teleological interpretations, B. Carter has stressed in a recent paper that he regrets not having used the term «cognizability principle» instead of «Anthropic

Principle».[21] So B.J. Carr has remarked that the AP is not to be used in the sense: «the universe does not exist, if we are not here to observe it», but only in the weaker sense «*if* we are here to observe it, the universe must be the way it is».[22]

The WAP can be totally divorced from metaphysical or even teleological overtones if couched in the language of a world ensemble. The hypothesis of the plurality of worlds has been looked on with suspicion by many scientists, but if formulated properly, it neither contains logical inconsistency - it is erroneously argued that the universe must be unique by definition - nor does it lead into the mystical realm of unfathomable entities. Foremost we have to distinguish between an ensemble of possible worlds (in the sense of Gibbsian statistical mechanics) and a plurality of real worlds (in the sense of the Everett-Wheeler-Graham interpretation of quantum mechanics).

It is ontologically non-committal to examine an ensemble of possible worlds as to their structure of features enabling them to generate observers or not. This is nothing but to establish a subset relation of the cognizable set in the set of all physically possible worlds. It is, of course, a *trivial* analytical truth that we inhabit a member of the cognizable subset. However, it is a *nontrivial* task to establish a probability measure on the frequency of cognizable worlds. This relative frequency is a contingent fact on the whole ensemble. Such an estimation of a probability measure was put forward in connection with the problem of isotropy. Barry Collins and Stephen Hawking, who have analysed this problem in 1973, made use of the hypothesis of a world ensemble. To reduce an extreme unlikely coincidence, they introduced a real set of many worlds endowed with every kind of initial data.[23]

Before looking closer at this special application of the WAP, we should stress that it is not illegitimate to enlarge the pertaining ontology of a scientific hypothesis, if it can be justified via a higher explanation. In any case, it is better to trespass Occam's principle of ontological parsimony than to make an explanatory use of the AP which violates highly confirmed long-term fundamental rules of epistemology. To grasp the crucial point just look at the above-mentioned example of Robert H. Dicke. Taken as a genuine explanation, Dicke's argument would amount to the assertion: $N_1 = N_2$, *because* we are here, which is preposterous, because it is back to front, that means, it violates our well-confirmed causal structure and asymmetry of time.

Metaphysical speculations on the plurality of worlds are as old as cosmology.[24] In modern science we find this hypothesis in the sense of the simultaneous existence of many spatiotemporal parts of an otherwise infinite universe, and we meet the hypothesis in form of a superspace of causally disjoint worlds spanning the range of all possible experimental results of a quantum mechanical measurement process.

Indeed, the notion of a plurality of worlds is not so an extravagant concept as it might seem at the first glance. It does not clash with the usual methodology of science. A world beyond our own is just a theoretical entity on a par with hypothesized, not directly observable elementary particles.

There is another conceptual distinction pertaining to the *extension* of the world ensemble.[25] Counterfactually, we can vary our known constants of nature like $\frac{e^2}{\hbar c}$, $\frac{m_N}{m_e}$, together with the dynamical traits of the universe like expansion rate, entropy content, matter and radiation density, to get the most comprehensive class of physically possible worlds. This is, of course, a subclass of the logically possible worlds which are defined so that contradiction is avoided.

It must be stressed that the general problem to assess the consequences of varying the dimensionless constants of nature is a difficult mathematical task, because we do not know the effect of compensatory changes in the values of other constants. One variation may exclude the existence of life, while another improves the situation. It is possible to estimate the consequences of a small change of say Sommerfeld's α, the strength of electromagnetic interaction, all other invariants of nature being fixed[26], but it is rather impossible to evaluate the consequences of variation in the whole network of constants pertaining to several fields of physics, although a few stability investigations have been made.

This indecisiveness also concerns Wheeler's reprocessing model of the universe according to which a finite universe will recollapse to a final singularity (big crunch). In every bounce, the fiery furnace of the cataclysmic singularity introduces a transmutation to the physical constants and the type of the expansion dynamics. It is difficult, however, to form an opinion on the measure of the cognizable combinations in the set of all combinations facing the infinite number of random oscillatory permutations of the constants.

If every physical basic feature - not just the particles, but also the fields and the geometry (including global topology) - are fossiles from the violent conditions encountered in the very early phase of the big bang[27], then it is difficult to estimate how often combinations favourable to life occur. One thing seems to be clear, however: If topology change is included and if, one day, the compact spacetime turns into a non-compact connectivity, then such a universe is bound to expand in the indefinite future and the constants are frozen in the cosmic structure forever. If, in that case, expansion were sufficiently violent, thereby preventing the formation of galaxies including planetary sites which are necessary for the origin of organisms, life would become extinct forever.

From a logical point of view, Jan Hacking has found a striking difference between the anthropic use of the reprocessing model put forward by John Wheeler and Brandon Carter's many world hypothesis. We can regard the sequences of universes like the rolls of dice. If the universes divorced by singularities have no memory as to their precedent history, that means that a universe has effectively no trace of its predecessors, then the sequence of the worlds is like the rolls of a die. A fair die does not remember how it fell last time. If Wheeler's model of succession, where all physical constants are frozen in the first stage of the big bang and then molten and reprocessed to new values, is taken in the sense of statistical independence among the different classes of values, then an explanation of our present orderly universe based on this model is a matter of the inverse gambler's fallacy.[28]

Even a long chain of universes of every physical construction does not explain the present existence of our orderly life-supporting universe. Our habitable universe is not getting more probable, if many previous universes have been reprocessed by a genuine chance device. That would be to argue like a gambler coming into a room and seeing a double six just rolling on a chance device. If he thinks, «because the double six is seldom, there must have been many rolls this evening» he is committing the inverse gambler's fallacy.[29]

If we retain the actual values of the constants of nature, we get a restricted set of physical worlds, namely those which are ruled by Einstein's theory of gravitation. This smaller ensemble is generated by varying the initial and boundary conditions of the field equations.

The advantage of such a stricter definition is easily to be seen, when we have a look at the answer of Barry Collins and Stephen Hawking to the seemingly innocent question: «Why is the universe isotropic?» Their starting point was the now well established isotropy of the universe which manifests itself most overtly in the fact that the temperature of the microwave background does not depend on the direction of measurement to the accuracy of $\frac{\Delta T}{T} \leq 10^{-4}$. The rotational symmetry of the 3 K background is even more baffling, when we realize that it extends beyond the cosmic event horizon. Parts of the sky that lie more than say 30° apart are equivalent in temperature, although the time available since the origin of the universe is too short to make this conformity understandable by a causal process.

Isotropy confronts us with the paradox of causality: Why are all the causally disconnected regions of the universe so meticulously fine-tuned? Do we need the idea of a command or a conspiration coming out of the singularity itself to resolve the horizon puzzle, which is equivalent to the causality riddle?

To find an explanation of the high isotropy Collins and Hawking[30] studied the asymptotic stability of the open FRW worlds under the action of spatially homogeneous anisotropy distortions. What happens if some deviation of anisotropy is put into the model at very early times? Do these worlds, more or less, approach regularity or are they becoming more irregular, as time goes on? The result of these highly technical investigations can be summarized as follows[31]:

If for the cosmological matter the two conditions $T^{00} > |T^{\alpha\beta}|$ and $\sum_{k=0}^{3} T^{kk} \geq 0$ stating that negative pressures never dominate the energy density and that the sum of the principle pressures is non-negative are satisfied, then the set of cosmological initial data giving rise to models which approach isotropy as $t \rightarrow \infty$ is of measure zero in the metaspace of all spatially homogeneous initial data. Largely a corresponding assertion can be made for the closed homogeneous universes. They do not isotropize at large times, e.g. the spatial 3-curvature does not become isotropic at the time of R_{max}. Only in the limiting case of the flat (k=0) model, when the expansion dynamics lies in the borderline between the hyperbolic and the spherical curvature - in Newtonian terminology that would mean zero binding energy - isotropization occurs at large times. Since this spatially flat model of the Einstein-de Sitter type[32] is of zero measure in the metaspace of all cosmological Cauchy data sets, and the astronomical data show the universe to be very close to 3-flatness, Collins and Hawking were confronted with the question[33], why something occurs for which theory supplies an almost zero probability.

In this situation, Collins and Hawking made use of the WAP. Since a counterfactual analysis shows that matter condensations like galaxies can only grow in a universe in which the rate of expansion is just on the borderline to avoid recollapse[34] the two authors subscribe to the WAP in the following manner: «... the isotropy of the Universe and our existence are both results of the fact that the Universe is expanding at just the critical rate. Since we could not observe the Universe to be different, if we were not here, one can say in a sense that the isotropy of the Universe is a consequence of our existence»[35]

As later investigations have shown, the conclusion of Collins and Hawking is not so straightforward as it looks at the first glance. It contains many hidden or suppressed astrophysical assumptions. Since we will be concerned more with the conceptual form of this novel type of argument, we shall take it in its original version.

4. Anthropocentricity, Teleology and Evolution

In modern science, teleological explanations had to give way to causal mechanistic explanations almost everywhere. Darwin's evolutionary theory expelled the question from the domain of science, whether the developmental trends of organic systems are goal-directed. Global teleology, the long-range purposeful development of nature, has been revealed as a myth. Local teleological behaviour, e.g. planned action or free voluntary decision, has been restricted to a rather small subclass of higher mammals. Goal-directed actions of higher animals are, of course, not at variance with a causal description of these processes as certain activations of the central nervous system. They can be reconstructed properly as emergent qualities of a higher level of complex organizations.

Given this unmistaken historical trend of natural science, it is more than surprising that even amidst pure physics (e.g. cosmology and astrophysics) apparent teleological explanations came to the fore. Collins and Hawking have been rather silent on the correct interpretation of the last sentence of their seminal paper: «... the answer to the question, why is the universe isotropic, is, because we are here.»[36] Other authors like R. Dicke and P.J.E. Peebles[37] are more explicit on the logics of the relationship between man and the universe within anthropic arguments: «Could it be ... that it is the presence of observers that determines the nature of the universe?» Does it really make sense to interpret the verb «determines» as pointing to a goal-directed process? Collins' and Hawking's argument suggests nothing like that.

It is a surprising fact that isotropy correlates strongly with the existence of intelligent life. That astonishment can be diminished, when the additional hypothesis is taken into account that realty consists not only of one universe, but of an infinite ensemble of universes with all possible initial conditions. Nearly each of these worlds becomes highly anisotropic at large times and therefore contains no observers. Hence, no teleological agent is needed (entelechia or Aristotelian steering tendency) to understand the amazing coincidence, it is just a selection effect.[38] It is not against, but a part of customary scientific methodology to be aware of selection effects that can bias our observation. If an objective determination of all galaxies in a fixed interval of brightness is intended, allowance has to be made for the effect that brighter galaxies are more easily to be seen and therefore tend to adulterate the result. In the same way, cosmological observations are biased by the selection effect due to the existence of observers simply. It should, however, be made entirely clear that this selection effect of «carbonaceous astronomers» may indeed reduce the surprise on the a priori improbable feature of nature, but it cannot substitute a causal explanation, *why* these properties of the universe prevail and just yield the necessary conditions for carbon-based life. This crucial point was also stressed by John Leslie in his recent comprehensive analysis of the Anthropic Principle: «An initial point to notice is that neither a Multiple Worlds explanation nor an explanation by reference to a Fine Tuner would supply a substitute for a long, scientifically very ordinary causal account of Life's evolution».[39]

It seems plausible on account of the universal validity of quantum mechanics to count an explanation as a *causal* one even if it includes probabilistic elements.

II. Four Types of the Anthropic Principle

WAP used in the right way can be regarded as a blank to be filled in by a causal connection of the pertinent facts. Already in 1974, B. Carter[40] was convinced that some coincidences can be interpreted only by involving the SAP: «The universe *must* be such as to admit the creation of observers within it at some stage». As indicated the term «must» is of crucial importance, it is open to a variety of meanings. Many readers of Carter's seminal paper took it implying coercion and since the latter points to an entity or agent that compels or forces the constants of nature, the teleological interpretation seems to be inescapable in order to get their values friendly to the observer. This is the meaning many have made out of the SAP: «In essence, it claims, that the universe is *tailor-made* for habitation and that both the laws of physics and the initial conditions *obligingly* arrange themselves in such a way that living organisms are subsequently assured of existence.»[41] Of course, it is not a logical consequence in the sense of the entailment relation to take SAP in this way, as we stressed earlier, and within the framework of naturalistic science it is difficult to make sense out of Davies' claim. Why *ought* a material universe to be tailor-made for habitation, what does it mean that the initial conditions are *obliged* to arrange themselves in a goal-directed way which points to the evolution of man? These moral terms are undefined before the advent of intelligent beings. They are predications devoid of physical meaning within the context of a naturalistic ontology. Barrow and Tipler have rightly remarked that the design version of the SAP can only be understood if couched in a theological language that transcends the limits of science. On the other side the earlier mentioned participatory Anthropic Principle is not in any way theologically impregnated, but inludes a strong anthropocentric element. The PAP is motivated clearly by quantum mechanics and by an extreme epistemological idealism. Taken at the face value, it says more than customary phenomenalism of the Machian type. Phenomenalism claims that the only object of our investigations can be the world of sense experiences, but not the way they are brought about by physical reality behind the phenomena. A causal theory of sense experiences is impossible accordingly. Taken verbally, the PAP states even more than phenomenalism, namely the causal relation between observers and the universe. The observer seems to engender or produce physical reality. Needless to say that we don't have the slightest hint to such a kind of retro-causal activity of intelligent organisms. Cognition as it is understood today is an activity of the neural network and is in no way exempted from customary time asymmetric causality. No physical model could be established within current scientific theories to fill in the mechaninism -responsible for a process of cognition-which, at cosmologically late times, atranges the constants and parameters of the early universe. That conjecture has trespassed the borderline between science and speculative metaphysics.

Barrow and Tipler have another version of the SAP on stock, the so-called «Final AP» (FAP) stating that intelligence has to come one day in cosmological history, and once in existence, will never die out. Since the FAP is more related to information theory and computer science, a subject not intended to deal with here, we will skip its discussion.

III. Are Anthropic Arguments Explanatory?

Given the four main varieties of the AP it is of central importance to realize which of them can be defended considering the approved rules of today's scientific methodology. Philosophers of science have uttered heavy criticism on the various APs foremost when they are put forward as explanations of a new anthropocentric type.[42] The patterns of explanation are thought to be well understood since C.G. Hempel analysed them as the deduction of an event E (explanandum) from a set of law statements together with some initial and boundary conditions. If we take the argument of Collins and Hawking as our main example, then the decisive boundary condition is that «intelligent life exists». Given the law of Einstein's gravitation theory, some additional astrophysical assumptions and some biological generalizations, we can construct a formally correct explanation of the isotropy of the universe. Nevertheless, its direction of explanation is wrong. The causal connection is one way, namely from the isotropy of the universe at late times to the existence of intelligent organisms, but not the other way round. The violation of the causal structure, a global property of spacetime, makes the *anthropic explanation* an illegitimate use of the AP. The methodological moral of this example is well-known, logic alone is not enough to analyse correct explanations. An explanatory argument has not only to be deductively correct, it needs to be supplemented by synthetic rules of a more general kind, pertaining to our concrete universe.

Anthropic arguments can, of course, point to novel hitherto unknown concatenations of organic systems with their cosmological embedding. If, however, answers to questions of the type

> «Why is it that 3-space is flat, or 3 K radiation is isotropic, or the universe is as old as a main sequence star (10^{10} y) or the universe is roughly 10^{10} light years in extent?»

are requested, then we will have to take care of the direction of explanation, resp. we need certain additional premises like the postulate of the world ensemble. If this enlargement of physical ontology is accepted, an answer as to why we observe something that is of infinitesimal probability can be given without inverting the causal structure of physical processes in any member of this ensemble. Therefore, I do not like to adopt the stance of D. Lewis, who argues[43] that the postulate of the world ensemble is a reason why we do not need an explanation at all. On the contrary, I hold that the ensemble hypothesis reduces the astonishment of improbability, but it cannot be taken as a substitute for the causal explanation why, in the individual worlds, the constants arrange themselves in the way they do.[44]

Some philosophers have taken the point of view that even extremely rare coincidences do not need an explanation at all, so why bother with improbable chance events? Since unlikely events occur sometimes, we should not be disturbed by the above-mentioned cosmological coincidences. John Leslie has rightly criticized this kind of shrugging one's shoulders.[45] Take an example: If, on a lonely beach, you come across an inscription several square feet in extent «Coca Cola», you wouldn't argue that considering the long geological times and the many possible combinations of sandgrains one time even the improbable combination «Coca Cola» has formed. Given the existence of humans everybody would tend to a causal explanation: A stroller on the beach must have written these two words, maybe as a joke. If, however, no organism were available for a dynamical explanation, a natural process would have to be found to make this contrived order of sandgrains comprehensible. I will give a sketch of this solution in the last chapter. John Leslie's own philosophical approach which he calls «extreme axiarchism» and which contains a creative efficacy of ethical requiredness[46] appears to be philosophically possible, but from a naturalistic point of view it seems much easier to defend the existence of many, even infinitely many, physical worlds than a Platonic existence of values. However, the philosophical problem whether the talk of Creative Values - as a piece of explanatory *metaphysics* in contradistinction to explanatory *physics* - may be included in scientific contexts of justification, cannot be pursued in this paper any further.

IV. ANTHROPIC AND DYNAMIC EXPLANATIONS

Some gentle reader might have got the impression, that pursuing the cosmic coincidences leads straight on into the dead end of speculative metaphysics without hope for a return to rational and empirically controlled physics. Is there no possibility in sight for a physical evaluation of life sustaining fine tuning of cosmic constants and parameters? My hope for an explanation of the cosmic coincidences that constrain the existence of intelligent life comes from the development of the Grand Unified Theories. This approach has the great advantage that it operates within the well-corroborated paradigm of science, without taking refuge to exotic modes of explanation. How this conservative strategy may lead to a causal understanding of the a priori improbable combination of parameters and constants, this can be grasped in the so-called inflationary cosmology built on the unification of strong, weak, electromagnetic and perhaps gravitational interaction.

The first step of this approach, the Glashow-Weinberg-Salam unification of the weak and the electromagnetic force, is now well-established by the discovery of the W^+, W^- and Z°, the vector bosons that mediate, together with the photon, the electroweak interaction. Obviously, it is not surprising that optimists among the theoreticians are full of enthusiasm that fundamental physics will come soon to its final goal, the unification of all physical interactions.[47]
At present, different theoretical schemes rival as candidates for the final comprehensive description: supergravity including the Kaluza-Klein model of higher dimensional spacetime, and, as a newcomer, superstring theory in which the fundamental entities are not point particles, but one-dimensional entities susceptible to vibrations, the so-called *strings*.

If the unification procedure does exclude gravity, we move towards the so-called GUT scheme, a unification of the forces relevant for microphysics up to the energy level 10^{15} GeV. Even in the GUT realm we can observe, how cosmology and high energy physics work together and yield causal processes apt to explain mysterious coincidences.

The standard model for the evolution of the universe is the hot big bang scenario which starts with a primordial fireball stage and continues with subsequent cooling by adiabatic expansion of 3-space. The standard hot big bang model is now well confirmed, but it includes a few drawbacks. A bundle of urgent questions cannot be answered by this model, therefore it is not wrong nor falsified, but it needs improvement. To mention just a few topics, where answers would be highly desirable:

i) the singularity problem

If general relativity is taken seriously up to extreme values of curvature, then the universe starts at a point which does not belong to the spacetime manifold. The initial singularity is an edge of Riemannian spacetime, but not a first event; it is not an event at all. What does it mean physically that spacetime does not exist at the origin of reality? Here something is missing.

ii) the horizon problem

If cognizance is taken of astrophysical observation, the spacetime of the universe is probably non-compact, in popular scientific diction called «open». How is it then possible that the many causal disconnected parts of the spatially infinite universe began by their expansion simultaneously at $t=0$? Is any hidden conspiration involved?

iii) the problem of homogeneity

Astrophysical fact and cosmological theory yield nearly perfect homogeneity of the observable part of 3-space; on a local scale, however, the universe contains irregular structures, from clusters of galaxies and large voids to planetary systems. The standard model handles homogeneity as a boundary condition which is fed in by hand into the model. Thereby the question of the *origin* of the overall homogeneity and of the germs of the local inhomogeneity is circumvented.

iv) the problems of contingency

All questions on the special value of the constants of nature, especially of the cosmological constant λ(= vacuum energy density), and the coupling constants (α, α_w, α_s, α_G) pertain to this group of problems. The reason for the 4-dimensionality of spacetime, totally ignored by physics up to recent times, belong to this category, too.

v) the flatness problem ($\Omega_0 = \frac{\rho_0}{\rho_c} = 1$)

When talking on the isotropy, we realized that only a spacetime which is flat in the 3-spatial dimensions has the unique property of isotropizing at large times. If we look, however, at such a universe at early cosmic time, e.g. at the Planck time ($t = 10^{-43}$ sec), we have to note that the expansion rate had to be fine-tuned to the accuracy of $(\frac{\rho - \rho_c}{\rho_c} t_P \; 10^{-57}$, in order to fulfill the observational fact $\Omega_0 = 1$ of today.[48] This extremely balanced behavior of the universe is highly incredible.

There are much more features of the universe which cannot be explained within the standard hot big bang model, for instance the monopol problem and the matter - antimatter asymmetry, but for our epistemological purpose it is sufficient to realize, how the gaps in the causal chain left by the AP can be filled in.

Within the last eight years, a series of scenarios have been worked out with changing success. In 1980, Alan Guth started the inflationary paradigm with a theoretical model that works in the context of the GUT. In the meantime, it has been amended by Andreas Albrecht and Paul Steinhardt[49], and the recent seemingly simplest version, the so-called chaotic inflationary scenario, has been launched by Andrei Linde.[50]

For our purpose only the central traits of the new symbiosis of cosmology and high energy physics are relevant. Mainly all inflationary models assume the existence of some stage of evolution at which the universe expands exponentially ($R \; \alpha \; e^{Ht}$). In this rather short interval of cosmic time, the universe is kept in a vacuum-like state containing some homogeneous classical fields, but no particles. The rapid exponential expansion is called inflation. At the end of the inflationary epoch, the inital vacuum state decays into particles. The interaction of these particles establishes thermodynamical equilibrium and only at this stage the universe becomes hot and can then be coupled to the standard big bang model ($R \propto t^{1/2}$ for a radiation-dominated FRW model).

The exponential expansion dominating the universe for a short period characterized by the de Sitter line element[51] is caused by the vacuum energy. It is the crucial ingredient of the inflationary paradigm.

We remember that the value $\Omega_0 = 1$ is a strong constraint for the origin and the evolution of intelligent life. In the frame of the inflationary scenario the present proximity of the universe to the critical density, that is the flatness problem, gets a solution easily comprehensible by a causal process. The huge exponential expansion caused the curvature to become negligible, like the surface of a balloon which is heavily inflated. In the same way, the horizon or causality paradox can be tackled with. The exponential de Sitter expansion led to a rapid widening of the cosmological horizon, the boundary of that area, where signals can propagate in accord with special relativity.

There is a possible answer to a special case of the contingency problem that acted a main part in the discussion on the AP, namely the dimensionality of spacetime. Already before the advent of the APs in the sense of Carter, G.J. Whitrow tried to elucidate the 3-dimensionality of space in an anthropic manner. He argued by means of a counterfactual analysis. If the dimension of space n would be n>3, there couldn't exist stable orbits for planetary sites which are possible habitations for humans. The dimensions n=1 and n=2 are excluded, because in such spaces a large number of nerve cells cannot be concatenated to form a complex neural network that, alone, to our knowledge, is able to produce mind generating ideas. It is no doubt that Whitrow uses the argument in an explanatory way, the existence of a «minding animal», among other premises, acting the part of an explanans for the 3-dimensionality of space.

«... that the number of dimensions of space is necessarily three, no more and no less, *because* it is the unique natural concomitant of the higher forms of terrestrial life».[52] Rightly, J.J.C. Smart has criticized this back to front explanatory use of an anthropic argument as preposterous.[53] However, there has been a move in recent unified theories to a renewal of an old idea of Th. Kaluza and O. Klein to introduce theories with spacetimes of more than four dimensions.[54] These extra dimensions of space - supergravity theories are favourably formulated in 10+1 dimensions, superstring theories in 9+1 dimensions - have been compactified, that means, shrunk into thin tubes of the order of the Planck length. According to such theories the universe had its full dimensionality, when $T \sim 10^{17}$ GeV, but all except our well-known 3 spatial dimensions have been confined to experimentally unfathomable small regions already 10^{-40} s after the singularity. Here a road opens to a much deeper understanding of the old question as to why space has exactly three and spacetime accordingly four dimensions. It is*because* - and here we can use this term in the normal explanatory sense - today our universe is cold that the higher dimensions of space have been frozen up.

A possible critique of this approach might argue that the problems have only been deferred to the adjacent question, why the 11 or 10 dimensions of spacetime existed at that early time. But this point seems to me irrelevant. Every explanation has to start somewhere, nothing at all can be inferred from the zero set of presuppositions. Therefore the compactification process is a step to a deeper understanding. Beyond that there are good math*ematical reasons[55] stemming from the symmetry group of the theories which give a rational defense as to why supergravity is properly defined in 11, and superstring theory in 10 dimensions.[56] On the other side, there are genuine questions on the Kaluza-Klein approach which cannot be put aside so easily.

We may wonder, why compactification stopped exactly with four dimensions of spacetime and not with any other number, and furthermore there should be a proof of uniqueness for a certain compactification scheme giving us the distinguished road from the full-dimensional high energy range down to our four dimensions and the low energy particle physics.

Even to these substantial questions a possible answer seems to have been found within the earlier mentioned scenario of chaotic inflation.[57] In this framework, the inflationary universe consists of many different smaller parts (mini-universes) in which all possible types of compactification produce every number of dimensionality. Our concrete four-dimensional realm and our special low energy particle spectrum are not in any way distinguished, but may be coupled to the preconditions of life. Although Linde may be right that there are certain distinguished compactification procedures that lead to kinds of mini-universes apt to contain life[58], he can dispense in no way with the specification of the causal mechanism that led to the pertinent contingent situation of our universe.

Most desirable, of course, would be a theory in which *only one type* of compactification can occur leading inevitably to our four dimensional inflationary universe together with the low energy physics we encounter in present experiments and observations.

To follow the path of dynamical explanations should not be taken as to devalue anthropic interconnections. The innumerable links that have been found between intelligent life and astrophysical and cosmological facts remind us, how strongly we are bound to our large-scale environment. Set in its proper stage, the APs are to be seen in the context of the unity of the universe.

It seems plausible to me to see the AP in an older tradition of ideas, namely the metaphysics of Leibniz' monadology which says roughly that every element of the universe contains in itself a mirror-like picture of the cosmic ground plan. A modern version of this microcosm-macrocosm speculation is G. Chew's bootstrap principle according to which «Nature is as it is because this is the only possible nature consistent with itself».[59] In the framework of the bootstrap picture the universe cannot contain purely accidental features. Since the APs connect the properties of organisms to certain traits of the cosmic environment, they can be called a partial realization of the bootstrap idea, or in the words of E. Harrison: «... the anthropic principle serves as a makeshift or poor man's bootstrap».[60]

However, since today the holistic bootstrap idea is far from yielding successful empirical physics - current elementary particle theories work more in the Newtonian-Cartesian idea of objects moving in a spacetime continuum - it is just a hope to be fulfilled by future research.[61]

The APs are not an invitation to turn the customary direction of explanation upside down, but to fill in the blank by dynamical considerations. Dynamical explanations built upon unified theories lead into deep philosophical waters. They contain surplus meaning in so far as they allow not only to deduce the anthropic constraints and numerical coincidences, but beyond that convey answers to questions physicists would have never imagined to bring into the domain of rational science. Therefore let us have at last a look on the singularity problem. Until quite recently, physicists have taken the stance that the genesis problem, that means the origin of physical reality itself, is beyond the grasp of scientific approach, because the creation of everything violates the logic of explanation. Paul Davies characterizes the former situation: «Because all physics breaks down there, the singularity does not belong to the physical universe, rather it marks the limits of the physical universe, it is the interface between the natural and the supernatural.»[62] The inflationary scenario opens a perspective that would come to grips with the singularity problem and the old question of supernatural borderlines of reality. Two possible solutions show up, depending on the type of the inflationary scenario chosen. Within the scheme proposed by Alan Guth and improved by Albrecht and Steinhardt non-conservation of baryon number is most pertinent to this problem. The violation of that conservation law in the GUTs allows to deduce an important feature of our physical world, namely the ratio of the number of baryons to entropy, $\frac{m_B}{s}$ (which is about $10^{-10\pm1}$). In the absence of baryon number conservation, it becomes *physically* possible that our universe emerged not only from no prior substance, but from nothing at all. This situation in which a physical and entirely causal explanation of the origin of all the matter, energy and entropy would be feasible, seduced A. Guth to the striking remark:

«I have often heard it said that there is no such thing as a free lunch. It now appears possible that the universe is a free lunch.»[63]

Another solution of the singularity problem comes in sight which lies more in the tradition of Lukrez' principle according to which nothing comes out of naught or goes into naught. If chaotic inflation is a correct scenario, then «it seems more likely that the universe is an eternally existing, self-producing entity, that it is divided into many mini-universes much larger than our observable portion and that the laws of low energy physics and even the dimensionality of spacetime may be different in each of these mini-universes».[64]

Of course, a perpetual existence of physical reality cuts off every question on its origin. It is the oldest answer in the history of philosophy of nature and perhaps the most rewarding for a scientist working in the rationalist paradigm.

NOTES

1. H. Bondi: *Cosmology*, Cambridge 2nd edition 1968, p. 13.

2. A.G. Walker: Completely symmetric spaces. *Journ. Lond. Math. Soc. 19* (1944) pp. 219-226.

3. S.W. Hawking & G.F.R. Ellis: *The large-scale structure of spacetime*. Cambridge 1973, p. 135.

4. This methodological point was a reason for E.P. Hubble to defend strongly the cosmological principle which expresses homogeneity and isotropy for 3-space. Cf. N.S. Hetherington: Philosophical values and observations in Edwin Hubble's choice of a model of the universe. *Hist. Stud. Phys. Sci. 13* (1983) pp. 41-67.

5. L. Boltzmann: Zu Herrn Zermelos Abhandlung: «Über die mechanische Erklärung irreversibler Vorgänge.» *Wiedemannsche Annalen 60* (1897) pp. 392-398.

6. K.R. Popper: Autobiography, in; P.A. Schilpp (ed.): *The Philosophy of Karl Popper*, LaSalle (Ill.) 1974, p. 126.

7. Cf. B. Kanitscheider: *Philosophie und moderne Physik*, Darmstadt 1979, p. 70.

8. J.D. Barrow / F. Tipler: The Anthropic Cosmological Principle. Oxford 1986, p. 230.

9 P.A.M. Dirac: The cosmological constants, *Nature 139* (1937) p. 923.

10. This sounds more revolutionary than it is, because in relativistic cosmology, too, total energy is not conserved: it decreases in an expanding universe and increases in a contracting universe. Energy conservation is confined to regions that do not partake in the expansion of the universe.

11. P. Jordan: Bemerkungen zur Kosmologie, *Ann. d. Phys. 36* (1939) p. 64.

12. A typical main sequence star has a mass of roughly $M_0 = 10^{60} m_n$, M_0 should therefore, according to Dirac's LNH, increase with $t^{3/2}$.

13. M. Bunge: Cosmology and Magic, *The Monist 47* (1962) pp. 116-141.

14. Time variations of coupling constants of electromagnetic, weak and strong interaction are constrained by

$$\frac{\dot{\alpha}}{\alpha} \lesssim 10^{-17} yr^{-1} \quad \frac{\dot{\alpha}_W}{\alpha_W} \lesssim 2 \cdot 10^{-12} yr^{-1}$$

Cf. A.J. Shlyakhter: Direct test of the constancy of fundamental nuclear constants. *Nature 264* (1976) p. 340.
The constancy of the gravitational constant has been checked by Reasenberg with the result G: $\frac{\dot{G}}{G} < 3 \cdot 10^{-11} yr^{-1}$ Cf. R.D. Reasenberg: The constancy of G and

other gravitational experiments. *Phil. Trans. R. Soc.* London A *310* (1983) pp. 227-238.

15. J.B.S. Haldane: Comment on E.A. Milne's paper: On the Origin of Laws of Nature. *Nature 139* (1937) Suppl. pp. 1003-4.

16. The field equations of the combined tensor - and scalarfield are

$$G_{\alpha\beta} = \frac{8\pi}{\phi}T_{\alpha\beta} + \frac{\omega}{\phi^2}\left(\phi_{,\alpha}\,\phi_{,\beta} - \frac{1}{2}g_{\alpha\beta}\,\phi_{,\mu}\,\phi^{,\mu}\right) + \frac{1}{\phi}\left(\phi_{;\alpha\beta} - g_{\alpha\beta}\,\Box\phi\right)$$

The term on the left side is the Einstein-tensor and the first term on the right side corresponds to the matter part of Einstein's field equations, the coupling of spacetime to matter by ϕ^{-1}. The second term on the left side is the energy-momentum tensor of the scalarfield, the third term has no intuitive meaning.

17. R.H. Dicke: Dirac's Cosmology and Mach's Principle, *Nature 192* (1961) pp. 440-441.

18. J. Barrow/F. Tipler: *The Anthropic Cosmological Principle*, Oxford 1986, p. 3.

19. B. Kanitscheider: The Anthropic Principle - Physical Constraints for the Evolution of Intelligent Life - An Epistemological Assessment, *International Journal on the Unity of the Sciences*, 2,2 (1989) pp. 273-297.

20. B. Carter: Large Number Coincidences and the Anthropic Principle in Cosmology, in: M.S. Longair (ed.): *Confrontation of Cosmological Theories with Cosmological Data*, IAU Symp. 63, Dordrecht 1974, pp. 291-298.

21. B. Carter: The Anthropic Principle and its Implications for Biology, *Phil. Trans. Roy. Soc.* London A *310* (1983) pp. 347-363.

22. B.J. Carr: The Anthropic Principle, *Acta Cosmologica 11* (1982) p. 143-151, esp. p. 150.

23. C.B. Collins/S.W. Hawking: Why is the Universe Isotropic? *Astrophys. Journ. 180* (1973) pp. 317-334.

24. Cf. e.g. Bernard de Fontenelle: Entretiens sur la pluralité du monde. *Oeuvres de Fontenelle*, Tome deuxième, Ire Partie, Paris 1818.

25. J.D. Barrow/F. Tipler: *The Anthropic Cosmological Principle*, loc. cit. p. 249.

26. This has been done by B. Carter: he could prove that one per cent lowering or raising of α will cause all stars to turn into red stars, resp. blue stars. In neither case, a normal main sequence star as the sun would be possible. Since life is coupled to a long-time burning of main sequence stars, it would probably never occur in a world brought about by this counter factual change. (Cf. Ch. Misner/K. Thorne/J.A. Wheeler: *Gravitation*, San Francisco 1973, Chapter 44).

27. J.A. Wheeler: From relativity to mutability, in: J. Mehra (ed.): *The physicist's conception of nature*, Dordrecht 1973, pp. 202-247.

28. J. Hacking: The Inverse Gambler's Fallacy: the Argument from Design. The Anthropic Principle Applied to Wheeler Universes. *Mind* 1987, pp. 331-40.

29. Hacking's argument has been contested however. See e.g. P.J. McGrath: The Inverse Gambler's Fallacy and Cosmology - A Reply to Hacking. *Mind 97*, No. 386, April 1988, pp. 265-68.

30. B. Collins/S.W. Hawking: Why is the Universe Isotropic? loc. cit. pp. 317-334.

31. For an updated version of the result cf. Barrow and Tipler, loc. cit. p. 425.

32. Here the scale function is $R \alpha t^{2/3}$.

33. Here the alternative is disregarded that the time interval since the beginning was too short for the fact that anisotropy could show effects on the 3 K radiation.

34. This is so because too strong an expansion, $\rho << \rho_c$, $\rho_c = \frac{3c^2 h^2}{8\pi G}$ destroys all germs of condensation, too weak an expansion $\rho_c >>$ c yields an early recollapse.

35. B. Collins/S.W. Hawking: Why is the Universe Isotropic? loc. cit. p. 317.

36. loc. cit. p. 334.

37. R.H. Dicke/P.J.E. Peebles: The Big Bang Cosmology - Enigmas and Nostrums, in: S.W. Hawking/W. Israel (eds.): *General Relativity. An Einstein Centenary Survey.* Cambridge 1979, pp. 504-517.

38. The assumption that everything that *can* occur according to the laws of nature *will* occur has its historical origin in the principle of plenitude (Cf. A. Lovejoy: *The Great Chain of Being*, Cambridge (Mass.) 1948). Carter's supposition that all logically possible universes consistent with relativistic cosmology actually coexist is a modern version of the principle of plenitude. Only causally sane deductive reasoning is involved therein. We as humans exist, because we are part of a possible universe, and all possible universes exist. Nothing is left to chance. The high state of orderliness of our universe is understood in such a way, that it is the only universe which supports our form of life. The principle of plenitude makes any resort to an intelligent designer superfluous. Carter's hypothesis is a piece of deductive reasoning within the realm of naturalistic ontology.

39. J. Leslie: London, 1989.

40. B. Carter: *Large Number Coincidences*, loc. cit., p. 294.

41. P.C.W. Davies: *The accidental universe.* Cambridge UP 1982, p.120.

42. J.J.C. Smart: Philosophical Problems of Cosmology, *Revue Int. Phil. 41*, 160 (1987) pp. 112-126.

43.D. Lewis: *On the plurality of worlds*, Oxford 1986, p. 132.

44. For an evaluation of the many worlds hypothesis cf. Qu. Smith: The Anthropic Principle and Many Worlds Cosmology. *Australasian Journ. Phil. 63*, 3 (1985) pp. 336-348.

45. J. Leslie: Anthropic Principles, world Ensemble, Design. *Am. Phil. Quat. 19* (1982) pp. 141-151.

46. J. Leslie: Cosmology, Probability and the Need to Explain Life, in: N. Rescher (ed.): *Scientific Explanation and Understanding.* UP of America 1983, pp. 53-82, esp. 56.
47. St. Hawking: *Is the End in Sight for Theoretical Physics?*, Cambridge, 1980.

48. A. Guth: Inflation Universes, a Possible Solution of the Horizon and Flatness Problem. *Phys. Rev.* D *23*,2 (1981) pp. 254-257.

49. A. Albrecht/P.J. Steinhardt: Cosmology for Grand Unified Theories with Radiatively Induced Symmetry Breaking, *Phys. Rev. Lett. 48*,17 (1982) pp. 1220-23.

50. A. Linde: Particle Physics and Inflationary Cosmology, *Physics Today*, Sept. 1987, pp. 61-68.

51. $ds^2 = -dt^2 + e^{2Ht} (dx^2+dy^2+dz^2)$

52. G.J. Whitrow: Why Physical Space has Three Dimensions. *Brit. Journ. Phil. Sci. 6* (1955) pp. 13-31.

53. B. Kanitscheider: Explanation in Physical Cosmology, *Erkenntnis 22* (1985) pp. 253-263.

54. Cf.P.G. Freund: Physics in 10 and 11 Dimensions. *Comments Nucl. Part. Phys. 15*,3 (1985) pp. 117-126.

55. B. Kanitscheider: Probleme und Grenzen einer geometrisierten Physik, in: Logic, Philosophy of Science and Epistemology, *Proc. 11th Internat Wittgenstein-Symposium*, Wien 1987, pp. 129-144.

56. M.B. Green: Unification of forces and particles in superstring theories. *Nature 314* (1985) pp. 409-414.

57. A. Linde: loc. cit. p. 68.

58 A. Linde: loc. cit. pp. 61-68.

59. G.F. Chew: "Bootstrap": a scientific idea? *Science 61* (1968) p. 762.

60. E. Harrison: *Cosmology*, Cambridge 1981, p. 115.

61. G.F. Chew: The Topological Bootstrap, in: A.H. Guth et al. (eds.): *Asymptotic Realms of Physics*, Cambridge 1983, pp. 49-50.

62. P.C.W. Davies: Quantum Gravity: A Unified Model of Existence? *Mitt. Astron. Ges.* Nr. 58, Hamburg 1983, p. 53.

63. A.H. Guth: Speculations on the Origin of the Matter, Energy and Entropy of the Universe, in: A.H. Guth (ed.): *Asymptotic Realms of Physics*, Cambridge 1983, p. 215.

64. A. Linde: loc.cit., p. 68.

Alberto Cordero

EVOLUTIONARY IDEAS AND CONTEMPORARY NATURALISM[1]

1. INTRODUCTION

As I was thinking about the rise of evolutionary ideas in twentieth century science, I remembered a passage in which the celebrated N.R. Campbell shows his attachment to an old metaphysics of time. It provides a telling measure of how bizarre and unexpected our present ideas must have initially been[2].

> Is it possible to find any judgement of sensation concerning which all sentient beings whose opinion can be ascertained are always and absolutely in agreement? ... I believe that it is possible to obtain absolutely universal agreement for judgements such as, the event A happened at the same time as B, or A happened between B and C. [N.R. Campbell, Foundations of Science, Cambridge, 1920, p. 29]

In this passage, written as late as 1920, the usually insightful Campbell is reassuring himself of the superiority of an absolute long gone sour. The idea of merely relative simultaneity is, he intimates, impossible; beyond human imagination. Already in 1905, however, Einstein's theory of special relativity had shown physicists how to claim that the time of an event is not the same for all observers.

E. Agazzi and A. Cordero (eds.), Philosophy and the Origin and Evolution of the Universe, 399–439.

That was only the beginning of our present evolutionary framework. Tampering with the alleged boundaries of imagination would soon become second nature to science. By the 1920s, conceptual revisionism in physics alone had surpassed everyone's expectations; there seemed to be nothing science would not eventually challenge. Scientists did not have to fear metaphysicians, for they had replaced them.

FORMAL FOUNDATIONALISM

Initially, the new scientific developments seemed in harmony with the anti-metaphysical views of the logical positivists, whose goal was to purify philosophy by removing its non-scientific elements and reconstituting the discipline with logic as its organon. They declared a piece of substantive or synthetic discourse meaningful or not on the basis of its logical form; it was meaningful if it displayed the form of scientific discourse and nonsensical or meaningless if it did not.

The new theories of space-time and matter did nothing but confirm the positivist claim that science is not constrained by *metaphysical foundations*. This was the positivists' grand negative thesis. As already indicated, however, they also had an affirmative claim about science. Science, they thought, was constrained by logical form, The gist of their approach was that one could distinguish between genuine knowledge and nonsense by keeping in mind certain logical difference between the languages of theory and observation, the analytic and the synthetic, the factual and the valuational, and the contexts of discovery and justification.

For over thirty years, logical positivists tried to characterize science in terms of form alone. By the mid-fifties, however, it had become clear that the dichotomies on which logical positivism had based its program could not be coherently articulated. Science simply did not have the clear and eternal form positivists had imagined it have.

Nor, for that matter, did it have any invariant form. If, earlier on, the radicalness of scientific innovation had blended smoothly with the formalist tenets of logical positivism, the mixture fell apart when historical research revealed that science changes, virtually at all levels, as its content develops. Science, it seemed, was not characterizable in essentialist terms of any kind.

FROM POSITIVISM TO NEGATIVISM

So, the very depth of the scientific innovations logical positivism had sought to authenticate was, in the end, one of the main sources of its downfall. The search for the timeless formal foundations of science turned out as wrongheaded as had the earlier search for its metaphysical foundations, and for similar reasons. This failure inevitably contributed to a thorough reappraisal of the foundationalist project in philosophy.

The terminology must be handled with care here. Following Quine, I shall employ the terms «foundationalism» and «first philosophy» as synonyms. Thus, in what follows «anti-foundationalist» will be a thinker for whom no standards of evaluation are independent of or better warranted than those current science endorses. He does not deny the existence of foundations as such, but only of foundations external to the practices of science. In his view, there is no science of justification which the philosopher practices but the scientist does not. Many different positions are compatible with this basic characterization of anti-foundationalism, as we shall see.

Even though the search for the essence of science failed, it was not easy to totally reject foundationalism. Insecurities abounded, and they came from all directions, especially from the perception that if science has no essential foundations, then it will lose its identity and identify with fiction.

With the demarcation between science and non-science blurred, relativism prospered in the relaxing atmosphere of the sixties. The reception of Thomas Kuhn's work as evidence of the existence within science of «revolutionary» argumentation in addition to «normal» argumentation, and the transformation of Paul Feyerabend from enthusiastic Popperean to defender of the pursuit of witchcraft in healthy society, are emblematic illustrations of the feelings of the time[3]. Romanticism was enjoying a sweeping revival. Could one reject foundationalism and yet avoid drowning in the perilous sea ahead? Philosophers who answered in the negative would encourage «neo-foundationalist» projects, many of which we still see today. Others refused to have anything to do with foundationalism; many of them abandoned rigorous philosophy and became «deconstructionists» or «literati» of one sort or another. Willard V.O. Quine, the patriarch of contemporary anti-foundationalism, was not one of these. There is natural science, he asserted, but no higher epistemological level from which to study it. Seeing that epistemology had a future only as science self-applied, he launched the project that would become the most influential brand of naturalism in philosophy today.

OVERVIEW

The above connection between the rise of our present evolutionary ideas and the fall of foundationalist conceptions of scientific knowledge will, I hope, give us some ground on which to stand when trying to place contemporary naturalism as a philosophy motivated by the impact of science on the fabric of knowledge. In the rest of this paper I want to critically discuss the surge of naturalist programs in recent philosophy of science. I have two main points in mind. First, I want to invite reflection on the reductive character of the most influential brand of naturalism. Here my principal targets will be Quine's conception of science and the elimination of reason many of his followers advocate. Second, I want to take up and defend a more scientific (and coherent) naturalism, one not linked to philosophical preconceptions about knowledge. Though I touch on a variety of points, my aim is to emphasize the unprecedented intellectual depth of contemporary science.

2.QUINEAN NATURALISM

The most influential naturalism today originates in Quine's attempt to explain scientifically how people come to hold elaborate theories about electrons, quantum fields and the like. In asking this and other epistemological questions from within science, Quine seeks to explain our belief in the science of which epistemology must be part. This is not the traditional way to do epistemology, but rather an anti-foundationalist proposal whose central note is precisely that there are no «upper-level» constraints on scientific inquiry.

Still, Quine's conception of science is, on the whole, conservative. He looks at the standards endorsed by current science primarily in logical terms. We have some «self- evident» truths, and we have a large (though *always* finite) collection of direct observations, but clearly these two sources of knowledge do not suffice to fully justify any scientific theory. Theories involve general claims and, so, from a strictly logical point of view, they can never be determined by any finite collection of data. Given any theory, it is always possible that a different theory exists which is equally compatible with the finite data at our disposal. As Duhem before him, Quine grants momentous epistemological importance to this difficulty. Since Quine, unlike Duhem, does not accept suggestions from «first philosophy», this encounter with logical possibility leads him to the view that scientific theories must ultimately be arbitrary constructions of the mind. In the last analysis, he says, what counts in favor of a particular explanation can only be specified in terms of such «virtues» as conceptual conservatism, logical modesty, simplicity, generality and testability. So, Quine is a naturalist who thinks that epistemology has a future only as science self-applied, but he is also a philosopher of the old guard who believes that mere logical possibilities matter epistemologically. Less explicitly, he is, in addition, a staunch empiricist.

Because Quine is a staunch empiricist, he regards belief-formation as a process well within the scope of logic and the psychology of *perception*. He thus takes for granted that the epistemological study of scientific ideas must begin at the level of neurological events in the sense organs (ultimately in «sensory irritations»), with additional contributions from various levels of pre-wiring and programming (he calls these «internal conditions»). Here is an example of his approach:

> We are studying how the human subject of our study posits bodies and projects his physics from his data, and we appreciate that our position in the world is just like his. Our very epistemological enterprise, therefore, and the psychology wherein it is a component chapter, and the whole of natural science wherein psychology is a component book-all this is our own construction or projection from stimulations like those we were meting out to our epistemological subject. (1969, p. 83).

Quine's first thesis is, thus, that theory formation is fundamentally a neuro-psychological topic. But, since psychological events are causally related to biological evolution, this basic thesis leads to a larger claim, namely, that human reasoning in general, and scientific inference in particular, must be fundamentally like induction, which, he adds, must in turn be like animal habit formation. Quine effectively claims that all our inferential capacities are not only continuous with but also *reducible* to those found in lower animals. I shall call this strong interpretation of the continuity of our cognitive capacities with the evolution of our bodies «reductive naturalism».

Science, then, is to Quine fundamentally the product of an organism that takes its perceptual input as the basis of simple inductive processes and is able to build up a complex linguistic system with the help of inductive methods fundamentally similar to those used in the acceptance of the most ordinary ideas about the world (1969, pp. 125-28).

For more than two decades, confidence in this type of naturalization has prevailed among most contemporary naturalists. There are, however, voices of discontent.

TWO BASIC REACTIONS

A typical attack is found in Siegel (1980). Siegel, a neo- foundationalist, warns that naturalized epistemology violates Reichenbach's distinction between context of discovery and context of justification, as well as other basic distinctions. The employment of science to investigate science, he declares, condemns Quine to circularity. He also complains that the goal of epistemology is not merely to describe, but also to prescribe. Psychology and epistemology differ, he says, in that the former studies the associations peculiar to specific individuals, whereas the latter focuses on those structures recognized by the relevant community to have justificatory force. Siegel's appeals to foundationalist distinctions beg the question (Roth 1983), but they raise a legitimate concern about the current irrelevancy of psychology to epistemology, at least with regard to the justification of claims in the mature sciences.

It is important to notice that the latter objection to Quine's epistemology comes also from the anti-foundationalist camp itself. Dudley Shapere is a case in point. He refuses to take at face value the Quinean claim that epistemology and the whole of science are, in the last analysis, a construction from sensory stimulations and pre-wiring alone (1987). Shapere also charges that Quinean concentration on the activation of sensory receptors as an ultimate term in the epistemological relation to be studied fails to deal with the real issue about the scientific process and its most relevant products, namely the justification of such exemplary theories as quantum field theories, the currently most advanced unification programs in physics and biology, and so on.

Like Siegel, Shapere particularly questions the relevance of Quinean psychologizing in epistemology. The life of experimental physics, he points out in his analysis of the solar neutrino experiment (1984, Ch. 16), makes it clear that psychological considerations play no role as reasons for raising specific problems, employing specific methods, or accepting specific beliefs in any advanced science.

CAMPBELL'S EVOLUTIONARY EPISTEMOLOGY

Attempts to adapt Quinean naturalism to satisfy the objections of its critics abound. A particularly extreme instance of this is found in Donald Campbell's evolutionary epistemology[4]. Campbell's copious production has been changing back and forth around a handful of evolutionary and realist slogans, but always maintaining the claim that selection theory is needed to replace every foundationalism, and always trying to make Quinean naturalism absorb, in one form or another, the positions of its critics. Selection theory, he explains, can be fruitfully employed not only for such developments as the evolution of the visual perceptual machinery of eye and brain, but also as a separate selection theory for the operation of vision in the generation of specific competent beliefs (Campbell 1988).

An important feature of Campbell's work is that it tries to flesh out Quine's abstract epistemology with real science. He maintains that a straightforward generalization of evolutionary biology suffices to articulate a full-fledged theory of the generation and development of conceptual change in science. Campbell also thinks that he can make his proposal realist by compounding evolutionary naturalism with the claim that competent reference is achieved by the referent's participation, however vicariously, in the selection of beliefs from actual alternatives[5].Campbell's view can be understood as saying that, eventually, we shall be able to explain conceptual change in science by way of detailed evolutionary accounts of the relevant mental processes, combined with causal accounts of the mechanisms of reference. As a complete explanation of scientific development, however, his approach seems unsatisfactory on at least two counts.

It is doubtful, to say the least, that contemporary natural science is universally prompted by means of anything like Campbell's «blind variation and selective retention»[6] (according to which scientific initiatives are generated by random variation of existing material followed by opportunistic selection of successful constructs). In the mature sciences, reasoning is rather tightly constrained by prior knowledge. Consider, for example, the case of physicists currently working on the unification of quarks and leptons. Their search for mathematical groups to use the Yang-Mills theory is so systematic that it sounds extravagant to suggest, as Campbell appears committed to do, that physicists have been led to the SU(5) group solely or even mainly by blind conceptual variation in any epistemologically relevant way.

Another source of dissatisfaction with Campbell's project is his «realism». It is a position made either vacuous or redundant by his talk of «competent reference». Once philosophical foundationalism is given up, the question arises, what can an appeal to the referent mean other than reference to whatever science compels us to believe there is? Surely Campbell is not suggesting that the referent of every physical theory is bound to be external to the cognitive subject, for then the very possibilities envisaged in the contemporary discussions of quantum mechanics would seem to falsify his theory[7]. It remains an open question, for example, whether or not, in the aftermath of the Bell experiments, the persistence of paradox about quantum superposition and entanglement will motivate scientists to settle, competently and fully, for something like the subjectivisms explored in Wigner (1963) or Putnam (1983). Many of us hope objectivism will prevail, but that is a different matter.

Also a different matter is the epistemological relevance of evolutionary ideas in general. My remarks about Campbell's approach are specific to his project and, as such, not necessarily applicable to other developments within the field generically known as «evolutionary epistemology». The latter is constituted by a wide family of proposals that relate the development of science in some non-trivial way to the process of biological evolution, with considerable variation regarding emphasis and depth, as shown, for example, by such antologies as Wuketits (1984). In particular, the foregoing objections do not question the seminal notion that biological evolution is continuous with cognitive individual and social development, but only the appropriateness of Campbell's own reductive proposal to the analysis of science.

QUINEANISM REVITALIZED

In the late seventies, many naturalists influenced by Quine decided to make their their philosophy more directly relevant to the study of actual science. The result of their efforts is evidenced in the production of such naturalists as Chris Cherniak (1986), Paul and Patricia Churchland (1979, 1984, 1986), Ronald Giere (1985), David Hull (1983, 1988), Philip Kitcher (1983), Eliot Sober (1984), Stephen Stich (1983, 1985), to mention but a few cases in point. To achieve the said end, these thinkers expanded the scientific horizon of naturalized epistemology and enriched their analyses with detailed applications of evolutionary theory, neurology and cognitive science, as some evolutionary biologists had begun to do. More importantly, they also directed their studies to actual conceptual change in science, which is something naturalism had failed to do to this point. Many, like Quine, would hold to strong reductive forms of naturalism. Unlike Quine, however, these new naturalists were not lacking first hand concern with science, nor were they nesting their hopes in merely logical and linguistic considerations.

In their writings we find important departures from Quine's conception of scientific theorizing. Many, for example, reject the notion that mere logical possibilities matter epistemologically, let alone constitute grounds for doubting specific theories. Following the practices of science, many point out that doubts about a theory arise only if and when specific results or arguments are produced against it.

Several naturalist approaches have been developed along these lines in the last decade. Not all the ongoing projects are reductionistic in Quine's sense; neither Cherniak nor Sober, for example, pursue strong reductionism. Still, most of the new naturalists continue to favor the elimination of the term «reason» from epistemology; I am referring to naturalists for whom reason and rationality are some sort of «second-class» properties which, like the secondary qualities of old age, are allegedly brought forth by some complex chain of cause and effect, originating in a world endowed just with the properties studied by basic physics and biology. Various reductionist stands can be distinguished in the literature. A particularly clear and strong position is found in the work of Ronald Giere.

STRONG REDUCTIVE NATURALISM

Giere wants to make naturalized epistemology relevant to the study of conceptual change and theory development in science; he does not want to leave us with a mere sketch of concept formation, as Quine and Campbell do. His goal is to offer significant answers to such questions as, what leads scientists to develop the models they do develop? and, how do their ideas come to be accepted? In order to achieve this goal, he puts the psychology of persuasion, cognitive science, and evolutionary biology at the center of his enlarged scientific platform. Giere also incorporates the reductive hypothesis. The result is a philosophy of science sharply focused on descriptive decision theory. Giere's account of science and theory choice starts, in fact, from a refined version of Quinean reductionism, which shows up in his project as two assumptions; one about the nature of our cognitive capacities, and the other about the foundations of theory choice.

The first assumption is the idea that our cognitive capacities are not only an evolutionary development of those possessed by the lower primates and other animals: doing science is one of our ways to interact with our environment. In Giere's view, the basic capacities of the lower animals must be the ones the naturalistic philosopher employs in attempting to study the scientific activities of his fellow humans.

That this approach, like many other theories from the Quinean camp, is radically against explanations of theory choice in terms of «reasons» is an important characteristic. As Giere (1985, p. 332) puts it, «Rationality is not a concept that can appear in a naturalistic theory of science-unless reduced to naturalistic terms.» He faces, then, the question of what these naturalistic terms are with the help of which we are to account for the ways scientists make their choices.

This leads to Giere's second reductive assumption, which is that descriptive decision-theory already contains some models of choice suitable to the explication of conceptual change in science. Descriptive decision-theory, let us recall, concentrates on the decision rule that, as a matter of fact, defines the best choice in a given situation as a payoff function. As such, DDT does not focus on rational agents, but on the behavior of what decision theorists call «satisficers», i.e., individuals whose choices are based on decision rules that can be grounded in explicit pay-off matrices.

How are we to select «rational» decision rules by looking at the characteristics of the strategies that are actually used in science? Giere is optimistic about the prospects. In his view, we just need to study scientific developments in a piece-meal way, looking for as much guidance as possible from the psychology and sociology of persuasion. That, he says, will direct us empirically to the best available strategies. All we must do then is clean up the findings and incorporate them our into a tentative theoretical model of human decision making.

We thus arrive to the core of Giere's theory of science: Scientists are satisficers; scientists, that is, follow something approximating a satisficing strategy when faced with the problem of choosing among scientific theories (1985, p. 348).

People functioning as mere satisfiers surely contribute to the direction and speed scientific developments actually take. But we knew that already from the sociology of knowledge. The question is whether or not scientific rationality transcends mere satisficing and, if so, how. Can we really do without asking, in a fundamental way, whether scientists are right or wrong in accepting certain theories and rejecting others? Giere thinks we can. As a case in point, he has analyzed the rise of the hypothesis of continental drift.

CONTINENTAL DRIFT

Similarity between the coastlines of South America and Africa on either side of the Atlantic led Alfred Wegener to propose a revolutionary hypothesis in 1915. The earth, he thought, must have developed by means of some form of continental drift. Wegener's hypothesis was at odds with the best knowledge then available, failed to convince scientists, and was largely ignored. In the 1950s, however, two developments changed the background of geological inquiry. First, large ridges were discovered in the ocean floors. Second, researchers found that the earth's magnetism has undergone several reversals of polarity in the past. These and other related developments were put together in a landmark theory by Harry Hess, which claimed that an ocean ridge is the result of molten lava rising from the interior of the planet and spreading from the ridge. Soon afterwards, two research teams, one led by F. Vine and D. Matthews, and the other by L. Morley, explored in detail the implications of Hess' hypothesis. They concluded that, if the theory was correct, then a fairly symmetric series of bands of opposite magnetic polarity should exist parallel to the center of every lava ridge. By 1966 the VMM prediction was verified beyond reasonable doubt. The ocean floor, it turned out, was actually on the move, and this was enough to convince most geologists and geophysicists that the drift hypothesis was basically right.

Giere (1985) discusses this case in revealing detail. His analysis centers on utility measures of the relevant agents desires (pp. 348-353). To him, the important aspect is not the intrinsic positive value «science» places on being objectively right, but rather the value of short-term professional payoffs at the individual level.

Giere's specific question is, why did most scientists rush to accept the drift hypothesis after 1966? His answer is that at in 1966 the payoff matrix for the decision became clear to about everyone. He grounds this explanation in two fundamental reflections about the scientific community:

(1).-The short-term professional payoffs of being «objectively right» were so high that all members of the community regarded being right as satisfactory, no matter what their own preferences might have been.

(2).-Most geologists who were not doing research in oceanography or paleomagnetism would have preferred that the traditional, static model of the earth had won, because their skills were attached to such views and switching carried the cost of gaining new knowledge.

On the basis of these considerations, Giere then constructs the following payoff matrix for the case (1985, p. 352):

	OPTION 1 Drift Models are Approximately Correct	OPTION 2 Static Models are Approximately Correct
Value of adoping Drift Mode	Satisfactory	Terrible
Value of Retaining Static Model	Bad	Excellent

Not unexpectedly, from the storms of individual greed and selfishness, the option that emerges with the best payoff turns out to be the one that follows the «golden rule» of standard epistemology for cases like this: Accept the outcome of the crucial test provided by the VMM prediction; i.e., if the VMM prediction is verified, accept a drift model; if not, continue accepting a static model.

Giere rejects the notion that all this could be better conceptualized in terms of rational acceptance. There is no need to do that, he says. From his reductionist perspective, the only important element is the efficiency with which the relevant community dealt with their total enviroment. Accordingly, Giere simply takes the satisficing strategies revealed de facto by this and other cases as part of his model of decision making, and then studies the circumstances in which humans fit the model. Nothing more, he believes, is required in order to provide a good scientific explanation of theory choice.

Reductive naturalism rightly calls attention to the role played by personal and social factors in the development of scientific ideas. But its reductive core pushes the study of science toward non-rational lines. The result is a socially exciting but seemingly distractive epistemology.

An objector may point out, for example, that the researchers in question would just have been bad scientists if they had failed to follow what I have called «the golden rule» for cases like this. They were in front of a sufficiently clear «modus tollens situation», no matter how strong other pressures on them might have been at the time. The geoscientists had available a rich context of reasons to guide and assess their choices, and these reasons were the only ones they had to keep in mind, if they were good scientists.

The objector, in short, could charge that talk about mere forces is distractive. Gossip about individual passion, greed and opportunism may understandably lead to explanations of the direction and pace of scientific developments at any given time, and so to good sociology, but not necessarily to good epistemology if, as a matter of fact, scientific practice is constrained by distinct critical standards.

So, isn't reductive naturalism blurring the point and missing it too?

REASONS

A more circumspect way of accounting for the development of the drift hypothesis is in terms of reasons, understood as considerations such that, to give reasons for a belief is to provide some basis for thinking that it is likely to be true. In anti-foundationist terms, this amounts to linking the acceptance of a hypothesis to the fallible assessment of its truth on the basis of such learned considerations as empirical adequacy, coherence, testability, fertility and unifying power.

On this approach, Wegener's hypothesis could not be sensibly pursued when it was first proposed, let alone accepted, simply because it was marred by a host of specific doubts. Apart from some geographical similarity, there was little else to go by; in particular, there was no evidence that the continents move. The alternate theory of the earth, the prevailing «static model», left the similarities between the coastlines of South America and Africa intriguing and mysterious, but was otherwise supported by the physics of the day. In addition, Wegener's proposal was difficult to structure in terms of prospective tests. His revolutionary theory was largely ad-hoc, the very kind of theory the practice of science has taught us to distrust. This state of affairs would change completely with the geological explorations of the 1950s.

By 1960, not only had static geology become suspect, but the drift hypothesis had been fleshed out physically and mathematically to the point that the VMM prediction gave scientists the opportunity for a crucial test. When, in 1966, the VMM prediction was verified, scientists were left with excellent reasons to accept the drift hypothesis and virtually no specific doubts about it. In spite of other professional and human pressures, rejecting the most plausible family of static theories had simply become the *right* thing to do.

In this second account, considerations related to truth assessment and rational norms, as opposed to mere satisfactions, are involved throughout. Reasons, however, are not presented as some occult feature of science, but rather as specific considerations that, as the story unfolded, compelled scientists to accept or refine their geophysical beliefs on the basis of current scientific knowledge. The result is an account of the rise of the drift hypothesis in the context of certain critical, normative, truth-seeking standards internal to science.

Notice that nothing in this account stands against the eventual naturalization of scientific rationality. The explanation just given of conceptual change simply avoids hasty reductionism, lest it miss the point. And yet, not missing the point has been a consistent preocupation among all reductive naturalists, from Quine to Giere. According to these theorists, however, the point is not just that science is a natural phenomenon and must be studied as such. As we have seen, to them the point is also that natural attributes do not transcend the level of subhuman species.

This way of looking at the point of reductive naturalism suggests two lines of criticism. There is, first of all, the question of how adequate it is as an account of decision making in science. Second, there is the question of the scientific status of its reductive assumption. We turn to each of these sources separately in the next section, after a few preliminaries.

3.PROBLEMS WITH REDUCTIVE NATURALISM

How adequate is reductive naturalism as an account of theory choice? Its followers have certainly tried to make Quine's abstract approach relevant to epistemological practice by embodying it with «real science». Campbell's detailed account of the development and functioning of sense organs and his epistemological generalization from that source attempts to do exactly this.

Giere's more cautious approach attempts moves in the same direction. Rather than making any wild generalization from chapters of science remote from epistemological practice like evolutionary theory or neurology, he uses them only as a general framework, which he then fills in with findings from the psychology of persuasion and descriptive decision theory. The result is an explanation of conceptual change that both closes the gap between philosophy and practice and is scientific through and through, at least at first sight.

As we have seen, however, strong reductionism seems to miss the point.

AN ADEQUATE EPISTEMOLOGY?

Let us try to make the latter claim precise by examining Giere's reductionism more closely. What seems wrong with his theory?

Giere's fundamental thesis is that scientists are satisficers. If by «satisficer» we mean a person who optimizes relevant pay off, then scientists are of course satisficers. So are all goal-oriented individuals: the philosophers of old Athens, gypsy cab drivers, Hollywood producers, and the chimpanzees of the zoo are satisficers. Scientists are just satisficers of a special kind, namely rational ones.

It seems clear that not all satisficers are scientists. Imagine a «community of imitators», say, a community A of people who simply repeat everything that the members of a certain community B of scientists do. The members of the two communities stand in a one-to-one correspondence with one another. The imitators (who need not know any science nor have any cognitive commitments) repeat the scientists' behavior because some excentric pays them good money to do so. All the individuals involved in this fantasy are satisficers, but not all of them are scientists; or are they?

If Giere's claim were that scientists are rational satisficers, his theory that scientists are satisficers would be both unproblematically correct and incomplete. It would be incomplete in the same way that the following explanation of conceptual change is incomplete because it fails to incorporate relevant information: «scientists are causalizers, i.e., their decisions follow causal chains».

It is important to see just how brave Giere's account is. He not only refuses to characterize scientists as rational satisficers, but the satisficing to which his theory refers involves the total natural enviroment of each and every scientist. This can be understood in at least two different ways.

One reading is that the rules of decision making scientists follow simply need not be «rational». They just have to be compelling to the relevant scientists, as in «Gosh, I better go along with the idea of subjecting dogs to psychoanalysis, because my boss is delighted with it.» On this interpretation, scientists do science just by virtue of implementing opportunistic strategies appropriate to the circumstances of each case. This, however, seems to make us unable to say, for instance, that throughout the 1940s and 1950s most of the «best» researchers at Lysenko's Institute of Genetics in the Soviet Union were doing bad science. The point is that implementing satisficing strategies is not by itself enough to guarantee scientificity. Satisficing strategies have no essential cognitive commitments and are implemented without necessary reference to any scientific knowledge.

So, read this way, Giere's explanation seems incomplete because the processes that define something as a satisficer are independent of any reference to a specific type of goal. Yet, the critic observes, science makes new theories acceptable or unacceptable, at least in part, by virtue of fairly specific norms and values.

A more promising reading of Giere's claim is that scientists are indeed rational satisficers, but that the capacities we associate with «rationality» are not just developed from, but also fully reducible to, those found in the lower primates. In this second interpretation, Giere's thesis is, at first sight, similar to other reductive hypotheses in science; for example, the biochemical hypothesis that living creatures are physico-chemical systems. There are «physiological processes», accepts the biochemist (meaning processes that deal with «purposes» and «functions»), but they are fully reducible to the properties and processes of plain physico-chemical systems.

This reading of reductive naturalism leads us to the second line of criticism mentioned at the beginning of this section. But, to understand that line effectively, we must notice a few things about physics and biology first.

ON NATURAL SCIENCE AND SCIENTIFIC REDUCTIONISM

Naturalists want to see reason reduced to the categories of the natural sciences. Nevertheless, the term «natural sciences» can cover many things and, to the extent that it does, be the basis of many different naturalisms.

To the reductive naturalist, the interesting conceptual core is provided by the set of well-established theories in physics, biology, plus all the relevant mathematics. This basis yields good models of the world of stars, rocks and birds according to current science, but not (at least not clearly) a direct picture of how belief functions in something like scientific theorizing.

As we have seen in connection with the drift hypothesis, there is another way of looking at science. According to this system, natural science involves concern for critical belief and, therefore, focuses not only on theories but also on such underlying aspects as thinking about the world and learning how to accept and reject theories about it. On this interpretation, a major constitutive element of science is *scientific rationality*.

Why is the reductive naturalist skeptical of the fundamentality of reasons in the context of hypothesis formation? In particular, if his rejection is scientific, how does the reductive naturalist hypothesis spring from contemporary science?

In a broad sense, the hard natural sciences are reductionistic. Molecular biology, atomic physics, the physics of the electroweak interaction, quantum chromodynamics, and all the major unification programs in fundamental physics are in some way reductionistic; each postulates a set of basic entities out of which all other entities are built. We must observe from the outset, however, that none of the disciplines just mentioned is a good example of nomological reductionism (strict reductionism with regard to properties and natural laws). The holistic peculiarities of some complex systems, the presence of seemingly irreducible many-body-forces, the features of quantum non-separability and quantum indeterminateness are some of the blocks that stand in the way of strict nomological reductionism. Still, it is undeniable that some fields pursue this form of reductionism. It is instructive to examine how one of them, biochemistry, works. (My aim here is not to equate biochemical reductionism with reductive naturalism, but simply to explore how a major reductionistic program in the hard sciences responds to practical and conceptual difficulties.

REDUCTIONISM IN SCIENCE

Biochemistry is one of the most outspoken reductionist disciplines in the history of science. Its study of organic functions from a strong physicalistic perspective is one of the great scientific achievements of our century. Biochemical analyses show how many characteristically biological attributes can be continuous with physico-chemical ones. These analyses explain, for example, how foods fit into the motions, electrical impulses, accumulation of nutrients in the cell, as well as the performance of many other organic functions beyond the cell. Robert Engel's paper in this volume gives splendid examples of biochemical reduction and of developmental continuity from molecules to life.

With regard to viral and celullar function, strong biochemical reductionism is made plausible both by the success of its analyses and by prior findings in chemistry, genetics and evolutionary biology. Nevertheless, the extension of this reductionistic program to other biological domains is limited by practical as well as by theoretical considerations.

Conspicuous among the disciplines that resist «complete» assimilation into biochemistry is the human mind. That our spirit is influenced by the physico-chemical structures and juices of the body is not in doubt; any wine lover knows that. But facts about the influence of matter on mind do not constitute adequate evidence for the thesis of *complete* assimilation. For one thing, the terms employed in our most illuminating descriptions of the mind do not function like the terms of good biochemistry.

On the practical side, the mind is simply too complex an object and current knowledge of even its simplest proto-structures is very poor. Methodologically, this circumstance makes it advisable to deal with mental phenomena primarily in mental rather than physicalist terms. It is a matter of convenience to avoid unnecessary distraction from the intended subject-matter. That is why scientifically robust conceptual studies of the mind do not fuss about nomological reductionism. For example, Paul and Patricia Churchland, two rather fierce reductionists in the field of cognitive studies and artificial intelligence, do not really pursue physicalistic reductionism, but concentrate instead on such things as the computational capacities of certain models of the mind[8]. That, incidentally, is already an excessive reductionistic move in the view of many[9]. The important point is that the claim that human beings may have natural properties which are not fundamentally physico-chemical is compatible with the Churchlands' approach.

On the conceptual side, the claim that mental states are identical with physical states is at odds with some contemporary views. One such view is that thinking and other mental properties need not be exclusively human. Future computers, whose physics and chemistry are expected to be very different from human beings, may also come to acquire these abilities.

TWO KINDS OF REDUCTIONISM

In practice the above puzzles do not worry biochemists, however. Biologists, in fact, seem to have little difficulty in shifting their reductionism from a strong, nomological, level (that the properties and laws of all living systems are completely reducible to those of mere physico-chemical systems) to a weaker *ontological* level (that the properties and laws of all living systems are just those of a creature made entirely of physical systems). Universal nomological reductionism remains, but only as «speculative» biochemistry, and not as part of the stuff a scientist is required to believe or pursue in order to do biochemistry, nor as part of what one finds in standard textbooks on the subject. The following difference between scientific and speculative reductionism thus emerges: while the former is meticulously guided and constrained by current scientific knowledge, the latter is not.

It does not follow, of course, that nomological reductionism in scientific biochemistry is ruled out in principle. Future neuroscience might show, for example, that all mental properties are strictly determined by the molecular properties of the brain. On the other hand, it might not.

What we see in biochemistry is a common phenomenon. The natural sciences present us time and again with this kind of property multilevel. On the whole, the picture that we get from the most reliable products of contemporary science shows a world of qualitative uniformity subdivided by qualitative variety, in which intrinsic properties, systemic properties and holistic properties carve up a multilevel ontology.

Like many other reductionistic claims, the issue of the status of reasons in science appears to be up for grabs, but the above results directly affect it in at least two ways. First, scientific reductive hypotheses are tightly constrained by current scientific knowledge. Second, the natural sciences do not presently endorse universal reductionism with regard to natural properties.

WEAK POSITIVE MOTIVATION

The most promising version of reductive naturalism, to return to our main point, agrees that scientific decision-making is constrained by «rational» elements, but insists that the capacities and values these elements involve are not different in kind from those available to subhuman species. According to the reductive naturalist, since our cognitive abilities are sufficiently similar to those of our primate cousins, naturalistic philosophy must attempt to study scientific activities employing only the capacities possessed by these and other animals. This is the cornerstone of the reductionist slogan that the term «reason» cannot appear in a naturalistic theory of science, unless reduced to naturalistic terms.

We are thus led to the question, what positive scientific motivations are there for this claim? As we have just seen, it is not that the natural sciences support universal reductionism, for they do not.In fact, only the most speculative and least scientifically established sociobiological views endorse something as radical as the reductive naturalist position.

Perhaps, then, the motivation is methodological. Popper and the most extreme hypothetico-deductivists believe that hypothesis formation in science amounts to irrational guessing, a process not constrained by current background knowledge. Science, according to this view of pedigree, can be analyzed as a succession of improbable hunches and critical tests. Is this what the reductive naturalist has in mind? Does he think that his hypothesis is scientific because it conforms to the rules of Popperian methodology? If so, the problem is that no methodology seems further from the truth or more refuted by the practices of advanced science than

Popperean methodology: mere hunches ceased to be well received in the mature disciplines long ago.

Hunch-oriented hypothetico-deductivism is still practiced in our culture, but mostly by philosophers and literati, not by scientists. The latter may occasionally indulge in hunches, but in the most mature contemporary sciences there is a genuinely bad feeling about wild, thoroughly ungrounded speculation. Reductive naturalism would, then, follow science in presenting itself as a hypothetico-experimental mode of inquiry. With regard to the grounding of its own central hypothesis, however, it would appear to be following old fashioned philosophy, not science.

The lesson from the above considerations, together with our exploration of reductionism in biochemistry, further articulates the initial feeling of uneasiness about reductive naturalism into a perception that such an approach is only weakly scientific, if at all. Doctrinaire in its methods, the central assumption of reductive naturalism lacks adequate independent support. And, judging from its treatment of the hypothesis of continental drift, the epistemological results are not very illuminating.

The reductive assumption leads one to ask some old questions of the sociology of knowledge and to keep in focus the behavior of flesh and blood scientists: to what extent social and political pressures influence the direction and rate of specific scientific innovations?, how exactly they do so?, why and how specific forms of indoctrination function in scientific communities?, and so on. These psychological and sociological considerations are no doubt relevant to a full understanding of the phenomenon of science. Nevertheless, as we have seen, values and valuations central to the scientific spirit are overlooked by approaches that focus exclusively on this restrictive type of naturalistic considerations.

Here, then, we develop doubts that weaken the scientificity of the reductive naturalist perspective. The latter does not seem to be supported by positive guidance from scientific practice. It is closer in spirit to speculative sociobiology than to scientific biochemistry. It appears, therefore, that, if strong reductive naturalism is to have any hope, it has to appeal to the inadmissibility of emergent properties in nature. Is there, perhaps, a *negative* existential or methodological argument in this direction?

A NEGATIVE MOTIVATION?

Perhaps what is at issue is not primarily naturalistic reductionism, but rather the implausibility of emergentism. Giere (1985, p. 343) says that if epistemology cannot be naturalized in reductive terms, then we have to accept that there are genuinely emergent properties. Somewhere along the evolutionary line from fishes to philosophers there would have emerged fundamentally irreducible properties that the models of evolutionary biology combined with experimental psychology, neurology and decision-theory alone cannot explain.

Giere is struck by how self-serving the search for differences between animals and humans can be. Emergence may be, however, a feature more widespread in nature than nomological reductionists are willing to anticipate.

The matter is currently a hot topic in connection with, for example, the so-called Bell experiments, which, increasingly, seem to support the conclusion that the compositionalist principle of classical separability is widely violated in nature. The properties of «two-body» systems, it now seems, may not always be nomologically reducible to those of their «component» subsystems; in more precise terms, the state of a complex quantum system is simply not always determined by the states of its constituent parts[10]. The debate here is by no means over, but one clear implication is that Giere's generic charge against the self-serving character of emergent properties in nature is unfounded.

GAP IN NEED OF A BRIDGE

To sum up, I am suggesting that the most encouraging reading of reductive naturalism leads to a position that imposes premature reductionist commitments to epistemology. My point is that the reductive thesis rests upon assumptions from continuity which do not really support the claim that the continuity of intellectual development is evidence for the elimination of «reasons» from epistemology. First, there is the question of conceptual distance. Second, there is the question of the legitimacy of emergent properties in natural science. I regard these two problems as current obstacles to strong reductive positions in naturalized epistemology.

My point is that there seems to be a gap between science and naturalistic reductionism. The reductive account of scientific change makes us think of an epistemology without reference to the «internal» life of science (scientific thought), as though this conceptual gap had already been filled or otherwise dismissed by appropriate background knowledge. It has not, and so rational epistemology should not yet be abandoned.

The issue is not whether scientific rationality can be naturalized, but rather how to advance epistemology in light of (1) the conceptual distance that now separates acceptable talk about animal behavior and acceptable talk about rational behavior, and (2) the best current scientific information about property levels in nature. How does this affect the prospects of naturalism as a contemporary philosophy?

4.REASON AND ANTI-FOUNDATIONALISM

At this point, it is instructive to step back and reconsider the background to Quine's project, specifically with regard to the legacy of the first American naturalists and the collapse of philosophical foundationalism. I shall maintain that the fate of reductive naturalism compromises neither the integrity of naturalism nor that of contemporary anti-foundationalism. My claim will be that one can agree with Quine and the reductive naturalists that methodological foundationalism is a hopeless program and that science self-applied is our only alternative, yet, at the same time, disagree with them about the role and place of emergent properties and reason in contemporary science.

NATURALISM

Naturalists attempt to explain everything by means of the categories of the natural sciences, but what they mean by this varies from one thinker to another.

In the days of Santayana, naturalism was a down-to-earth response to mechanical materialism (the view that all phenomena can be reduced to matter in motion). Its central goal was to avoid both supernaturalism and the reductive fallacies of materialism. F.J.E. Woodbridge, John Dewey, Morris Cohen, Ernest Nagel and Herbert Feigl are remembered as naturalists. Their naturalisms blended smoothly with the anti-foundationalism of the great pragmatists. Dewey, in particular, defended the view that the categories of explanation must always remain this-wordly and was skeptical of doctrines of justification which allegedly philosophers practice but scientists do not.

In more recent times, pluralistic naturalism is defended, for example, by Mario Bunge. Reasoning that the physics and biology of complex systems make emergentist and holistic advances, as do neurology, psychology and sociology, he advocates a pluralistic ontological framework[11]. Like earlier naturalists, Bunge excludes from the universe only supernatural agents. His naturalism is one of concrete entities and qualitative variety that allows for ontological levels.

The figures I am using as paradigms of anti-foundationalist naturalism have nothing to do with nomological reductionism. Much of their work may be rooted in a bygone era, but their naturalistic preoccupations remain alive in the work of many contemporary philosophers.

RATIONAL HISTORY

In the 1960s, some thinkers, motivated by the collapse of foundationalism, began to study scientific theories from the perspective of the intellectual goals and circumstances that propelled their development, the context in which specific theories existed, and their history. Some, like Kuhn, encouraged reductionist naturalization. Others, like Stephen Toulmin and June Goodfield, urged us to pay attention to the developing character of science.

Focusing on the importance of reasons for the dynamics of scientific change, Toulmin and Goodfield attempted to indicate why the different problems of the natural sciences came to be tackled in the order in which they did, insisting on the importance of studying the strong points of the systems of belief that our own scientific ideas have come to displace[12]. Later on Toulmin embarked on a more doctrinaire and, judging from the reactions, less successful investigation of the general problem of human understanding[13], but this in no way lessens the merits of his earlier rationalist anti-foundationalism. His judicious anti-foundationalist attitude has been furthered by able successors.

EPISTEMOLOGY INTERNALIZED

Dudley Shapere, in particular, is the author of a particularly robust merger of anti-foundationalism and rationalism in contemporary philosophy. We have already encountered him in connection with his critique of Quine's project. Shapere sees science, including scientific rationality, as a «give-and-take interaction between the methods with which it approaches nature and what it learns about nature» (1984, p. xxxiii). His position may thus be called «scientifically internalized philosophy»[14] («internalism» for short).

Shapere maintains that traditional epistemology is bankrupt, not because science has failed to yield knowledge on the basis of reasoning, but simply because philosophy has failed to get at what reasoning and knowledge are all about. In his view, recent mainstream philosophies -including Quinean naturalism- are simply unable to account for the enterprise of science[15]. Sterile in their methods, barren in their results, we do not need these philosophies, he advises; science itself tells us everything that is relevant to the question about the scope and limits of knowledge.

According to Shapere, it is science that tells us, for example, what it is for something to be «real» in the only sense presently available to us with any clarity, namely the «scientific». In advanced disciplines like fundamental physics, the sorts of considerations that are relevant to the formulation of problems, the introduction of new ideas and the reformulation of old ones, and to the acceptance and rejection of theories, are considerations that have been scientifically found to be well-grounded and relevant. It is such considerations, he says, together with the data given in scientific observation (as opposed to merely perceptual observation) that count as reasons in science. Shapere's paper in this volume gives us a detailed picture of his method and positive contribution to the understanding of conceptual change, so it is not necessary for me to describe these elements further.

Shapere's critics are understandably restless. They ask, where in his work do we find a good explication of the concept of 'a reason'? They see vagueness in his epistemology. Unlike reductive naturalism, it is not crisp.

Shapere's position, however, seems to be that rendering complex subject-matters prematurely «crisp» is dangerous. He invites us to look at what has been going on in history and philosophy of science since the 1960s. One of the deepest lessons from the development of science, he maintains, is that we actually learn how to think, and that it is as part of this process that we have learned to move away from the search for essences and philosophical «levels» of inquiry. Hence Shapere's insistence that we must not trivialize the conception of scientific rationality. Reason, in his view, is fundamentally a historical organon whose individuation can only be genealogical and circularly rational, the individuation of a reason undergoing a process of rational change. It is precisely because scientific rationality involves a give-and-take interaction between thought and learning, because it has this peculiar historical character, that talk about it need not be about something that transcends science in its actual historical, theoretical and pragmatic context.

So, in this approach, we cannot understand scientific rationality without first becoming aware of the rational past and present of science. Once this is done, we may try to understand the use of the contemporary concept of «a reason» by attending to the ways in which claims, problems, possibilities, methods, goals, etc. are accounted successful and doubt-free and how such ideas come to be judged relevant to specific problems.

Shapere's approach is also criticized as being at most a methodology, something unsuitable for solving philosophical problems, not a philosophical theory (Leplin 1987, 1988). His account of good reasons is said to open the door to relativism (Doppelt 1987). Illuminating responses to these charges exist (Shapere 1988, 1989c; Nersessian 1989). The debate is far from over, of course, and Shapere's approach is well represented by his paper in this volume. Here, therefore, I would like to add only two points. First, the internalist approach already has yielded such fruits as Shapere's powerful analysis of the concept of observation (1984, Ch. 16) and Nersessian's detailed study of the development of electromagnetic concepts (1984), to mention but a few achievements. So, it is certainly not not a mere «program». Second, so far the critics of internalism appear to miss its radical flavor, for they either simply reject its historicity or presume that there is some non-scientific account of validation left, thus begging the question.

What, then, of the impact of evolutionary ideas on the constitution of contemporary naturalism?

5. NATURAL SCIENCE, NATURE, AND NATURALIZATION

I have discussed three major movements influenced by the rise of evolutionary thought in contemporary natural science: Quinean naturalization, multi-level naturalism, and scientifically internalized anti-foundationalism. We must now consider the question: what is the *positive* legacy of these movements to naturalism?

In spite of the criticisms presented in this paper, the positive legacy of Quine to future naturalism is considerable. This is particularly true with regard to his anti-foundationalist stand, his role as the great twentieth century propagandist of the notion that there is no science of justification which the philosopher practices but the scientist does not.

Philosophically, I think naturalism inherits, first of all, a debate about the character and status of reasons in science. According to internalists, the effect of assesing the acceptance of a theory exclusively in terms of non-rational satisficing relations moves the theory from the category of belief to be explained in terms of reasons given as evidence to the category of a happening to be explained in social and psychological terms, i.e., in terms of social pressures, upbringing, individual anxieites, and so on.

Internalists do not deny that individual scientists often postulate hypotheses on bases that irreducibly reflect biological and psycho-social structures. Indeed, within the framework of evolutionary biology, it seems both easy and appropriate for naturalists to conceive the initial conditions of what we call human knowledge. Particularly insightful and seminal in this regard is, I think, Konrad Lorenz's interpretation of the Kantian categories in terms of biological adaptation, according to which what is a priori for the individual is a posteriori for the species (1975). Internalists seem right in insisting, however, that to say, as most Quineans do, that strong naturalistic reduction is all there is to understanding science is a claim to be adequately justified, not one to be begged[16].

Partly because of their desire to maximize objectivity, the reductive naturalists stick to categories and structures made respectable by mainstream positivism and pragmatism in the 1950s, tending to endorse various forms of semantic reductionism, operationism, behaviorism, etc. More at ease with the tentative character of human knowledge at all levels, the internalists appreciate the role of foundationalist structures but think it important to generalize and revise them in the light and spirit of contemporary knowledge, effectively subjecting such concepts as theory, observation, reality, etc. to a process of scientific internalization[17].

Still, from Giere and the second generation of Quinean epistemologists naturalism gets a sense of the importance of the work of historians of science. Also from these philospheres naturalism inherits an enhanced awareness that animal, individualistic and social factors do play a role in the generation of scientific ideas. The philosophical relevance of historically detailed and scientifically comprehensive case studies cannot be exaggerated.

From Bunge and many earlier naturalists, philosophy of science receives a worldly conception of the qualitative diversity of nature, as well as a scientific sense of tolerance to multi-level ontologies.

If my previous commentaries are correct, however, the most fertile legacy to future naturalism comes from the scientific internalization of epistemology. The position of Shapere and the internalists encourages attention to the goals and circumstances of naturalism, the intellectual context in which it exists, and its history.

These are, I think, some starting points for the next generation of naturalist projects in the philosophy of science. Through them, epistemology may become science-self-applied, as Quine says he wants. I have argued, however, that future anti-foundationalism, particularly naturalism, will be better off without the doctrinaire reductionism apparent in the works of many of Quine's successors, not to mention Quine's own conception of science.

DIFFERENCES AND PROSPECTS

Reductive naturalists may conceivably tell internalists that their appeal to reasons takes them into the realm of the occult and beyond what is the legitimate starting point of good science, namely, public, objectively accessible reality. Internalists may conceivably respond by pointing out that it is doubtful that any acknowledged scientist has ever managed to do any science without guidance from reasons[18]. The internalists' point is that reductive naturalists invite us to set all this aside in favor of a project that denies rationality a fundamental status in epistemology, but do so on the basis of what internalists claim to be an outmoded and mistaken view about the experimental basis of science, thus begging the question. In the interpretation proposed in this paper, internalism and reductive naturalism end up closely related cousins, yet not equally scientific.

THE CRITIQUE OF NATURALIZATION

The challenge for the contemporary naturalist is to produce a philosophy of science capable of harmonizing our scientific undersanding of mankind, knowledge and the world. It is sometimes objected, however, that the whole project of naturalization is unsound «in principle». Some critics of Quinean naturalism, Hilary Putnam (1982) most particularly, point out that epistemology and the philosophy of science cannot be naturalized in principle, because science presupposes reason and this, in turn, requires both reference and values, neither of which, he insists, can be naturalized.

Putnam's critique takes the form of a transcendental argument. Although he has Quinean naturalism as its primary target, his objection is derived from a larger argument purporting to show that the notion that there is exactly one true and complete description of the world (he calls this view metaphysical realism) is incoherent (1981). Putnam's general argument is complex and extremely problematic[19]; here I want to restrict myself to its significance for the project of contemporary naturalism. Two points strike me as relevant.

The first has to do with metaphysical naturalism, i.e., naturalism as a purely metaphysical thesis with no immediate epistemological connection, one involving no claims about the relationship between «real nature» and the present state of scientific knowledge. Whether or not Putnam's general argument is right about metaphysical realism, this seemstobe of no consequence to a scientifically internalized project that explicitly couples naturalism to the body of scientific knowledge. From the point of view of the latter, the complexity and precision of the partial theories and data already accepted renders frivolous the claim that, if there is one true and complete description of the world, then there must be another which is non-trivially different from the first. Nor is the suggestion that there is only one correct description of the world intrinsically incoherent within contemporary scientific thought. On one major program in fundamental physics, for example, the constraints on correct description are already as specific as to require a unified non-Abelian gauge theory of physical action.

My second point has to do with the critique of reductive naturalism. Giere 's suggestion (1985) that Putnam's complaint about naturalization is ultimately a defense of the existence of emergent properties seems right. If so, however, that part of Putnam's attack is correct. Had the natural sciences successfully challenged the existence of emergent properties, there would be a plausible naturalistic basis for restricting the ontology of naturalism accordingly,; I have argued, however, that current science does nothing of the kind. It could come to do it, though, and so my denial of nomological reductionism should not be construed as a «philosophical» reaction but just as a *contingent* by-product of current science.

What, then, of Putnam's transcendentalist argument against naturalization? If we keep our conceptions of science and nature close to current scientific thought and current scientific practice, then it is by no means clear that the naturalization of reason and value is beyond the scope of science.

Can, then, reason be naturalized or not, and if so how? I think that depends, in part, on how we understand the terms «natural science» and «nature».

NATURAL SCIENCE AND NATURE

Science provides us with theoretical models relevant to the understanding of natural learning, as Giere and the reductive naturalists correctly stress. Just as importantly, however, science provides us with a style of thinking which has evolved along with the rational development of those models and now pervades them, as Shapere and the internalists stress.

To understand science and human reason from the point of view of epistemology, then, we need the best «theoretical models» furnished by natural science. But we also need to appreciate the ways of scientific thinking used by scientists in the course of the development of those «models». Naturalists, in short, must think scientifically, and introduce epistemological hypotheses the way scientists have learned to introduce proposals in the most mature sciences.

An equally anti-foundationalist handling of the idea of nature seems also in order. If nature is what the natural sciences tell us it is, then its best representation today is in terms of a multilevel ontology. The picture of nature we get from the most consolidated scientific descriptions to date is «onion-like», the picture of a world of qualitative diversity in which the entities of more external level have the properties of those below them, but the reverse is not the case. Contemporary science thus provides us with a robust anti-foundationalist framework, but not one hostile in principle to emergent properties.

THE SEARCH FOR A CONTEMPORARY NATURALISM

The story I have been telling in this paper leaves many questions open, but I hope to have defined with some precision the vital office of contemporary naturalism. Let us take stock and summarize the main points I have argued for:

1. The rise of contemporary evolutionary ideas is a standing warning to those who would downplay either the radical novelty or the revisionary depth of conceptual change in natural science.

2. Twentieth century naturalism is a development prepared by successive projects of naturalization, each focused on a different aspect of science. Although I have argued that the most influential project today, reductive naturalism, is seriously flawed, I have also attempted to show that a more scientific and promising variety of naturalism is in the making.

3. Contemporary naturalism has many acknowledged sources; I have argued, however, that its best grounds are in contemporary science itself, including scientific thought. If the story presented in this paper is correct, the plausibility of naturalizing the philosophy of science results primarily from the rise of scientific evolutionary ideas.

4. Throughout the twentieth century, evolutionary ideas have cornered the project of old philosophical foundationalism into increasingly weak and obscure positions. Simultaneously, the unprecedented level of reliability attained by some theories of physics and biology has contributed to the rise of naturalism in two ways, first by reasserting the possibility of scientific knowledge, and second by pointing the way toward the constitution of scientific foundations for the knowledge-seeking enterprise.

5. So, in spite of what the collapse of foundationalism might be doing to other intellectual enterprises, naturalism does not lack strength, because its force originates in the predictive and explanatory power of science, which is now at a historical peak.

6. It is a matter of fact that contemporary science increasingly pictures nature as a world closed under causal relations, yet a world of uniformities subdivided by variety, in which intrinsic properties, system properties and holistic properties carve up a universe of concrete entities distributed in a multilevel ontology. So, the reference assumptions and the growing predictive and explanatory success of the most fundamental theories of contemporary natural science give increasing credibility to ontological naturalism, but not (yet) to nomological reductionism. What gains credibility is the notion that the universe is a causally closed system. This rules out disembodied spirits and supernatural powers, but not natural emergent properties.

7. And so, a contemporary naturalist can coherently claim that the ultimate basis of his position is found in the conceptual and ontological commitments of current science, not in formal or universal essences. An important corollary is this: for as long as the elimination of «reasons» from epistemology lacks scientific support, rational epistemology should not be abandoned.

These openings can be elaborated indefinitely, but the bottom-line is, I think, simple enough. The fate of philosophy of science, whether in the end it is naturalized or not, will probably be determined by the results of contingent science, not by any necessary or transcendental truths about naturalization.

Queens College, City University of N.Y., USA

NOTES

1. The presentation of this paper was supported by National Science Foundation Grant INT-8905411. I would like to thank Dudley Shapere, Ernest Sosa and Efrain Kristal for many helpful comments.

2.Quoted by M.B. Hesse in *The Structure of Scientific Inference*, London, 1974, Chapter 1.

3.See, for example, Feyerabend (1975). Feyerabend, P. (1975). «How to Defend Society Against Science.» *Radical Philosophy*, 11: 3.

4.Campbell (1974), (1987). He now calls his position «selection- theory epistemology» (1988).

5.Campbell proposes the following «dogma» (his own characterization): all instances of achieving fit of belief to referent require a «blind-variation-and selective-retention component, *however indirect or presumptive*» (1988, my emphasis).

6.This was a central issue in a recent Conference on Scientific Creativity, University of North Carolina at Greensboro, March 1989; proceedings expected 1991, J. Leplin (ed.).

7.The philosophical consequences of the Bell experiments are discussed, for examplen in Cushing & McMullin (1989).

8.Paul M. Churchland (1979), Patricia S. Churchland (1986).

9.See, for example, Searle (1984).

10.A discussion is found, for example, in Cordero (1988).

11.Bunge's multi-level ontology is presented most accessibly in Bunge (1977).

12.Toulmin & Goodfield (1961), (1962).

13.The approach in Toulmin (1972) follows a form of historical explanation that is a generalization of Darwin's populational theory of variation and selection. Toulmin's ambitious effort is still inconclusive, stopped by many severe difficulties in the conception of his program. For a perceptive review, see Suppe (1977, pp. 670-681). Another thinker whose approach may also be read as a contribution to a freer naturalism is Arthur Fine. His book *The Shaky Game* (1986) conceives of rationality as based on topic specific principles. His thought, however, seems to undergoing much change of late. Both the concluding chapter of the book mentioned and Fine (1989) express pronounced doubts regarding the interpretative powers of scientific reason, as opposed to the powers of personal and social constraints.

14.For an overview, see Shapere (1984)

15.Shapere (1984, 1987, 1988, 1989a, 1989b).

16.See, for example, McMullin (1983).

17.For a detailed example of the scientific internalization of allegedly metascientific concepts, see Shapere 1987 on observation.

18.This argument against the reductive naturalist paraphrases Giere's reply to the skeptic (1985).

19.See, for example, Devitt (1984).

REFERENCES

Bunge, M. (1977), *Ontology I: The Furniture of the World*, Vol 3 of *Treatise on Basic Philosophy*. Dordrecht: Reidel.

Campbell, D.T. (1974), «Evolutionary Epistemology». In P.A.Schilpp (ed.), *The philosophy of Karl R. Popper,* La Salle, IL: Open Court:413.

(1987), «Neurological embodiments of belief and the gaps in the fit of phenomena to noumena». In A. Shimony & D. Nails (eds.), Naturalistic Epistemology, Dordrecht, Holland: Reidel: 165.

(1988), «The Epistemological Role of Selection Theory».Conference on «Evolutionary Epistemology», Center for Phil. Sc., Univ. of Pittsburgh, Dec. 1988.

Cherniak, C. (1986), *Minimal Rationality*. Cambridge, MA: The MIT press.

Churchland, P.M. (1979), *Scientific Realism and the Plasticity of Mind.* Cambridge: Cambridge U.P.

(1984), *Matter and Consciousness*. Cambridge, MA: The MIT Press.

Churchland, P.S. (1986), Neurophilosophy: Toward a Unified Understanding of the Mind/Brain. Cambridge: The MIT Press.

Cherniak, C. (1986), *Minimal Rationality*, Cambridge, MA: The MIT Press.

Cordero, A. (1988), «Probability and the Mystery of Quantum Mechanics». In E. Agazzi (ed.), *Probability in the Sciences*: 217. Dordrecht: Kluwer.

Cushing, J.T. & McMullin, E. (1989), *Philosophical Consequences of Quantum Theory*. Notre Dame: Univ. of Notre Dame Press.

Devitt, M. (1984), *Realism and Truth*. Oxford: Basil Blackwell.

Doppelt, G. (1988), «The Philosophical Requirements for an Adequate Conception of Scientific Rationality», *Philosophy of Science* 55: 104.

Fine, A. (1986), *The Shaky Game*, Chicago: Univ. of Chicago Press.

(1988), «Interpreting Science», in A. Fine & J. Leplin (eds.),PSA 1988, Vol. Two: 3. East Lansing: PSA.

Giere, R.N. (1985), «Philosophy of Science Naturalized»,*Philosophy of Science* 52: 331.

Hull, D. (1983), «Conceptual Evolution and the Eye of the Octopus»,in *Proceedings of the 7th International Congress of Logic, Methodology, and Philosophy of Science*.

(1988), *Science as a Process*. Chicago: University of Chicago Press.

Kitcher, P. (1983), *The Nature of Mathematical Knowledge*. Oxford Oxford U.P.

Leplin, J. (1987), «Truth and Scientific Progress». In J. Leplin(ed.), *Scientific Realism*, Berkeley: Univ. of California Press: 193.

(1988) «Is Essentialism Unscientific?», *Philosophy of Science* 55: 493.

Lorenz, K. (1975).«Kant's Doctrine of the A Priori in the Light of Contemporary Biology», in R.I. Evans (ed.), Konrad Lorenz: *The Man and his Ideas*: 181-217. New York: Harcourt Brace Jovanovich.

McMullin, E. (1983), «Values in Science». In P.D. Asquith & T.Nickles, PSA 1982, volume two: 3. East Lansing: Phil. Sc. Assoc.

Nersessian, N. (1984), *Faraday to Einstein: Constructing Meaningin Scientific Theories*, Dordrecht: Kluwer.

(1989), «The Method to Meaning». Forthcoming in *Philosophy of Science.*

Putnam, H. (1981), *Reason, Truth and History*, Cambridge: Cambridge U.P.

(1982), «Why Reason Can't Be Naturalized?», Synthese 52: 3.

(1983), «Quantum Mechanics and the Observer». In *Realism and Reason, Philosophical Papers*, Vol. 3: 248. Cambridge: Cambridge UP.

Quine, W.V. (1953), *From a Logical Point of View*, Cambridge:Harvard UP.

(1969), *Ontological Relativity and Other Essays*, NY:Columbia UP.

(1977), *The Ways of Paradox and Other Essays*, Cambridge:Harvard UP.

Roth, P.A. (1983), «Siegel on Naturalized Epistemology and Natural Science», *Philosophy of Science* 50: 482.

Searle, J.R. (1984), *Minds, Brains, and Science.* Cambridge:Harvard U.P.

Shapere, D. (1984), *Reason and the Search for Knowledge*,Dordrecht: Reidel.

(1987), «Method in the Philosophy of Science and Epistemology». In N. Nersessian (ed.), *The Process of Science*, Dordrecht: Nijhoff: 1.

(1988), «Doppelt Crossed», *Philosophy of Science*, 55:134.

(1989a), «Modern Physics and the Philosophy of Science». In A. Fine & J. Leplin (eds.), PSA 988: 201. East Lansing: Phil. Sc. Assoc.

(1989b), «On the Introduction of New Hypotheses in Science».Forthcoming in J. Leplin (ed.), *Proceedings of Conference on Scientific Creativity*, Univ. of North Carolina at Greensboro, March 1989.

(1989c), «Leplin Leapt». Forthcoming in *Philosophy of Science*

Siegel, H. (1980), «Justification, Discovery and the Naturalizing of Epistemology», *Philosophy of Science* 47: 297.

Sober, E. (1984), *The Nature of Selection.* Cambridge, MA: The MIT Press.

Stich, S. (1983), *From Folk Psychology to Cognitive Science*: The Case Against Belief Cambridge, MA: The MIT Press.

(1985), «Could Man Be an Irrational Animal?». *Synthese* 64:115

Suppe, F. (1977), «Afterword», in F. Suppe (ed.)The Structure of Scientific Theories, 2nd edition, Urbana: Univ. of Illinois Press.

Toulmin, S. & Goodfield J. (1961), *The Fabric of the Heavens*,NY: Harper & Brothers.

(1962), *The Architecture of Matter*, NY: Harper & Row.

Toulmin, S. (1972), *Human Understanding,* Vol I, Princeton, N.J.:Princeton U.P.; Vols. II and III still unpublished.

Wuketits, F.M. (ed.) (1984). «Concepts and Approaches» in *Evolutionary Epistemology*. Dordrecht: Reidel.

Francisco Miró Quesada

ORIGIN AND EVOLUTION OF THE UNIVERSE AND MANKIND

FACING THE ORIGIN

To get an idea of the scientific progress that has been made over the past few decades, it is enough to think that, nowadays, it is possible to produce (via particle accelerators) states similar to those that existed a billionth of a second (a nanosecond) after the very birth of the Universe. Although these results have been obtained with methods of measurement and theoretical frames far superior to the classical ones, they seem incredible. It is difficult to believe that we know which kind of matter existed in those early moments. Nonetheless, we know it (at least with a good probability). Before the Big Bang, the theory becomes speculative. After the Big Bang, some empirical corroboration is possible. For instance, as a consequence of the Big Bang there ought to exist a homogeneous radiation through the whole Universe, with a temperature of 3 K. The existence of this radiation was predicted in 1940 by Gamow, and it was verified in 1965 by Penzias and Wilson (quite independent of the prediction).

SCIENCE OR METAPHYSICS

Although our knowledge before the Big Bang becomes speculative, it is of great interest. The era of inflation can explain

E. Agazzi and A. Cordero (eds.), Philosophy and the Origin and Evolution of the Universe, 441–456.

not only the Big Bang but many other aspects of the Universe that seemed quite baffling before (for example, the way energy originated). Some empirically-minded scientists reject any attempt to explore the Universe before the Big Bang because they consider this exploration to be metaphysical. Since the decisive analyses of Kant regarding the structure of scientific knowledge, it is common to consider a theory that cannot be empirically corroborated as metaphysical. According to this criterion, the theories about the evolution of the Universe beyond the first nanosecond would not be scientific. Though perhaps it is impossible to avoid metaphysics, if we want to grasp the birth of the Universe, we can advance very near to doing so. The theories that follow this path have certain traits which are not present in the metaphysical ones. They have been forged through the application of relativity and quantum theories, that is, the same theories that led to our current understanding of the Big Bang. So, in a way, they include these theories. The former theories have been able to reconstruct the evolution of the Universe from the Big Bang to the present. They are very powerful theories indeed, and this power is being utilized by some physicists to approach the beginning of the Universe by an incredibly tiny amount of time.

So, we have the following situation. The same theories that can be, in some acceptable ways, corroborated through experience, enable us to go beyond experience. The new theory includes the former corroborable ones. The impossibility of corroboration of the new results are not due to metaphysical vagueness, but rather to the fact that we do not possess sufficiently strong accelerators to reproduce the temperatures existing in the pre-Big Bang eras. (To recreate the temperature of the Planck era we ought to build an accelerator as large as the Milky Way.)

NOTHINGNESS AND BEING

Human reason is all-including. It is never satisfied with partial results. Once it has conceived, even if in a speculative manner, the Planck era, it needs to go farther. It must reach the very beginning of the Universe, the precise instant at which the cosmos was born. But, when scientific speculation is pushed to the limit, we face the risk of falling into metaphysics. Nevertheless, human reason is unable to resist the temptation, and nowadays there are several competing theories on this issue. Let's take a look at them.

A) The Great Explosion (Standard Model). According to this model, originated with the work of Gamow, the Universe begins with the Big Bang. At the very instant of birth, the cosmos is only a mathematical point with infinite energy. This energy produces the great explosion from which the whole Universe, as we know it today, originated. This theory, which is an extrapolation of the general theory of relativity, presents a grave difficulty. The point of origin is a singularity, that is, energy becomes infinite at this point and space-time must have an infinite curvature. Infinite quantities make no physical sense. We must therefore discard this theory.

B) A World Without Boundaries. Stephen Hawking has proposed a remarkable theory. Utilizing imaginary time in the formulas concerning the evolution of the Universe, he has come to the conclusion that the cosmos has no origin because it is self-closed with respect to space and time. By utilizing imaginary time, he eliminates all differences between space and time. A consequence of this elimination is that the Universe has no boundaries. The question about the origin becomes meaningless.

This theory, although heavily speculative, has some advantages. It reduces the theoretical possibility that many Universes co-exist. It can also explain some established facts, such as the isotropic property of the expanding Universe and the Big Bang. According to its author, the no-boundary condition, together with the uncertainty principles of quantum mechanics, could explain all the complicated structures that we see in the Universe (Hawking, 1988).

C) Free Lunch. If the laws of quantum physics are applied to the analysis of the origin of the Universe, we are led to the amazing conclusion that the world has emerged from nothingness

(Vilenban, Coleman). These laws show that in the sub-atomic world there can be facts without cause. The cause-effect relationship, as conceived by classical physics, cannot be defined in a precise way in the micro-world. We can only know the probability of an event which follows from another event. This probability can be very small, so small that the classical concept of cause makes no sense.

This theory has been called the "free lunch" theory because if matter and energy emerge from sheer nothingness, then they are gratuitous. They owe nothing to antecedent causes. The theory is quite attractive, but it has unsurmountable conceptual difficulties because it supposes that there was a time before the birth of the Universe. Of course, supporters of the theory say that "before this moment" makes no sense at all. Thought of in this way, the problem of what kind of time there was before the Universe disappears. This reply is very similar to the one given by medieval theologians concerning the moment of creation. The circularity cannot be avoided, however. It is impossible to think about the beginning of time because the concept of "beginning" can only be applied if time already exists. Nevertheless, it is perfectly compatible with the laws of physics because it is true that energy and matter can jump into existence out of nothingness (this jump being given a temporal frame in our Universe).

D) The Eternal Return. Prigogine has proposed a quite different theory. Through a very deep study of entropy that led him to reformulate important aspects of relativity and quantum mechanics, Prigogine reached the conclusion that the arrow of time is irreversible. This claim is contrary to the tradition that began with Galileo and Newton and culminated with Einstein. As Prigogine has shown, this is a tradition that embraces even quantum mechanics.

For Prigogine's theory of entropy, it follows that time has no beginning. Therefore, the circularity that appears when we try to derive the temporal from the intemporal is avoided. It also follows that the Universe evolves through time, probably with periodic expansions and retractions. This leads to an interesting coincidence with the myth of eternal return, so common in ancient cultures, that was defended by Nietzsche. There is a fundamental difference, however. According to Prigogine, every return is different from all its predecessors. It is unique. With poetic insight, Prigogine compares the successive returns to the states of the Universe to the returns of Spring. Every Spring is similar to the others, but is also different. Every leaf of grass, every flower, every hour is different; they are unrepeatable (Prigogine, 1988).

EVOLUTION

A theory of the origin must be completed with a theory of evolution. If we are unable to know the ways things evolved after they came into existence, then we shall be unable to understand our present situation. As far as cosmology is concerned, we have gone from the evolution of the Universe to the origin. We have gone against the current of time. We have been motivated by the intellectual drive to understand why our world is as it is. To do this, we had to go back in time. Every period had to be understood by means of the antecedent one. When we came to the final step we therefore already had a vision of the cosmological evolutionary process.

The advances concerning the evolution of the Universe are indeed amazing. We shall not consider the details of an evolution wich have already been examined by other contributors in this volume. It is sufficient to say that from a certain era up to the present, the theory of the evolution of the stars is connected in several ways with empirical corroboration. We know why the stars have such a tremendous amount of energy. We know how long they live and that their life span depends on their size. The origin of things as strange as novae, neutron and white stars, pulsars and quasars is rather well-established by theory and, in some cases, the theory is supported by indirect verification. We have made great progress concerning the nature, origin and evolution of galaxies. We know that there are different types of galaxies, that there are clusters of galaxies and clusters of clusters of galaxies[1]. We also know that new stars are always being born. Although the formation of planets presents a deep problem, in recent years we have been able to develop a new theory about their origin and to gain much understanding about their evolution.

Our knowledge of the evolution of the Universe is complemented by our knowledge of the ways life originated and its subsequent evolution. There are gaps which are difficult to fill, but, in spite of these gaps, the fact of evolution is well-established. As the planets began to cool, new conditions appeared that were favorable for life to emerge. It is supposed that in our galaxy, although those conditions are rare (at least regarding the prevailing ones on Earth), there are many thousands of planets in which life, similar to the terrestrial is possible. The theory of the genetic code makes it possible to conceive a contingent origin of life. Through electrical discharges on the seas, which contained the substances necessary for the synthesis of nucleotides, some cells emerged that were able to duplicate themselves. From these cellular complexes were formed, and, finally, organic life appeared.

The theory of evolution had a tremendous impact on Western civilization and a decisive influence in the advent of a new conception of man. From the time it was created until today, the Darwinian theory of evolution has received a growing acceptance. However, it does present some difficulties that cannot be eliminated without making theoretical adjustments. New theories have therefore been created which explain away the difficulties. The development of the theory of the genetic code awards a much stronger explicative power to the evolutionary thesis. Neo-Darwinism is the old Darwinian theory supplemented by the new science of genetics. It is called the "synthetic theory" because it combines Darwinism with the recent advances in genetics. The new discoveries about the genetic code offer a new perspective concerning evolution because they offer the possibility of understanding the relationship between macro and micro-evolution.

The synthetic theory is widely accepted in the scientific community, but there are other theories that try to improve it or that plainly reject it. A rival theory that is gaining acceptance is the theory of "punctuated equilibrium". According to this theory, every species presents two different evolutionary stages. In one stage they remain a long time without change (stasis), whereas in the other they experience very rapid transformation (crisis). In this second stage, a species changes by bifurcation. From the perspective of geological time this transformation is practically instantaneous. Another conception, called the "neutral theory", claims that the variation of most specific traits is indifferent to natural selection. Only a few variations are utilized for survival. As we can see, there are still active discussions on many issues (Gayon, 1989). Some theorists are interested in ontological problems. They claim that the species have the same ontological status as individuals.

But, in spite of these and other disagreements, the universal acceptance that life evolves and that man is a product of an evolutionary process is solidly established. (It is even accepted by Christian philosophers.) It is also widely accepted that man appeared in central or perhaps southern Africa and that, from there, he expanded throughout the world. This expansion remains unexplained, but there is strong evidence that in the beginning there were several types of hominides and that only Cromagnon man was able to survive.

The origin of the Universe, the origin of life and the origin of man have been great enigmas that seemed to defy any possibility of solution. However, as we have seen, we have made enormous progress. The claim made by Scheler (1938) in *The Place of Man in the Cosmos*, a book that was very popular in the forties, namely, that man had never been so enigmatic concerning himself as nowadays, cannot be supported anymore. The progress in our knowledge of the origin and evolution of the Universe, and also concerning the origin of life and of man, has created a new vision of the place of man in our cosmos.

But, as soon as man appears on Earth, a new kind of evolution begins. It is no longer cosmological nor biological; it is social evolution. Apparently, to know the social evolution of man seems to be a simple matter. We just have to look at history. However, it is a problem because this evolution takes place through different civilizations or cultures, and to understand it we must be able to understand the symbolism created by man in different historical situations. We must grasp the deep sense of different Weltanschauungen. And, to understand the multiple creations of a culture from the perspective of a different culture can be very difficult indeed. Think, for instance, about the problems that must face philosophers and philologists to understand pre-socratic thought.

These are not the only considerations, however. During the latter centuries Western civilization dominates history and invades, deforming them, the other civilizations. The present world is the result of this invasion. The difficulty is not only to understand the evolution of the other cultures under the impact of the Western one but, also, to understand the modifications and deformations that the invader culture has suffered through the reflux of the invaded civilizations. On the other hand, Western civilization is characterized, especially in later times, by its vertiginous dynamism. The whole structure of this civilization is rapidly transforming itself. This transformation is social, economical and political, it includes value systems, religious belief, art, and even the ways science and philosophy look at themselves.

Science and philosophy are the most characteristic creations of the modern world. That is, amid the myriad aspects that constitute a civilization, science and philosophy are the ones that distinguish Western civilization from other cultures to the highest degree. Thanks to science and philosophy, modern man has been able to see himself as he does. And, of course, thanks to technology, which is a consequence of scientific progress, but which enables, in its turn, this same progress.

REASON AND SCIENCE

Science and philosophy are a creation of human reason. So, if we want to understand the evolution of man through his long historical trajectory, it is necessary to elucidate the evolving dynamism of rational knowledge through these two disciplines. This evolution is very complicated, and it would be too long to analyze it in the present text. Because science has contributed directly to the understanding of the origin and evolution of the Universe and of life, we shall limit ourselves to analyze, very briefly, the evolution of rational thought concerning them. We are fully conscious that our analysis will be unilateral because the evolution of science is related to the evolution of philosophy, and vice versa.

The Hindus, the Chinese, the Sumerians and the Egyptians had mathematical knowledge. But, this knowledge was elementary and -this is what really matters- they did not proceed to develop mathematics by means of the axiomatic method[2]. A theorem is a universal knowledge which refers to all possible objects of a certain kind. This universality is essential to scientific knowledge. It confers to it its descriptive, explicative and predictive power.

The Greeks knew that mathematical knowledge could be applied to the study of nature. (Let's think, for instance, of Archimedes and the School of Alexandria). But, due to historical conditions that would be too long to describe here, they could not go very far in this direction. In the Middle Ages, there are many symptoms that show the way Western science would follow in the later centuries. Modern science was really born during the Renaissance and from this time on it begins to influence history. The great scientists, like Leonardo and Galileo, were convinced that nature can be rationally known and that mathematics is its language. Through mathematics, human reason can contemplate what really happens at the core of physical reality. From Newton onwards most physicists and philosophers of science thought that physical science had conquered definitive knowledge and that future natural science would only increase the perfection of acquired knowledge and incorporate new knowledge within the invariable frames created by Newton. Positivism, in a way, changed this view, claiming that the only physical objects we can know are just phenomena and that general laws of physics are but generalizations of phenomenic sequences. Even Mach, the most radical of Positivists, accepted that the laws of nature, although a consequence of induction, followed the Newtonian paradigm.

It is only with Einstein that the rational mechanism that characterizes scientific knowledge begins to be understood. Induction plays a limited role; it can function as a hint to the creative power of rational thought. Physical theory is an intellectual creation, not a description of observable facts. It is connected with phenomena through corroboration. If the corroboration fails, then the theory must be readjusted or changed. Popper led this conception to the limit. This change of the conception of modern man regarding science shows that the ways of reason are unpredictable. Some modern philosophers and historians of science, like Shapere, claim that science (at least natural science) has a strong drive towards autonomy concerning external influences, even philosophical ones. He calls this process "internalization of considerations". The process of internalization develops itself deeply connected with its domain of knowledge and it incorporates the beliefs that have been successful in establishing this knowledge (Shapere, 1983, XXI). According to this conception of science, founded on a rigorous analysis of physical science, especially relativity and quantum mechanics, the evolution of science cannot be forecast. As science advances it progressively incorporates new concepts and throws away old beliefs. There are beliefs, however, that are retained because they are considered relevant to previously acquired knowledge and there are no reasons to have doubts about them.

The internalization process leads, according to Shapere, to a position that could be dubbed "historicist". It is not a relativism, because the way science evolves is far from arbitrary and it does not depend (at least in its constitutive aspects) on variable external factors. It depends on its proper dynamism and this dynamism consists of the "reasons" scientists have to accept some beliefs and reject others. It is a historicism because science is permanently evolving and this evolution includes all its aspects, even its principles.

It is worthwhile to compare Ortega's historicism with Shapere's. The historicism of Ortega is external. The variation of science through time depends on cultural dynamism; the changing social conditions, political, economic, etc. influence science and impose a change of principles. Shapere does not deny that there can be some external influence, but he claims that the principal changes are due to an internal process. Scientific knowledge of nature is founded in "reasons" and these reasons are not a priori norms because they are manifest through the development of the internalization process. Every aspect of scientific knowledge can vary, but this variation is always due to "reasons". Even logic and mathematics can vary if there are sufficiently convincing reasons. Because what is considered a "reason" in a determinate period of the evolution of science can be different in another period (Shapere, 1983 XXVII, XXXIII), the principles and methods of science are historical.

There is in this proposition a fundamental truth: Human reason, through which scientific knowledge is justified, functions according to very wide guidelines. In the era of the great Rationalism, scientific knowledge is seen as a definitive conquest that has been acquired through the application of logical, mathematical and physical normative principles. But, in the process of internalization that characterizes scientific evolution, the normativity of principles suffers, in many cases, a transformation or even disappears. In this variation, however, we find the true meaning of scientific rationality. And this rationality is the process of internalization of scientific knowledge, its drive to total autonomy, although this fundamental goal is probably only an ideal of reason (Shapere, 1983, a, 23).We think Shapere is, in essence, correct. But, his conception of "being a reason" is not sufficiently clear. And, until a rigorous analysis of the mechanism through which in the constitution of scientific knowledge, a part of it is considered a "reason", is not performed we shall not be able to understand scientific knowledge. If we go deep enough in this analysis we can have unexpected surprises. For instance, in Logic there exist several non-compatible systems. But, when we meditate on the reasons that have led to these systems, we find, at least in many cases, that they concern universal principles. Generally speaking, the constitution of a system of logic corresponds to some abstract structure whose knowledge can only be acquired by means of that system.

It can be seen through strong argumentation that there exists, on an a priori basis, a connection between the types of structure and the corresponding logical systems. This is not the case in the sciences of nature, but we cannot forget that these sciences cannot exist without logic and mathematics. This fact shows that although it is true that science evolves through a process of self-construction and that it becomes increasinely independent, this self-construction is only possible through rational dynamisms that, in some of their aspects, are invariant[3].

The ways science has evolved concerning its scope and meaning are remarkable. For instance, what nowadays is considered as observable would have been seen as non-observable a few decades ago. A probabilistic explanation would have been considered as nonsense. Deterministic causality, which for centuries was considered fundamental for scientific knowledge, is now seen as a mere approximation in the world of molar objects. Nevertheless, if we analyze the processes which have produced the evolution of science, we discover that they present invariant aspects, aspects on which the evolution is based. It is true that the conquest of its autonomy and self-sufficiency are essential components of the rationality of science, but this rationality is far from being contingent. If the concept of observed "object" has changed with time, it is not because scientists have arbitrarily decided to change the meaning of "observable" but because they dispose now of sufficient information and they can handle ultraprecise and refined instruments which reveal certain kinds of objects, at least through their effects in the laboratory (e.g. the neutrino). On the other hand, today, as in earlier times, direct sense observation is the ultimate criterion of corroboration. And, although it is true that the concept of explanation has been broadened, there must always be a deductive component. Even in the probabilistic explanations, there is no explanation nor prediction without logic. Quantum logic is frequently presented as a proof that in the evolutionary processes of science anything can vary, even logic. But, all of quantum theory has been developed within the frame of classical logic. It is true that orthomodular lattices, as extensions of Hilbert subspaces, can be considered as models of certain kinds of propositions of quantum mechanics. But, although following this route it is possible to describe quantum phenomena, quantum logic is unnecessary to develop the whole theory. In any case, if some day, due to new observable evidence, quantum logic would become a necessity, it is clear that this necessity would come from the fact that the mathematical structure utilized to describe sub-atomic reality would impose some kind of non-classical logic. And this imposition, as we have already seen, would mean that there exists a non-contingent relationship between the mathematical structure and logic that would be needed to describe it. However, in the present condition of the theory, this does not seem to be the case.

TOWARDS A NEW SOCIETY

If we pass now from the evolution of science, and the concept of reason in its cognitive function, to Western civilization, we find that there exists a similar process. The so-called "primitive societies" are ruled by rigid principles. The social relationships are, in these societies, rigorously pre-determined. Different human groups are strongly stratified. In Athens we find that a process towards liberation begins to oppose to rigidity. But, there are still slaves.

During the Middle Ages there are some important advances which become wider and more apparent in the Renaissance. But, in spite of varied and rapid changes there are still very strong rigidities concerning social position and hierarchy of power[4]. It is in the 17th and 18th centuries that the rigidity begins to dissolve with respect to socio-cultural patterns. This process has no precedents. In other civilizations, like those of the Chinese and the Hindu we find a clear social and cultural critique, but it is less deep and systematic[2]. The result of the historical progress of Western civilization is nowadays clear. There are no rigid principles that impose a predetermined human behavior. So power cannot be rationally justified but through consensus. The only reason to impose a collective behavior is the preservation of the freedom of all members of the collective. The rational analysis of the justification leads to an inescapable conclusion: the only limit of personal behavior is the freedom of the behavior of others. Rationality, justice and liberty are thus closely related.

Of course, we are only making an outline of the whole situation. The history of the last three centuries has not developed in a linear way. There have been advances and retrogression. But, in spite of the zig zags, some of which have been tragic, the advancement has been effective and seems to be irreversible. The march towards a free society is supported by technological development. Rational technology is, of course, neutral and can be utilized for domination as easily as liberation. When it is applied indiscriminately, and only for economical pursuits, it generates the danger of the destruction of nature. But, an analysis of the present historical tendencies shows that we can have the hope, almost the certainty, that the modern techniques will be applied for the good of mankind.

The wide scope of Western civilization has made possible the understanding that, as is the case in the Cosmos, in the universe of civilizations none is privileged. As a consequence of cultural self-criticism, Western man has become aware that there are points of view different from his own, and that he must attempt to understand them and learn from their most important values. Once he has liberated himself from the bondages that oppressed him, he is then in a position to understand the meaning of human differences. From this process begins to emerge a unified mankind within the frame of cultural diversity, a mankind of incomparable worth.

Science has played a key role in the evolution of modern society. The radical changes in the self-awareness of mankind have been, in most aspects, triggered by the great astronomical, physical and biological discoveries. The critical analysis of society has advanced side-by-side with scientific progress and, in many aspects, it has been originated by this progress. We think, therefore, that colloquia like this one are very important indeed. Because they contribute to the liberation of human thought and to finding historical orientation. In scientific knowledge we find the most characteristic trait of the world we have begun to forge. To spread it throughout the world is to enhance the fundamental project of man: the creation of a just and free society in which all men can live like brethren.

Lima University, Peru

NOTES

1) There are, of course, some baffling aspects of our cosmos that up to now have not found any explanation (for example the T-Tauri stars that have anomalous spectra). But, concerning the whole cosmic population our knowledge of stars and galaxies embeds a landscape of remarkable scope.

2) In China and India there are philosophical systems that, in many respects, are similar to the Greek ones. But, there is an essential difference. Greek philosophy was created together with Greek mathematics and astronomy, and all the other sciences originated from philosophy. For the Greeks, philosophy is the first step to science. This situation is not found in other civilizations. Why is this so? We don't have any good explanations for this wondrous fact. As a working hypothesis, we remark that it is possible that this situation could be due to the discursive character of Greek philosophy. To be created, philosophy must be necessarily communicable. "Logos", for the Greeks, meant reason and also word, discourse. We think it is probable that this relationship between reason and language contributed to the creation of science. To express with the utmost rigor their philosophical discoveries, they tried to formulate them in a language without ambiguity. When they succeeded with respect to certain fields of philosophical knowledge, a new science began to emerge.

3) As a matter of fact, it is sufficient to employ a first order logic to perform the usual deductions that are made in science. However, if one tries to formalize all the expressions that are used in the language of science and all the deductions that are performed not only within the theory but also about the theory, then first order logic is not sufficient and it is necessary to employ at least some sort of modal logic and, in some cases, even relevant logic. To express in a formal way a theorem of impossibility, for instance, we must use modal logic. In some cases

we could avoid this by means of first order logic, but the statements and deductions that must be made to avoid this logic are artificial and complex.

4) Of course, total rigidity does not exist in every historical period. Although this rigidity is found in many of the so-called primitive cultures, in many historical civilizations rigidity is never total. For instance, in Athens during the regime of Pericles and in Rome (in some cases during the Republic and in many cases during the Empire), there was a wide margin of freedom. The basic rigidity of those societies continued to exist, however. Slavery was never abolished neither in Greece nor in Rome.

REFERENCES

Davies, Paul (1984). *Superforce.* New York: Simon and Schuster.

Einstein, Albert (1921). *La Theorie de la Relativite.* Paris:Gauthier-Villars.

Gayon, Jean (1989). *Critics and Criticisms of the Synthesis*, manuscript.

Hawking, Stephen (1988). *A Brief History of Time.* New York:Bantam Press.

Miro-Quesada, Francisco (1988). *Proceedings of theInternational Congress of Philosophy,* Cordoba.

Oparin, A.I. (1957). *The Origin of Life on Earth.* New York: Academic Press.

Pagel, Heinz (1982). *The Cosmic Code.* New York: Bantam Press.

(1985). *Perfect Symmetry.* New York: Bantam Press.

Popper, Karl (1959). *The Logic of Scientific Discovery.* NewYork: Basic Books.

Prigogine, Ilya (1988). *Entre le Temps et l'Eternité.* Paris.

Shapere, Dudley (1983). *Reason and the Search for Knowledge.* Dordrecht: Reidel.

Scheler, Max (1938). *Die Stellung des Menschen in Kosmos.* Darmstadt.

INDEX OF NAMES